COHERENT OPTICAL COMMUNICATIONS AND PHOTONIC SWITCHING

COHERENT OPTICAL COMMUNICATIONS AND PHOTONIC SWITCHING

Proceedings of the Fourth Tirrenia International Workshop on Digital Communications
Tirrenia, Italy, September 19–23, 1989

edited by

G. PRATI

Instituto di Scienze dell'Ingegneria
Università degli Studi de Parma
Parma, Italy

1990

ELSEVIER
Amsterdam – Oxford – New York – Tokyo

ELSEVIER SCIENCE PUBLISHERS B.V.
Sara Burgerhartstraat 25
P.O. Box 211, 1000 AE Amsterdam, The Netherlands

Distributors for the United States and Canada:
ELSEVIER SCIENCE PUBLISHING COMPANY, INC.
655 Avenue of the Americas
New York, N.Y. 10010, U.S.A.

Library of Congress Cataloging-in-Publication Data

Tirrenia International Workshop on Digital Communications (4th : 1989)
Coherent optical communications and photonic switching : proceedings of the Fourth Tirrenia International Workshop on Digital Communications, Tirrenia, Italy, September 19-23, 1989 / edited by G. Prati.
p. cm.
Includes bibliographical references.
ISBN 0-444-88412-2 (U.S.)
1. Digital communications--Congresses. 2. Optical communications--Congresses. 3. Switching circuits--Congresses. I. Prati, Giancarlo. II. Title.
TK5103.7.T57 1989
621.382'7--dc20 89-71503
CIP

ISBN: 0 444 88412 2

Printed in The Netherlands

FOREWORD

These are very exciting times for those of us who work with coherent optical communications, photonic switching, and advanced photonic technologies in general. With the successful deployment of the first trans-oceanic fiber optic transmission systems, and the wide acceptance of fiber for telecommunications trunking applications, we can all look forward to what I believe is the ultimate challenge: bringing fiber to the end user - with the concurrent broadband services that fiber can support. This application will greatly increase the demands for transmission capacity and the need for advanced high-speed switching technologies. A challenge we face is in the use of advanced wavelength multiplexing technologies (like coherent communications) to make the delivery of broadband services to the end-user economical and flexible to implement. Another opportunity that lies within this context is the utilization of photonic technologies to facilitate the implementation of practical broadband switching and networking techniques. This includes both centralized switches at various types of communication network nodes, as well as various forms of distributed switching.

The use of coherent communications (or alternatively direct detection, dense wavelength multiplexing techniques) provides increased flexibility in the design of broadband distribution networks. The extraordinarily large bandwidth of fiber can be exploited without the need for electronics operating beyond the data rates associated with the user services. Thus, fiber can be shared conveniently amongst several or many users - an important factor in making fiber-to-the-customer cost effective. In addition, the passive nature of many components that can be used in wavelength multiplexed networks - e.g., splitters- is attractive when one considers the need to remotely locate multiplexing and branching points. The recent upsurge in interest, and the corresponding results, on optical amplifiers adds yet another dimension of flexibility and opportunity that is synergistic with coherent communications and other forms of dense wavelength multiplexing.

Thus, it is with great hopes and expectations that the community of researchers working on coherent communications techniques and applications moves forward into the next decade.

To capture the whole realm of possibilities for the utilization of photonic technologies within switching and networking applications we shall use the designation: photonic switching. However, as the later papers will explain, this does not imply the direct replacement of photonic devices into existing equipment and network architectures as a substitute for electronic devices.

Photonic switching will be used here to capture the general opportunity to utilize existing and new photonic devices in conjunction with existing and new electronic devices to realize new or improved switching and networking capabilities. It includes such things as simple mechanically activated optical switches used for network protection switching and network reconfiguration, where an entire optical signal is redirected as a unit. It includes the use of optical interconnections within equipment to facilitate internal high speed and high density interconnects. It includes the use of tunable transmitters and receivers at the core of a high speed packet or circuit switch, with large quantities of electronics providing supporting interface and control functions. In principle it includes the concept of an all-optical switch-whose architecture and component technologies have yet to be invented.

I am happy to say that in the past several years materials and device scientists have been coming together more and more with systems architects to uncover the real opportunity areas for coherent communications and photonic switching. In the early days of fiber optics (circa 1970) it was somewhat easier to visualize the application (e.g. the architecture) which represented the target for a practical realization of the technology. Basically, the target everyone was pursuing was a point-to-point fiber optic link with higher capacity, longer repeater spacing, and thus lower cost than 1.5-2 Mb/s digital transmission over copper wire pairs. One could calculate, relatively easily what the requirements were to break even against the incumbent technology. It is unlikely that coherent communications will be used only as another alternative for increasing the performance or reducing the cost of point-to-point links. It is unlikely that an all-optical switch will emerge which can be directly substituted for today's electronic switches - except perhaps in some relatively simple applications such as the protection switching application mentioned above. Therefore, one has to simultaneously consider new network applications, new network architectures, and a variety of hybrid photonic-electronic possibilities in order to uncover the applications likely to be most attractive.

It is my hope that this book will shed some light on the current thinking regarding where the applications are, what the technology status is, and where the opportunities are for future research.

Steward D. Personick

Network Technology Research
Bell Communications Research, Inc.
Red Bank, New Jersey USA

TABLE OF CONTENTS

SPONSORING ASSOCIATIONS, ORGANIZATIONS AND INSTITUTIONS

Consiglio Nazionale delle Ricerche
EUREL
IEEE, Middle and South Italy Section
Università di Parma, Parma, Italy
Università di Pisa, Pisa, Italy

ACKNOWLEDGEMENTS

The editor is indebted and wishes to express his thanks to the workshop technical program committee, namely Professor Mario Armenise, Dr. Ian Garrett, Professor Ivo Montrosset, Professor Takanori Okoshi, Dr. Stewart D. Personick, Professor Paul R. Prucnal, Dr. David W. Smith, whose cooperation has made it possible to bring the workshop and this book into being.

The support of the following organizations and institutions is also gratefully acknowledged:

Alcatel-Face
Italtel
Marconi Italiana
Siemens Telecommunicazioni
Telettra

PART 1

COHERENT OPTICAL COMMUNICATION TECHNIQUES AND TECHNOLOGIES

Recent Status of Coherent Lightwave Technologies for High-Speed Long-Haul and FDM Transmission

Toshihiko SUGIE and Kiyoshi NOSU

NTT Transmission Systems Laboratories
Yokosuka-shi, Kanagawa, Japan

The recent progress of coherent lightwave technology is described. Emphasis is placed on experimental optical heterodyne / homodyne detection and optical FDM (frequency division multiplexing) techniques, including newly developed optical devices.

I. Introduction

Coherent lightwave techniques, which make use of optical frequencies or phase, are having significant impact on communication systems. Because of the possibility of an improvement in receiver-sensitivity up to 20-30 dB, research and development activities have been accelerated. However, to realize coherent lightwave systems many technical difficulties have to be overcome ; such as frequency stabilization, laser-spectrum purification, insensitive polarization states of the fibers, and modulation/demodulation techniques. With the development of coherent light sources, sophisticated communication technologies that have very large bandwidth potential of the carriers will become feasible [1]~[3].

The present research trends of coherent lightwave transmission are summarized as follows ; (1) high-speed, (2) long-haul transmission, and (3) dense optical multiplexing. A key technology for high-speed and long-haul transmission is optical heterodyne/homodyne detection and that for dense optical multiplexing is optical FDM (frequency division multiplexing). This paper reviews the current state of coherent lightwave technology on these points.

II. High-speed and Long-haul transmission

II-1 Fundamental technique

Heterodyne/homodyne detection has a lot of advantagies. With FSK or PSK modulation, heterodyne detection permits significant improvement in receiver sensitivity, and it allows signal processing of the electrical intermediate frequency band, such as, equalization of optical signal waveform distortion which is caused by the fiber chromatic dispersion. These merits and applications of heterodyne/homodyne detection are summarized in Table I . A system configuration for heterodyne/homodyne detection is shown in Fig.1. Transmitter

---

The authers are with NTT Transmission Systems Laboratories, 1-2356, Take, Yokosuka-shi, Kanagawa, 238-03, Japan.

Tel. +81 - 468 - 59 - 3041, Fax. +81 - 468 - 59 - 3396

Table I Merits and applications of heterodyne / homodyne detection

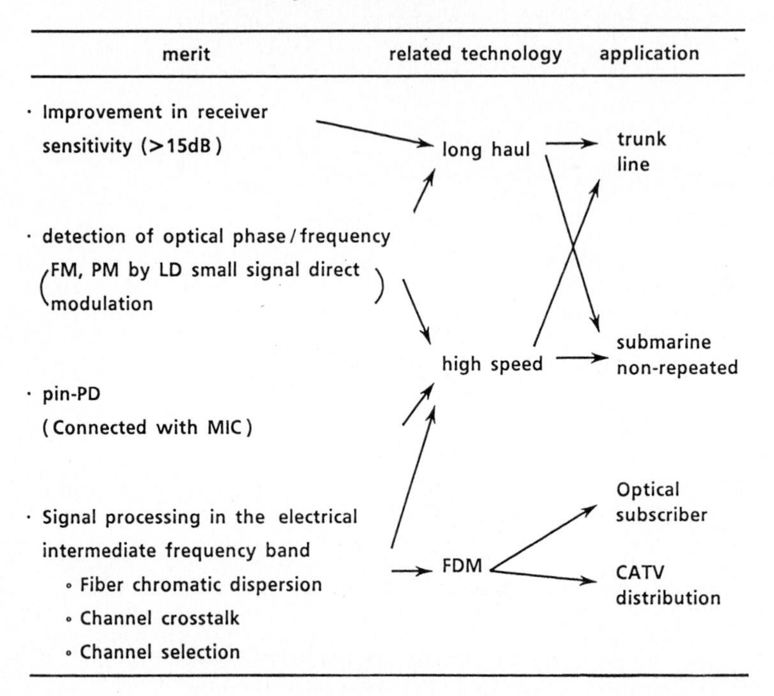

configurations are classified into direct modulation and modulation by an external modulator. Technical difficulties must be overcome to realize transmission systems, such as narrow linewidth [4][5] and frequency stabilization of laser diodes (LD), insensitivity to polarization states of the fibers [6] and modulation/demodulation techniques. These difficulties and the techniques recently proposed are summarized in Table II as well as the related experiments.

(1) Transmitter

A transmitter consists of an LD, a modulation circuit and a stabilization circuit. Both narrow LD linewidth and modulation characteristics are most important, because they determine a modulation format which can be applied.

(i) Linewidth : Narrow linewidth in LD must be realized to reduce the required IF bandwidth and excess penalty. Although required linewidth is determined by the modulation format, a linewidth less than 1 MHz must be achieved for several gigabit transmission, for example $10^{-3} \times$ bit rate for FSK-heterodyne differential detection. This has been resently achieved by long cavity DFB-LDs, MQW-DFB-LDs, and LDs with an external cavity, by several organizations.

(ii) Modulation technique : Frequency modulation (FM) is easily carried out by using LD frequency chirping that corresponds to the injection current. This means that CP-FSK is achieved by direct modulation of the LD injection current. However, an external modulator must be used for ASK and PSK. A wideband

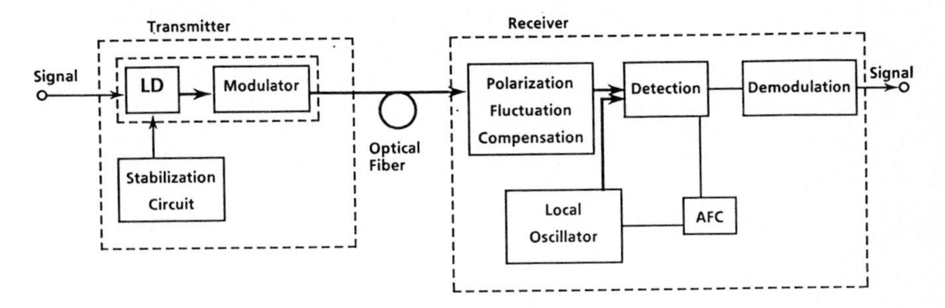

Fig 1. A system configuration for heterodyne/homodyne detection

Table II Technical problems of high-speed, long-haul transmission

Item	proposed technique	Feature	Reference
[Transmitter] narrow linewidth-LD	· Long cavity DFB-LD · MQW-DFB-LD · LD with external cavity	760 KHz 700 KHz 70 KHz	(7), (8), (11) (9) (10) ···
flat FM response	· Multielectrode-DFB-LD · equalization · cording	 ~4 GHz AMI,bipolon FSK	(7) (11)~(14) (15) (16) (17)
external modulator	· $LiNbo_3$-modulator · MQW-modulator	10 GHz 20 GHz	(18) (19) (20)
[Optical fiber] polarization insensitivity	· polarization controller · polarization diversity (conventional two IF method frequency conversion method) · polarization-maintaining fiber · polarization scrambling	fiber, $LiNbO_3$, YIG	(21)~(23) (6), (24) (25) — —
improvement of dispersion effect	· IF equalization	202 km, 4 Gb/s, 8 Gb/s	(26) (27)
nonlinear effect	· SBS	less than + 10 dBm	(28)
[receiver] preamplifier	· Inductor peaking · twin-pin, HEMT(balanced receiver)	2~8 GHz 8.7 pa/√Hz	(29) (30)
IF amplifier	· HEMT (HIC) · Inductor peaking	~20 GHz 20 dB	(31)
frequency tracking	· differential detector with 90° hybrid	± 30 MHz (1.8 Gb/s)	(33)
phase diversity	· optical hybrid		(32)

frequency response is required in these devices. A flat FM response for an LD has been achieved for FSK transmission by using a multielectrode DFB-LD. Fig.2 shows typical configuration and characteristics for this multielectrode DFB-LD. Narrow linewidths of less than 1 MHz and with a flat FM response up to several GHz are achieved . Codes such as AMI and bipolar FSK have been proposed for direct modulation to avoid LD non-flat FM response effect. An equalization circuit for FM response is employed to drive the LD and this can improve the response by about 1~2 times as compared to an LD without equalization circuit.

Wideband external modulators of more than 10 GHz have been developed for ASK, and PSK. These modulators were fabricated by $LiNbO_3$, and MQW. A low drive voltage (half wavelength) of less than 10 V and a low insertion loss of 3 dB are achieved by using the coplanar waveguide as a traveling-waveguide.

(2) Optical fiber

The polarization state fluctuation and chromatic dispersion of fibers attribute to the penalty increase in heterodyne/homodyne detections. A lot of techniques have been proposed to overcome these problems.

(i) Polarization insensitivity : The polarization state of the optical fiber fluctuates, when LD light injected into the fiber propagates. However, the polarization state of the LD signal propagated along an optical fiber must coincide with that of the local LD light. There are several typical methods; polarization-maintaining fiber, polarization controller, polarization diversity and polarization scrambling. In these technique, sensitivity degradation, response time and excess loss are important. These must be reduced to negligible. At present, various types of polarization controllers are proposed such as coiled fiber, opt-electric crystal, and Faraday rotator among others. Takeing account of response time, polarization diversity is most attractive because of stable operation that is independent of oplarization fluctuation, although the two orthogonal polarization components are separately detected, A polarization diversity experiment is shown in Fig.3 [24]. Degradation due to polarization fluctuation is estimated to be less than 0.3 dB for this diversity technique. Recently, a polarization diversity optical receiver that uses polarization frequency conversion technique with two local oscillators has been proposed [25].

(ii) Dispersion : Fiber dispersion attributes to distortion of the waveform transmitted along a long fiber. Although degradation by fiber dispersion

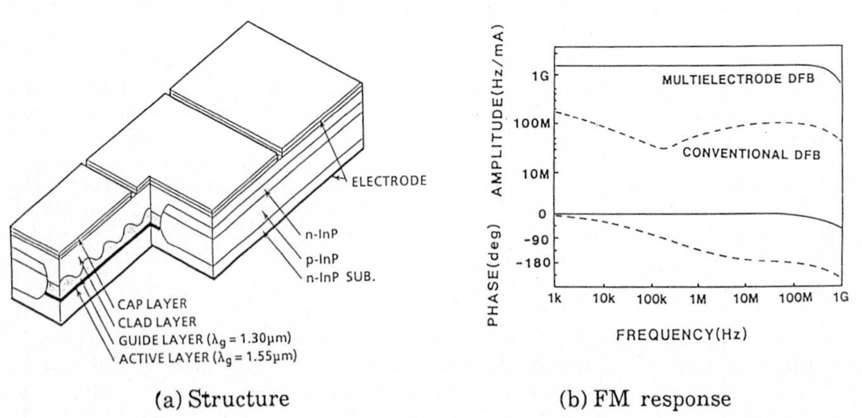

(a) Structure (b) FM response

Fig 2. Multielectrode DFB-LD

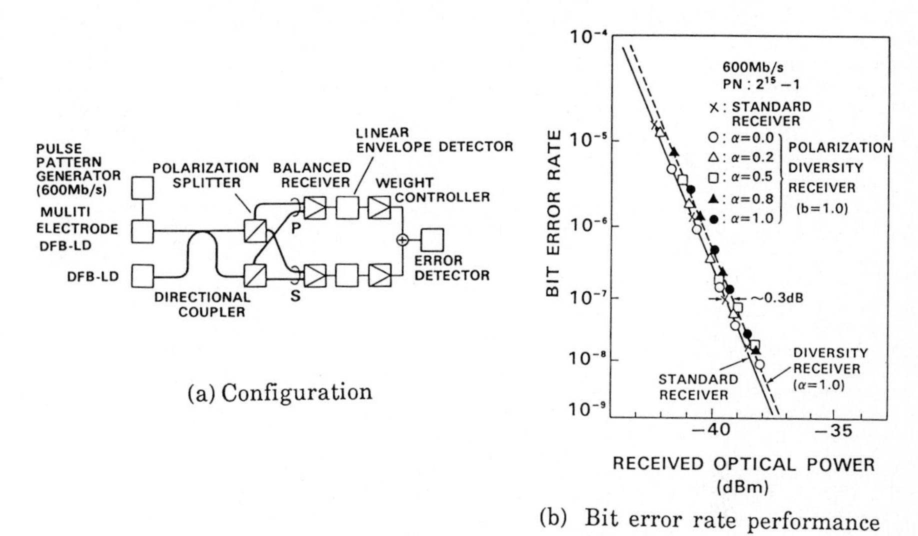

(a) Configuration

(b) Bit error rate performance

Fig 3. Polarization diversity

depends on the modulation/demodulation format, waveform distortion is compensated for by IF equalization in heterodyne detection, i.e. signal processing in the IF band. A 202 km long fiber transmission experiment is demonstrated at 4 and 8 Gb/s using a microstrip line delay equalizer [26] [27]. Frequencies for signal and local oscillator LDs are allocated by the delay characteristics of the transmission fiber and the IF equalization circuit.

(iii) Non-linear effect : Among the non-linear effects in a fiber, transmitted power that is limited by stimulated Brillouin scattering (SBS) is considered to be problem to long haul transmissions. Although several experiments and theoretical considerations have been carried out from the viewpoints of modulation format, and modulation speed [28], this fatal effect on long haul transmissions can not be reported on at present. This is becoming a serious problem as higher optical power is being injected into fibers.

(3) Receiver

(i) Optical receiver and IF amplifier : A heterodyne/homodyne receiver consists of pin-photo diodes, low noise amplifiers, a local oscillator and a frequency/phase tracking circuit. A lot of effort has been made to achieve both wide bandwidth and low noise characteristics for each component and the receiver configuration. A balanced receiver that consists of dual pin-PDs is proposed to suppress local LD relative intensity noise (RIN) by more than 10 dB. IF HIC amplifiers employing HEMT have also been developed with a averaged Noise Figure of 4.1 dB in the range of from 100 MHz to 18 GHz [31]. These two opints are inevitable to achieve long haul transmission [29].

(ii) Phase diversity : As bit rate increases, the required bandwidth of the IF amplifier, instead of LD linewidth, increases in the heterodyne detection. Although homodyne detection has advantages for the required amplifier bandwidth and detection sensitivity, optical PLL must be developed. Experiments have been carried out but problems still remain to be solved. However, phase diversity is a promising technique to overcome these problems for

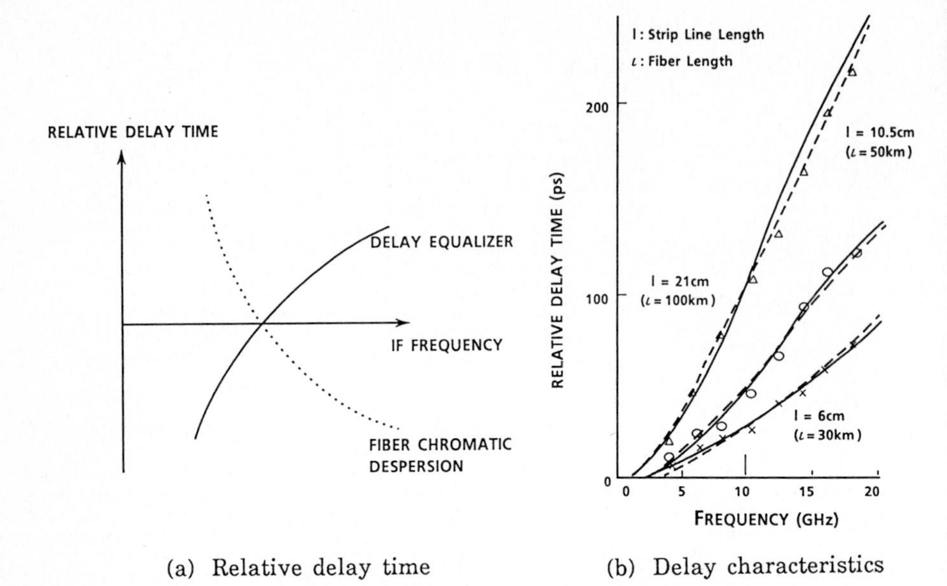

(a) Relative delay time (b) Delay characteristics

Fig 4. Delay characteristics

high-speed transmissions [32]. Receiver bandwidth is the same as that of homodyne detection and optical PLL is not required in phase diversity, although LD frequency must be stabilized. This has been recently studied.

(iii) Frequency tracking : Frequency tracking systems must be insensitive to both the fluctuations in mark density and polarization. The IF frequency average for mark and space can only be used when the mark density is constant and simply combined IF signals may disappear in polarization diversity when two constituent signals are of equal power. A newly developed tracking system, which consists of both differential detectors and a 90° hybrid, can reduce tracking error to ±30 MHz at 1.8 Gb/s when the mark density fluctuates from 1/8 to 7/8 [33].

II-2 Transmission Experiment

A number of current and proposed experiments are summarized in Table III. The important technical problems that are shown in Table-II are overcome in these experiments. Degradation due to fiber chromatic dispersion is successfully equalized by a newly developed microstrip delay line equalizer. The 8 Gb/s, 202 km optical CPFSK transmission experiment was conducted at 1.55 μm using a non-dispersion shifted fiber [27]. Delay characteristics of the optical fiber and the microstrip line are shown in Fig.4. Experimental results are shown in Fig. 5 as well as the experimental configuration. Penalty degradation is reduced to 4.5 dB at 10^{-9}. By developing a low noise balanced receiver of 100 photons/bit receiver sensitivity and a narrow linewidth multielectrode DFB-LD, a 290 km transmission has been achieved at 2.5 Gb/s in CPFSK [29]. This receiver consists of dual pin-PDs and a low noise HEMT trans-impedance preamplifier, which uses an inductor peaking technique. Response and noise are flat over 9 GHz and the averaged input equivalent noise current is reduced to 8.7 pA/√Hz in a range from 2 GHz to 8 GHz, which is shown in Fig.6. MQW-LDs with a linewidth of less than

TABLE III
HETERODYNE / HOMODYNE DETECTION SYSTEM EXPERIMENTS

NO	A - 1	A - 2	A - 3	A - 4	A - 5	A - 6
Bit rate	8 Gb/s	2.4 Gb/s	2 Gb/s	1.2 Gb/s	150 Mb/s	150 Mb/s
Span	202 km	290 km	204 km	201 km		
System	CPFSK heterodyne	CPFSK heterodyne	CPFSK heterodyne	DPSK heterodyne	CPFSK heterodyne	CPFSK heterodyne
Int. Freq.	5.4 GHz	5.0 GHz	3.4 GHz	2.4 GHz		900 MHz
Technical Significance and feature	Multi electrode -DFB-LD Equalization of optical signal wave form	Multi electrode -DFB-LD Low noise receiver	MQW-DFB-LD	Fiber-external-cavity laser DBOR-HEMT (Twin-pin)	External-Cavity laser polarization diversity	AMI code polarization diversity
Result	Reduction of Penalty increase due to fiber dispersion	High bitrate, Long span	First MQW-DFB-LD experiment	Stabilization, AFC (doubling)	Under sea test of polarization diversity	code for LD non-flat FM responce
Year month	1989 Jul.	1989 Jul.	1988 Sept.	1988 May	1988 Mar.	1988 Sept.
Journal, Conf.	IOOC'89	IOOC'89	ECOC'88	Electron. Lett	Electron. Lett	IEICEJ National Conference
Reference	(27)	(29)	(16)	(34)	(35)	(17)

1 MHz [16] and fiber-external-cavity lasers with a linewidth of 70 kHz [34] are also used for the transmission experiments over 200 km, in CPFSK and DPSK, respectively. The degradation due to linewidth is negligible in these LDs. Transmission distance limited by fiber loss and chromatic dispersion is calculated in Fig. 7 as a function of bit rate in the case of CPFSK. The transmission experiments are also indicated to clarify the technical significance. Fibers with a 0.17 dB/km loss and a 17 ps/km/nm dispersion as well as fibers with a 0.2 dB/km loss and a 2.5 ps/km/nm dispersion are assumed. A signal power of +10 dBm and a receiver sensitivity of 20 photon/bit are also assumed.

II-3 Important techniques for further development

Taking account of recent progress in optical amplifiers and modulation/demodulation techniques, sophisticated optical transmission is expected. Besides a balanaced receiver, an external modulator and a narrow linewidth light source with high-speed and high-power, the following techniques are important.

(1) Optical amplifier : Especially, optical amplifier repeaters are attractive for long-haul transmissions.

(2) OE-IC for a polarization diversity and a phase diversity : An uniform characteristics of components as well as compactness will be achieved.

(3) Low loss fiber, such as fluoride fiber in mid-infrarared region : This will intrinsically improve transmission distances.

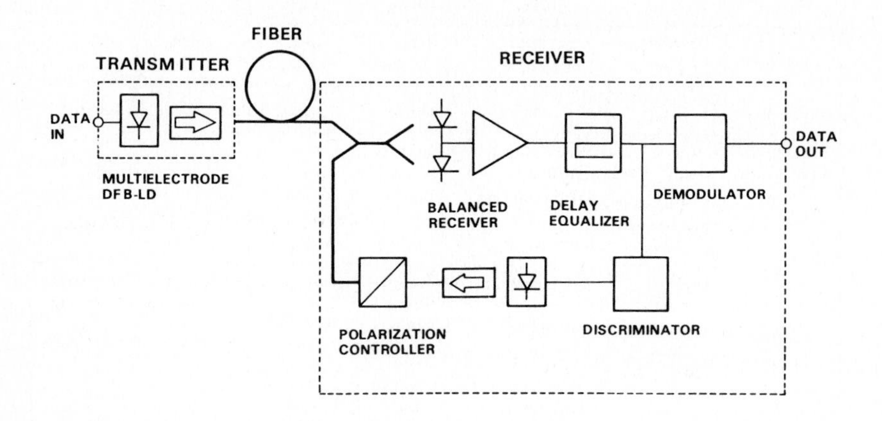

(a) Experimental configuration

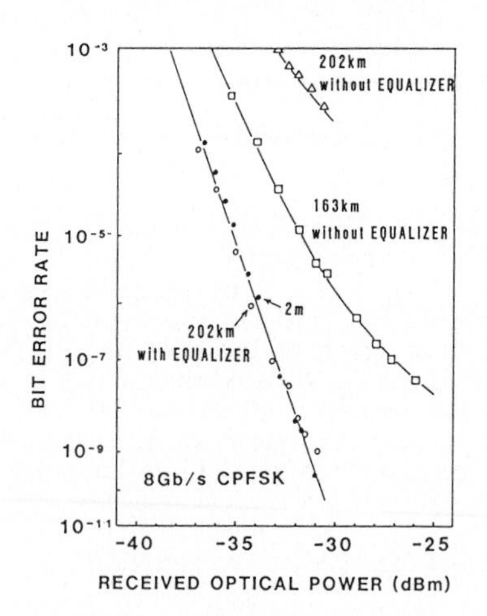

(b) Bit error rate performance

Fig 5. Fiber chromatic Dispersion Equalization Experiment

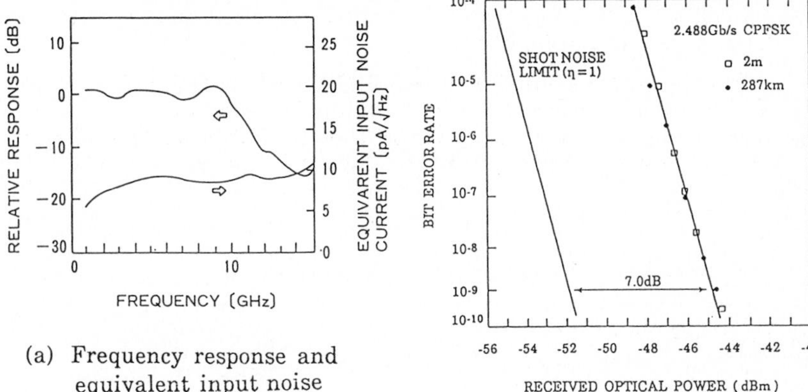

(a) Frequency response and equivalent input noise current of the double balanced receiver

(b) Bit error rate performance

Fig. 6 100 photons / bit receiver characteristics

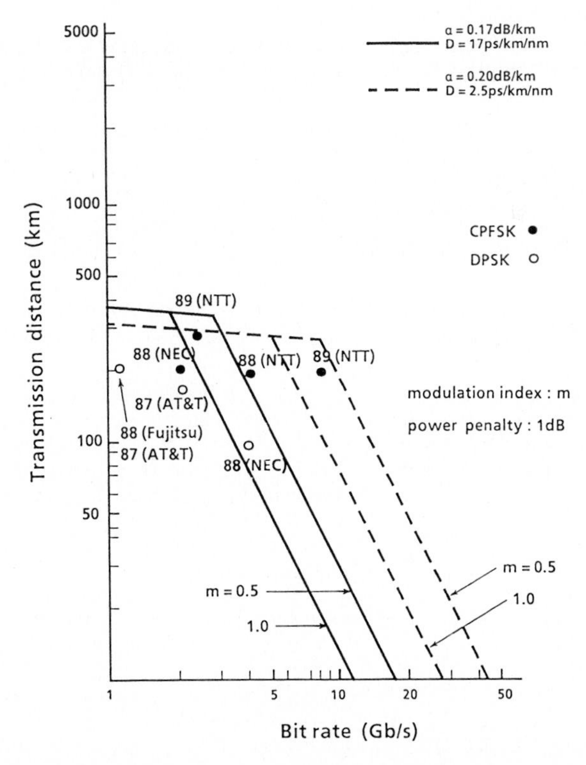

Fig 7. Transmission distance

III. Optical FDM

III-1 Fundamental technique

Optical FDM systems make full use of the fiber low loss range from 1.5 to 1.6 μm, and can transmit more than 1000 channels stimutaneously. Optical frequencies are allotted for services and a receiver can select one or more signals from those distributed at several GHz intervals. This technology is expected to be used in future broadband information distribution systems, as well as in ultra large capacity intra-city trunks.

There are two optical multiplexing configurations. One uses an optical filter as a combiner, and the other uses a M×N star coupler, which works simutaneously as a distributor as well as a combiner. In the former configuration, optical carriers are multiplexed by an optical filter, such as a periodic filter multiplexer at the head end. After transmission, they are distributed by a power divider, such as a 1×N star coupler. In the latter, system configuration for optical FDM multiplexed and distributed by a star coupler is shown in Fig.8. From the viewpoint of subscriber distribution, this configuration is suitable when subscribers are scattered around the head end, whereas the configuration multiplexed by an optical filter is suitable when subscribers are at some distance from the center and distributed in packed groups [36]. Technical difficulties in FDM transmissions are also summarized in Table IV.

(1) Head End (Transmitter)

Stabilized multicarrier frequencies must be multiplexed by an optical FDM transmitter. This implies that carrier frequency stabilization and multiplexing techniques are inevitable.

(i) Multicarrier frequency stabilization : The following techniques are proposed to stabilize the multiplexed carrier frequency. An LD frequency sweep method that uses heterodyne detection, an optical resonator or filter method such as Mach-Zehnder interferometer or others. By using a ring resonator, multiplexed 16 ch carriers are stabilized in 5 GHz spacing [37]. Output fluctuation of the ring resonator was fed back to the bias current of each LD through the PID.

(ii) Multiplexing : There are two optical multiplexing methods ; one is an optical filter method and the other is a power coupler method. In principle, the

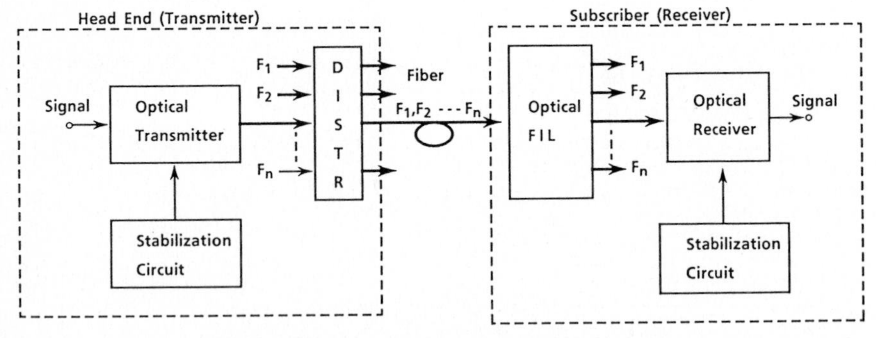

Fig 8. A system configuration for optical FDM

TABLE IV

Technical problems of optical FDM

Item	Proposed technique	Feature	Reference
[Head End (Transmitter)]			
multicarrier frequency stabilization	· optical ring resonator	16 ch, 5GHz (30 MHz)	(37)
multiplexing	· optical filter	periodic filter	(38)
	· star coupler	16 × 16 coupler	(36)
[Optical fiber]			
non-linear effect	· Fourwave mixing	4 dB degradation (2 Gb/s, CPFSK)	(39)~(41)
polarization insensitivity	· polarization controller		(21)~(23)
	· polarization diversity		(6) (24) (25)
[Subscriber (Receiver)]			
channel selection	· optical periodic filter	5 GHz-channel spacing, 16ch	(36) (42)
	· tunable LD	DBR-LD : 4nm	(45)
		DFB-LD : 2nm	(7) (11) (44)
image rejection	· polarization insensitive image rejection	optical/electrical 90° hybrid 560 Mb/s	(46)

former can combine optical signals without loss, while the latter has an inherent loss due to multiplexing. However, because of the loss increase in optical filters fabricated in serial connections, the latter can be used mainly for local information distribution. At present, a 16×16 star coupler with a low loss of 13.5 dB has been developed [36].

(2) Optical fiber

As shown in the previous section, the polarization insensitive technique is also important in optical FDM. Still more, degradation caused by the four-wave mixing process will appear when the output power of each LD is on the order of mW and fiber length is more than several km [39]. This nonlinear effect in FDM is determined by the channel spacing, channel number, input power, transmission distance and fiber dispersion. Fig.9 shows calculated interchannel crosstalk for the case of a non-dispersion shifted fiber and a 1.55 μm dispersion shifted fiber [40]. For example, when 100 carries spaced at 5 HGz, for 1.55 μm, and with a launched power of 0.2 mW/channel are transmitted through a 10 km long dispersion shifted fiber, the interchannel crosstalk will be more than -20 dB. Frequency spacing must be expanded to suppress this crosstalk. Non-dispersion shifted fibers are more suitable for FDM systems when LD oscillation at 1.55 μm is used. Receiver sensitivity degradation due to four-wave mixing in 2 Gb/s CPFSK is also measured at about 4 dB in the dispersion shifted fiber [41].

(3) Subscriber (receiver)

Because narrowly spaced optical signals must be selected in the receiver, a channel selection or a demultiplexing techniques are inevitable. An image-rejection technique is also important to reduce the adjacent channel effects.

(i) Channel selection : There are two kinds of channel selection methods for narrowly spaced optical signals. One is optical filtering method using a passive optical filter such as a periodic filter [42]. The other is electrical IF band filtering method in optical heterodyne detection [43]. In the optical filtering method, either direct detection or heterodyne detection can be utilized. At present, periodic filter construction is the same as that of the Mach-Zender interferometer. It consists of two 3 dB directional couplers and two waveguides with different

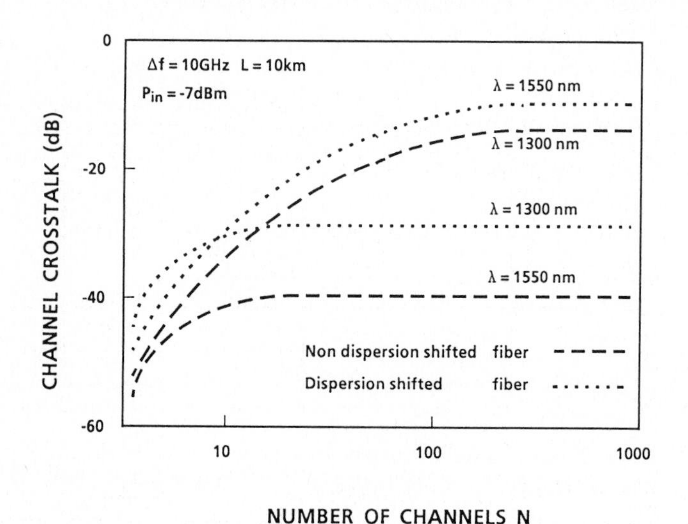

Fig 9. Interchannel crosstalk due to four-wave mixing process in FDM

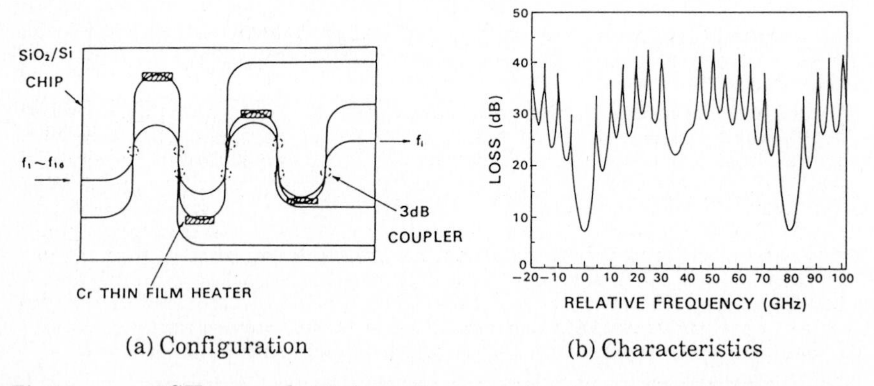

(a) Configuration (b) Characteristics

Fig 10. 5GHz spaced, 16 channel waveguide frequency selection switch lengths connecting them. A channel spacing of 5 GHz with 16 ch is developed [36]. Configuration and filtering characteristics for this channel waveguide frequency selection switch are shown in Fig. 10. On the other hand, electrical filtering by heterodyne detection has the advantages of high sensitivity, large numbers of channels and long transmission distances, over optical filtering/direct detection. However, the heterodyne scheme needs a wideband tunable laser diode to select one optical carrier from the spread multicarrier. Recently, DBR-LDs with a 4 nm continuous tunable range and a linewidth of less then 20 MHz has been developed [45].

(ii) Image-Rejection : The optical heterodyne image-rejection technique is attractive because it removes the image spectrum of the adjacent channel and this enables a narrow channel spacing. A polarization-insentive image-rejection receiver is proposed and the effectiveness of its image-rejection technique has been demonstrated for a two channel 560 Mb/s DPSK system [46].

III-2 Experiment

Table V summarizes the present state of FDM technology. Recently, a 16 channel optical FDM distribution/transmission experiment at 622 Mb/s has been demonstrated [36]. 16 DFB-LDs could be stabilized to within 30 MHz with a frequency spacing of 5 GHz by using a waveguide ring resonator. Frequency

TABLE V
FDM TECHNOLOGY SYSTEM EXPERIMENTS

	B-1	B-2	B-3
Nunber of channel	16ch	10ch	4ch
Channel spacing	5GHz	8GHz	3GHz
Bitrate	622Mb/s	400Mb/s	560Mb/s
System	IM	FSK	DPSK
Technical significance and feature	tunable optical periodic filter type demultip lexer	tunable DBR-LD polarization diversity randam access AFC	Fiber-external cavity DFB-LD DBOR-HEMT (Twin-Pin) polarization diversity
Result	Optial FDM distribution transmission experiment	Coherent CATV transmission demonstration	Coherent transmission of HDTV signal demonstration
Year month	1989 Jul.	1989 Jan.	1988 Sept.
Journal Conf.	Iooc'89	OFC'89	IEICEJ National Conference / JLT
Reference	(36)	(46)	(47)

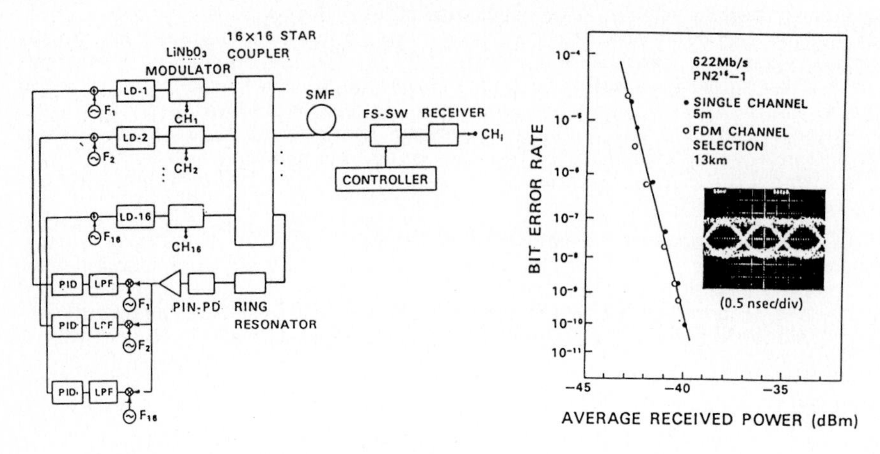

(a) Experimental configuration (b) Bit error rate performance

Fig 11. 16channel optical FDM distribution / transmission experiment at 622Mb/s

selection for multiplexed carriers has been successively carried out using a waveguide selection switch with crosstalk of less than 20 dB per channel. Configuration of the FDM distribution/transmission experiment and bit error rate characteristics are shown in Fig. 11.

Channel selection using tunable DBR-LDs and polarization diversity configuration have been demonstrated [46] [47]. Also, a simultaneous amplification experiment utilizing an Er^{3+}-doped fiber amplifier is demonstrated for a 16 channel FDM system [48]. The received power penalty of 1.1 dB that is associated with 16-channel transmission appeared, mainly caused by SNR degradation. A signal gain of more than 15 dB per channel was achieved in this experiment. This suggests that 1×32 star couplers can be additionally connected to the 16×16 star coupler to expand the number of subscribers.

III-3 Important techniques for further development

Optical FDM is expected to be applied to new sophisticated networks, such as optical routing techniques and crossconnect switching nodes. The following techniques will be important.

(1) Absolute frequency control technique : This will be inevitable to select the narrowly spaced signals.

(2) LD arrays and OE-IC for multiplexing/demultiplexing : Compactness and low comsuption power will be achieved.
These will lead to the increase of multiplexing channels.

IV. Future trends

The present research trend of coherent lightwave transmission is shown in Fig.12. A lot of effort has been made to develop fundamental technologies for high-speed long-haul and FDM transmission. They have been experimentally confirmed as shown in the previous section, although several technical problems still remain to be solved.

For the next generation of coherent lightwave transmissions, advanced technologies will be developed that are applicable to areas from subscriber loops to large capacity and long-haul trunks. Still more attempts will be made to use lightwaves in such a way as to get the same performance as radio waves.

V. Conclusion

Recent progress in coherent lightwave technology has been reviewed with emphasis on the experiments, from the viewpoint of high-speed, long-haul transmission and dense optical multiplexing. Present technological problems and future trends have also been discussed. For the next generation, advanced technologies for system use must be developed, along with the fundamental technology.

Acknowledgements

The authors would like to express their thanks to Dr. S. Shimada, Dr. H. Kimura and Dr. T. Ito, of the NTT Transmission systems Laboratories, for their guidance and encouragement. They would also like to express their thanks to the members of the Lightwave Communications Laboratory for their valuable advice and discussions.

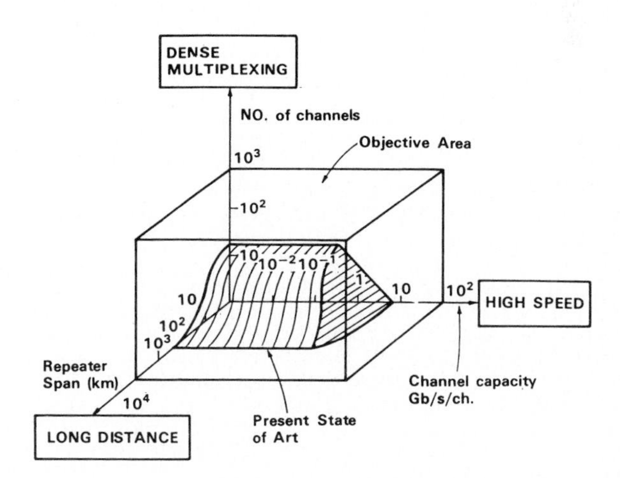

Fig 12. Present research trend of coherent lightwave transmission

References

[1] Yamamoto,Y., and Kimura, T., IEEE J. Quantum Electron., QE-17, (1981) 919.
[2] Okoshi, T., IEEE J. Lightwave Technol., LT-2, (1985) 341.
[3] Midwinter, J.E., IEEE J. Lightwave Technol., LT-3, (1985) 927.
[4] Garrett, I., and Jacobson, G., IEEE J. Lightwave Technol., LT-5, (1987) 551.
[5] Hodgkinson, T.G., IEEE J. Lightwave Technol., LT-5, (1987) 573.
[6] Okoshi, T., et. al., IOOC'83, Tokyo, June 27-30 No. 30C 3-2, (1983) 386.
[7] Fukuda, M., et. al., IEEE Photon, Technol. Lett., vol.1, (1989) 6.
[8] Ogita, S., et. al., Electron. Lett., vol. 24, (1988) 613.
[9] Takano, S., et. al., 11th LD conf., Q1 Boston, (1988).
[10] Onaka, H., et. al., CLEO'88, TUC4, (1988).
[11] Kotaki, Y., et. al., 7 th IOOC'87., Kobe, 19A2-4 (1989).
[12] Yoshikuni, Y., IEEE J. Lightwave Technol., vol. LT-5, (1987) 516.
[13] Iwashita, K., et. al., Electron. Lett., vol. 23, (1987) 1022.
[14] Yamazaki, S., et. al., Electron. Lett., vol. 21, (1985) 283.
[15] Iwashita, K., and Takachio,N., OFC'88 Post deadline paper No. (1988).
[16] Yamazaki, S., et. al., ECOC'88, (1988) 467.
[17] Tsushima, H., et. al., IEICEJ. National Conference B-429, Sept. (1988).
[18] Kawano, K., et. al., IEEE Photon. Technol. Lett., vol. 1, (1989) 33.
[19] Mekada, N., et. al., MOC'87, H4 (1987).
[20] Kotaka, I., et. al., IEEE Photon. Technol. Lett., vol. 1, (1989) 100.
[21] Matsumoto, T., et. al., Electronics and Communications in Japan Part 2, vol. 71, No. 6, (1988) 36.
[22] Kubota, M., et. al., Electron. Lett., vol. 16, No.15, (1980) 573.
[23] Okoshi, T., et. al., Electron. Lett., vol. 21, No. 18, (1985) 787.
[24] Imai, T., ECOC'88 (1988) 159.
[25] Tsushima, H., and Sasaki, S., Electron. Lett., vol. 25, No.8., (1989) 539.
[26] Iwashita, K., and Takachio, N., Electron. Lett., Vol. 24, (1988) 759.
[27] Takachio, N., et. al., IOOC'89 Kobe, Post deadline Paper 20 PDA-13 (1989).
[28] Aoki, Y., et. al., IEEE, J. Lightwave Technol., vol. LT-6, No.5,(1989) 710.
[29] Ichihashi, Y., et. al., IOOC'89, Kobe., Post deadline Paper 20PDA-12 (1989).
[30] Wada, O., et. al., Electron. Lett., vol. 24, (1988) 514.
[31] Ohkawa, N., Electron. Lett. vol.24, No.17, (1988) 1061.
[32] Ishida, O., and Okoshi, T., ECOC'88 Brighton, vol. 1, Sept, (1988) 155.
[33] Imai, T., and Iwashita, K., IOOC'89, Kobe, 18C2-6 (1989).
[34] Chikama, T., et. al., Electron. Lett. vol. 24, No. 10 (1988) 636.
[35] Ryu, S., et. al., Electron. Lett., vol. 24, No.7, (1988) 399.
[36] Toba, H., et al., ICC'89 Boston, 14.5, (1989).
[37] Toba, H., et. al., Electron. Lett., vol. 25, No. 9, (1989) 574.
[38] Nosu, K., and Iwashita, K., IEEE Lightwave Technol., vol. LT-6, No.5, (1988) 686.
[39] Shibata, N., et. al., IEEE J. Quantum Electron. vol. QE-23, (1987) 1205.
[40] Azuma, Y., et. al., IEICEJ, National Conference, Sept. B-435 (1988). Shibata, N., et. al., GLOBCOME'89 (1989).
[41] Shibata, N., et. al., IOOC'89, Kobe, 18C1-3 (1989).
[42] Toba, H., et. al., IEEE J.Selected Areas Commun., vol. SAC-4, (1986) 686.
[43] Shibutani, M., et. al., OFC'89, THC2 (1989).
[44] Kitamura, M., et. al., IOOC'89, Kobe, 19A2-2 (1989).
[45] Murata, S., et. al., Electron. Lett. vol. 24, No. 9, (1988) 577.
[46] Naito, T., et. al., OFC,89 ThC-3, 141, (1989).
[47] Chikama, T., et. al., IEICEJ, National Conference B-408 Sept. (1988). Kuwahara, H., et. al., IEEE Lightwave Technol. to be Published (1989).
[48] Toba, H., et. al., Electron. Lett., vol.25, No. 14, (1989) 885.

MODULATION AND DEMODULATION TECHNIQUES IN COHERENT LIGHTWAVE COMMUNICATIONS

Hideo Kuwahara, Terumi Chikama, Shigeki Watanabe and Hiroshi Onaka

Fujitsu Laboratories Ltd.
1015, Kamikodanaka, Nakahara-ku, Kawasaki, 211, Japan

Among the various modulation and demodulation schemes, PSK/DPSK and CP-FSK schemes are the most attractive for long distance transmission. The advantages of these schemes are summarized. DPSK transmission is described exclusively, including its bit error rate characteristics at 1.2 Gb/s and 560 Mb/s and its degradation mechanism, long span transmission over 200km, polarization insensitive reception using polarization diversity scheme, laser phase noise canceling by carrier recovery synchronous detection and its extension to multi-value modulation. Multi-channel HDTV signal transmission and image rejection reception technique are also demonstrated for optical frequency division multiplexed transmission. New device technology to realize a narrow linewidth and flat FM response is shown for a large capacity CP-FSK scheme.

1. INTRODUCTION

Coherent lightwave transmission is promising for future communications networks. Its advantages include its 10 dB or greater receiver sensitivity over conventional intensity modulation and direct detection (IM/DD) systems, and its enhanced frequency selectivity in optical frequency division multiplexing (OFDM) systems. A number of experiments and several field trials have already been conducted [1,2].

Coherent lightwave communication's various modulation and demodulation formats are similar to those of traditional radio frequency communications. Each modulation and demodulation scheme has its advantages and disadvantages and the most suitable must be selected for a given application. It is also possible to use different modulation signals, such as FSK and PSK signals, in the same network simultaneously due to the independet nature of each carrier. This is also another advantage of coherent lightwave transmission.

Figure 1 compares the mod/demod schemes by receiver sensitivity. Each mod/demod scheme also has requirements especially for the linewidth of the laser. Homodyne are the most highly coherent schemes and have the most stringent requirements for the linewidth/bitrate ratio (about $5x10^{-4}$). ASK or FSK envelope/filter detection are the lowest coherent shemes. They require linewidths of only 0.1- 1 times the bitrate, which is important for realization of low cost systems for subscriber systems [3]. In these systems, however, the receiver sensitivity improvement from IM/DD is not so large, and tuning capability in coherent lightwave transmission is important. Synchronous heterodyne detection and delayed demodulation schemes such as DPSK and continuous phase (CP-) FSK are intermediate coherent schemes, requiring a modest linewidth (about $3x10^{-3}$) while realizing a high sensitivity. They are most attractive for trunk line applications [4,5].

In this paper we will describe the intermediate coherent schemes. First the advantages of PSK/DPSK and CP-FSK schemes are summarized and several modulation and demodulation schemes are compared. We investigated the transmission characteristics of 1.2Gb/s and 560 Mb/s DPSK systems and the DPSK degradation mechanism, polarization insensitive reception using polarization diversity, laser phase noise cancellation by carrier recovery synchronous detection, optical frequency division multiplexing, and image rejection reception. We will also describe new device technologies for realizing a large capacity CP-FSK scheme.

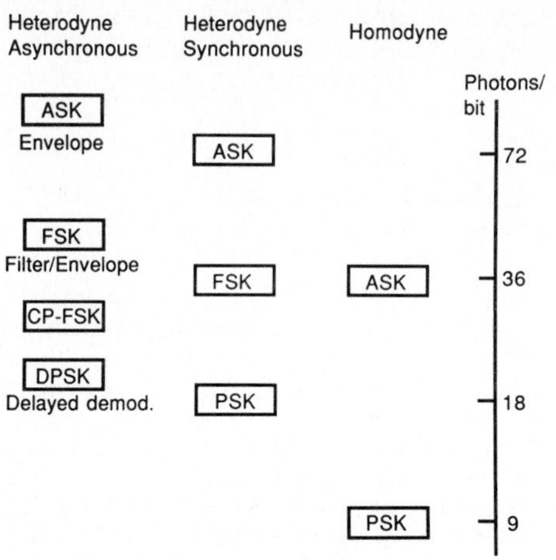

Fig. 1 Comparison of mod/demod schemes in terms of receiver sensitivity.

2. ADVANTAGES OF PSK/DPSK AND CP-FSK

Among modulation schemes, heterodyne PSK/DPSK and CP-FSK have been investigated most intensively. Figure 2 summarizes the advantages of PSK/DPSK and CP-FSK modulation. The first is the high receiver sensitivity. As shown in Fig.1, PSK synchronous demodulation has a receiver sensitivity of about 18 photons/bit. The receiver sensitivity difference in PSK synchronous detection and DPSK delayed demodulation is about 0.5 dB, and is caused by the noise increase in the multiplying process in the demodulation. CP-FSK receiver sensitivity is theoretically 1.5 to 2 dB lower than DPSK because its modulation and demodulation processes have band limiting characteristics.

The second advantage is compatibility with the optical amplifier. When we attempt transmission over ultra-long spans, for example, over 1000 km of single mode fiber, optical amplification is promising. The constant envelope characteristics of PSK/DPSK and CP-FSK are important when using an optical amplifier, especially a semiconductor laser amplifier, because the carrier density in the amplifier is affected by the power envelope. In ASK the fluctuation in carrier density causes the pattern effect.

The third advantage is the increase in the nonlinear threshold level. The most harmful effect in the long distance trunk line system would be stimulated Brillouin scattering because the threshold level of SBS is several milliwatts in single mode fiber. This level is mainly determined by the power spectrum density in a Brillouin bandwidth of several tens MHz. Therefore, ASK, where half of the transmitting power is concentrated in the carrier, has power density disadvantages. PSK/DPSK and CP-FSK with small modulation indexes have a rather flat power spectrum, and there is no power concentration. This will become important when the power of the transmitter laser increases to more than 10 dBm.

For PSK/DPSK an important advantage is the relaxed requirements for the transmitter laser diode. PSK/DPSK are severely affected by the phase noise of the light sources [6]. However, recent active development in narrow linewidth lasers and phase noise canceling receivers will overcome this problem, especially in high bit rate systems. In PSK/DPSK, external modulation techniques are commonly used, and this allows the use of lasers with narrow linewidth characteristics using various device structures and module configurations.

Another advantage of PSK/DPSK scheme is it is less affected by chromatic dispersion than other schemes [7]. It is an important point especially in long distance transmission.

Advantages of PSK/DPSK and CP-FSK schemes

* High receiver sensitivity
* Constant envelope - Compatibility with optical amplifier
  (ASK - pattern effect)
* High threshold of nonlinear effect

PSK/DPSK

Easier requirements for transmitter laser
Less affected by fiber dispersion
Multi-value modulation
Easy carrier recovery
(Useful for synchronous demod. and AFC)

CP-FSK

Direct modulation of semiconductor laser
Compact spectrum

Fig. 2 Advantages of PSK/DPSK and CP-FSK schemes.

The next advantage of PSK/DPSK is its easier application to multi-value modulation. Multi-value modulation will be important in coherent lightwave transmission to effectively utilize the receiver bandwidth or to increase the transmission capacity. PSK requires the simplest modulation and demodulation techniques for transmitting multi-value signals.

Another additional advantage of the PSK/DPSK signal is the ease of recovery of the carrier component from the received signal. An ordinary binary PSK signal is modulated in the 0 and π phase. The modulation components can be cancelled by doubling the IF signal. This capability is important in synchronous demodulation and also in extracting stable IF signal for the AFC loop.

CP-FSK is attractive because modulation can be done by simple direct modulation of the semiconductor laser current. Direct modulation has great advantages in practical implementation such as drive power for the modulator and the insertion loss of the external modulator. CP-FSK can be powerful, provided that lasers with a wide and flat FM response as well as a narrow linewidth can be realized [8]. CP-FSK has a compact spectrum, which is also advantageous in multi-channel OFDM systems. As for the dispersion caused by the frequency modulation, equalization compensation in IF region [4] is effective in CP-FSK.

3. PSK MODULATION/DEMODULATION SCHEMES

3.1. PSK Modulation

The Z-cut Ti:$LiNbO_3$ straight-line phase modulator is effectively used [9] for external phase modulation. The refractive index change of the waveguide modulates the phase of the light passing through the waveguide. Figure 3 shows the configuration of this type of travelling wave modulator for wide band modulation. The modulator in our experiment has about 2 dB of insertion loss from the input polarization maintaining fiber to the conventional single mode output fiber. This configuration would benefit from the development of a modulator operating with a smaller drive voltage.

The Mach-Zehnder modulator used as an intensity modulator in multi-gigabit optical transmission systems, can also be used as a phase modulator. Figure 4 (a) shows the waveguide structure of the Mach-Zehnder modulator. It consists of two $LiNbO_3$ waveguides with Y-branches. Figure 4 (b) shows the principle of operation as a phase modulator. By applying a voltage difference between the two electrodes, the optical phase in waveguides I and II rotate in opposite directions. If the modulation efficiency of the two waveguides are equal, the phase of the combined optical signal changes from 0 to π abruptly. Optical output power at both the 0-phase and the π-phase are the same. Moreover, this suppresses the

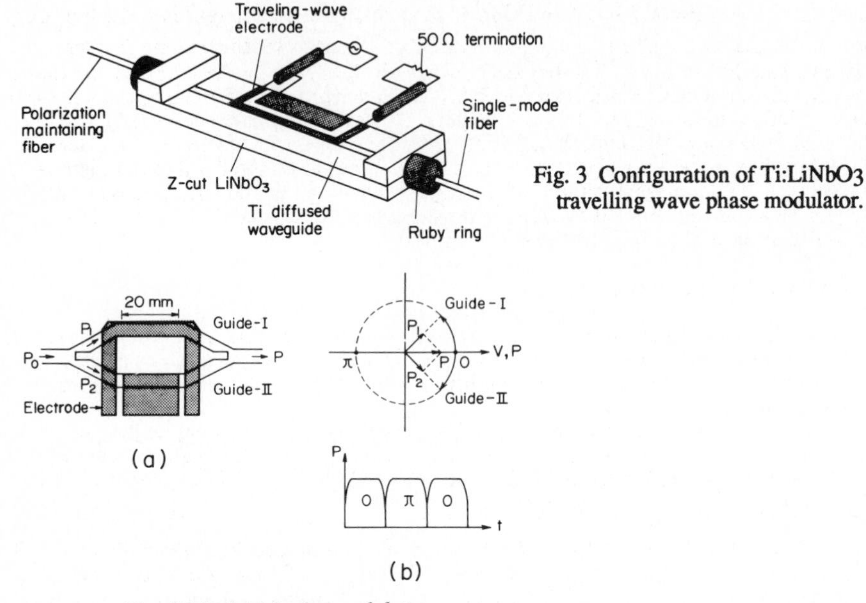

Fig. 3 Configuration of Ti:LiNbO3 travelling wave phase modulator.

Fig. 4 Mach-Zehnder modulator
(a) Waveguide structure.
(b) Operation principle of the phase modulator.

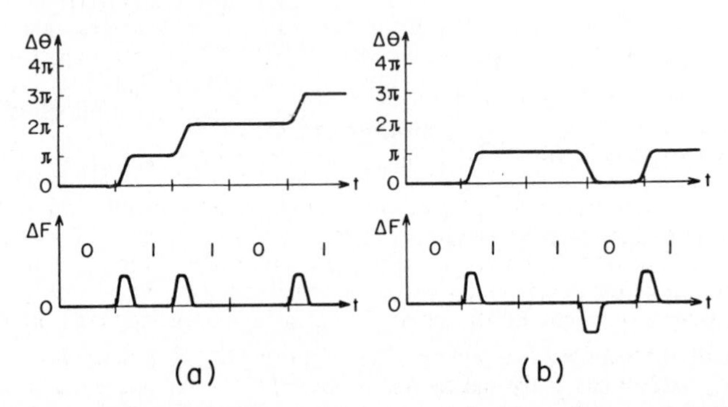

Fig. 5 Phase shift and modulation current in direct phase modulation.
(a) Uni-polar pulse modulation, (b) Bi-polar pulse modulation.

influence of imperfect phase modulation at the rising and falling edge of the pulses, where output power is reduced. Using this type of modulation the power penalty from the chromatic dispersion of the fiber can be expected to be small, because power is small at rising and falling edge where the chirping is ocurring. In section 4.1 we will discuss 1.2 Gb/s DPSK transmission using a Mach-Zehnder modulator.

Direct PSK modulation, as well as CP-FSK modulation, can also be achieved by current modulation of a laser diode [10,11]. Figures 5 (a) and 5 (b) show the relation between the modulation current and resultant phase shift. In the case of (a), a return-to-zero current pulse corresponding to the mark bit is applied to the laser, and the optical frequency is modulated. The pulse width and the height is adjusted so that the integration of the frequency shift pulse, that is, the phase shift becomes π. This modulation scheme requires a laser with a flat and wide FM response. For the case in which the duty cycle of the RZ pulse becomes 1, this modulation scheme is equivalent to MSK. Demodulation is performed by one-bit delay demodulation. In the case of (b), positive pulses corresponding to the rising edges of mark bits and negative pulses corresponding to the falling edges are applied to the laser. This modulation current exactly simulates the derivative of the phase modulated optical signal with finite rise and fall times. This scheme also requires a laser with a flat FM response, while the requirements in the low frequency region are not so stringent because of the bipolar modulation current.

3.2. PSK Demodulation

In this section, we discuss two typical demodulation techniques and their features; delay-line demodulation and carrier recovery demodulation. Configurations of the delay-line demodulator and the carrier recovery synchronous PSK demodulator are shown in Figs. 6 (a) and 6 (b), respectively.

In the delay-line demodulator, the IF signal is divided into two arms, and signal in one arm is delayed one bit period T. The delayed and undelayed signals are then multiplied by a mixer which acts as a phase detector. The demodulated signal g(t) is given as

$$g(t) = \cos\{2\pi f_{if} T + \theta(t) - \theta(t-T) + \Delta\phi(t)\} \tag{1}$$

where f_{if} denotes the IF center frequency, and θ and $\Delta\phi$ are the modulated phase and the phase noise, respectively. The phase difference $\theta(t) - \theta(t-T)$ is detected. A delay-line demodulator is easy to implement because it does not need a PLL at the receiver. In addition, delay-line demodulator is less sensitive to the laser phase noise than synchronous demodulation because the laser phase noise accumulates in one period only. Assuming that $\Delta\phi$ has a zero-mean Gaussian probability distribution, the variance of the phase deviation $\Delta\sigma_D{}^2$ can be expressed as

$$\Delta\sigma_D{}^2 = 2\pi\, \Delta\nu_{if}\, T \tag{2}$$

where $\Delta\nu_{if}$ denotes the linewidth of the beat spectrum. Delay-line demodulation is more advantageous in high bit rate systems having a small T.

In the synchronous demodulators using carrier recovery (b), the recovered IF carrier is multiplied with the IF signal to demodulate the data signal. For PSK modulation, synchronous demodulation using PLLs provide the best receiver sensitivity. This requires, however, a very narrow laser linewidth and is not practical at present. We proposed a novel PSK synchronous demodulation scheme using carrier recovery at the IF stage, in which the differences in delay time beween the signal and carrier recovery routes are kept short [12]. In a carrier recovery route, first the IF signal is doubled by a frequency doubler (FD) to cancel the (0, π) phase modulated component [13]. The resultant wave of twice the IF center frequency $2f_{if}$ is frequency divided by the following frequency halver (FH) to recover the reference carrier wave. The recovered carrier and signal are multiplied by a mixer, and the

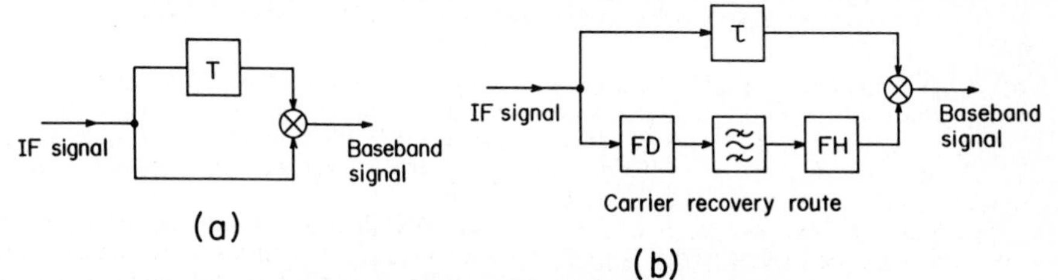

Fig. 6 (a) Configuration of the delay line demodulator.
(b) Carrier recovery synchronous demodulator.

demodulator output h(t), is given as

$$h(t) = \cos\{2\pi f_{if}\Delta\tau - \theta(t-\tau) + \Delta\psi(t)\} \quad (3)$$

where $\Delta\tau$ denotes the delay time difference between the IF signal route and the carrier recovery route, and $\Delta\psi$ the phase noise. The variance of the phase deviation for this case, σ_{CR}^2, can be expressed as

$$\sigma_{CR}^2 = 2\pi\, \nu_{if}\, \Delta\tau\,. \quad (4)$$

According to Eq. (4), the phase noise can be decreased if $\Delta\tau$ is small. For the special case of $\Delta\tau = 0$, it is possible to completely suppress the influence of the phase noise. Moreover, from Eq. (3), it is not necessary to set f_{if} to a specific value, and the power penalty caused by the IF center deviation can also be suppressed, provided that $\Delta\tau$ is small. In conventional DPSK, f_{if} must be set to a multiple of half the bit rate (see Eq.(1)), and the reference signal is delayed for a 1-bit duration T. A comparison of the phase noise of DPSK and CR-PSK (Eqs. (2) and (4)) shows that a great improvement in insensitivity to phase noise can be expected for CR-PSK. Experimental results are shown in section 5.

4 DPSK HETERODYNE TRANSMISSION EXPERIMENTS

4.1. DPSK Transmission Experiment

The performances of 1.2Gb/s and 560Mb/s DPSK heterodyne transmission [8] were intensively investigated to clarify the performance limitations of the receiver sensitivity to system parameters, such as laser phase noise and modulation depth. Figure 7 shows the configuration of the experiment. We developed and investigated the performance of two key optical components, that is, an external fiber cavity (EFC) DFB-LD module and a dual detector balanced receiver (DBOR).

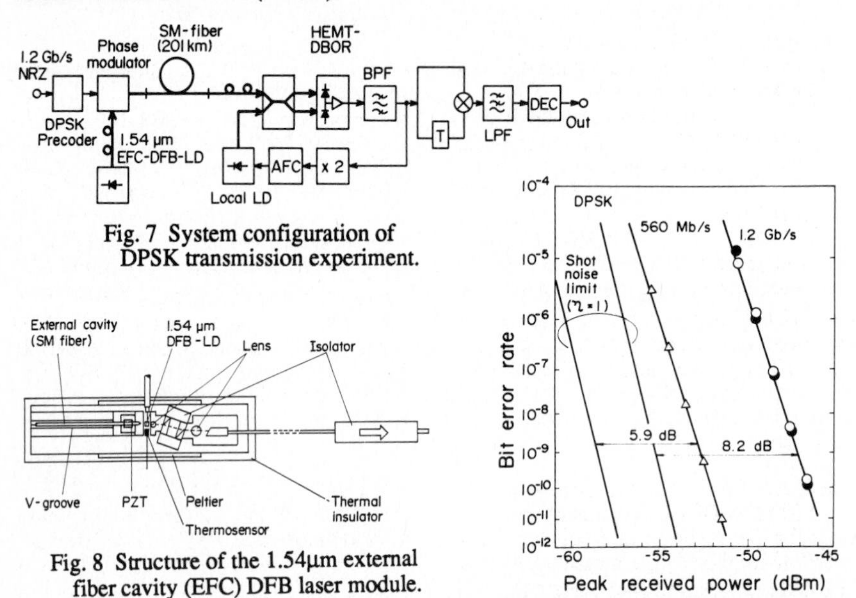

Fig. 7 System configuration of DPSK transmission experiment.

Fig. 8 Structure of the 1.54μm external fiber cavity (EFC) DFB laser module.

Fig. 9 Bit error rate characteristics in 1.2 Gb/s and 560 Mb/s DPSK transmission.

A. Key Components

PSK/DPSK schemes require light sources with linewidths of less than about 0.1 % of the bitrate. An external cavity configuration is one of the most practical techniques because it can use well-established DFB-LDs. Compact EFC DFB-LD modules have been developed and used as signal and local oscillator sources, and exhibit stable operation with a narrow linewidth [14]. Figure 8 is a schematic diagram of the module which contains an external fiber cavity and optical isolators. A piece of fiber several cm long is weakly coupled to the cleaved facet of a 1.54-μm InGaAsP FBH-DFB laser chip. The far end of the fiber is reflectively coated 80 %, and the near end is processed into a tapered, hemispherical end to reduce reflection. The phase of the feedback light from the cavity is controlled by a piezoelectric translator (PZT). The temperature of the module was stabilized within 0.01 deg. by a Peltier element and a PID controller. The total isolation of two-stage isolators was more than 70 dB, sufficient to eliminate the influence of the reflected light. The linewidths of the DFB-LDs are effectively reduced from original widths of about 5 MHz to less than several hundred kHz by loading the external fiber cavity.

A dual-detector balanced optical receiver (DBOR), consisting of two photodiodes in a balanced mixer configuration, is promising as a coherent receiver because: (1) efficient use of local laser power, (2) excess intensity noise suppression of the local laser, (3) immunity to DC drift in the homodyne receiver, and (4) suppression of baseband modulation and adjacent channels interference in the OFDM. We developed and used a monolithic GaInAs twin-PIN photodiode with a 20μm diameter photosensitive area, followed by a high electron mobility transistor (HEMT) used as a front-end high-impedance amplifier. To make a balanced receiver operating in GHz region, the signal delay time, the quantum efficiency, and the capacitance of the two photodiodes must be well balanced. The monolithic integration of two photodiodes is a powerful technique to realize a good balance. Developed twin-PIN shows wideband response up to about 16 GHz even at a high power operation of several mA photo current [15]. Microlenses for optical coupling is also integrated monolithically.

B. 1.2Gb/s DPSK Transmission

The signal light was modulated by the straight line travelling-wave type Ti:$LiNbO_3$ phase modulator with a differentially coded 1.2 Gb/s NRZ (PN:2^{15}-1) signal. The modulated optical signal was transmitted through 201.4 km of conventional single mode fiber (0.23 dB/km) and detected by the DBOR. The polarization state of the transmitted signal light and local laser light was adjusted manually using a fiber polarization controller. The IF signal was amplified, filtered by a band pass filter (1.1 - 3.8 GHz), and demodulated by the following delay-line demodulator. The IF center frequency was stabilized by an AFC to within 3 MHz. The demodulated baseband signal was filtered, then fed to a Si-bipolar LSI decision circuit .

The experimental bit error rate (BER) is shown in Fig. 9. A receiver sensitivity of -46.9 dBm (132 photons/bit) at a BER of 10^{-9} was achieved over 201.4 km transmission. No degradation due to fiber transmission was observed. The degradation from the shot noise limit (-55.1 dBm: $\eta = 1$, B = 1.2 GHz) is estimated to be 8.2 dB, and can be attributed to the quantum efficiency of the photodetector (2.6 dB), the thermal noise or insufficient local laser power (0.9 dB), filter and amplifier imperfections (4.2 dB), and imbalance in the DBOR (0.3 dB). The power penalty due to the laser phase noise was less than 0.2 dB, because the beat spectrum linewidth in the experiment was less than 700 kHz. A span loss margin of 48.3 dB was obtained.

C. 560Mb/s DPSK Transmission

560 Mb/s DPSK experiment was also performed using almost the same configuration except for the intermediate frequency (2.24 GHz), an optimized BPF (1.6 - 2.8 GHz), and the LPF (360 MHz). The measured BER is also shown in Fig. 9. A receiver sensitivity of -52.5 dBm (78 photons/bit) for a BER of 10^{-9} was achieved. Degradation from the shot noise limit (-58.4 dBm) is estimated to be 5.9 dB. This power penalty can be attributed to the quantum efficiency of the photodetectors (1.6 dB), the thermal noise (0.7 dB), the laser

phase noise (0.4 dB), and imperfections in the amplifiers and filters (2.9 dB). Compared to the 1.2 Gb/s system, the electrical imperfection reduction reflects the smaller required bandwidth. The power penalty from the front-end is reduced by replacing the DBOR with one of higher quantum efficiency and the same thermal noise.

D. Impact of Phase Noise and Imperfect Modulation

We experimentally investigated the BER dependence on the normalized IF linewidth at 1.2 Gb/s and compared it with the theory [16]. The IF beat linewidth is changed by controlling the bias current and temperature of the EFC-LD module. Figure 10 (a) shows the BER degradation. Results coincide well with the theoretical values. The required beat linewidth was 1.2 MHz, and 600 kHz for the laser modules. Our EFC-LD module satisfies these requirements.

We also evaluated the degradation in sensitivity due to the deviation from 180 degree phase shift modulation caused by drive voltage limitation and device aging. This imperfection leads to intersymbol interference and deviation from the optimum threshold. We measured BERs at 1.2 Gb/s by changing the drive voltage. The measured power penalties, shown in Fig. 10 (b), agree well with theoretical prediction [17]. The allowable deviation for a 0.5 dB power penalty is 20 degrees, which is not so severe in practical implementation.

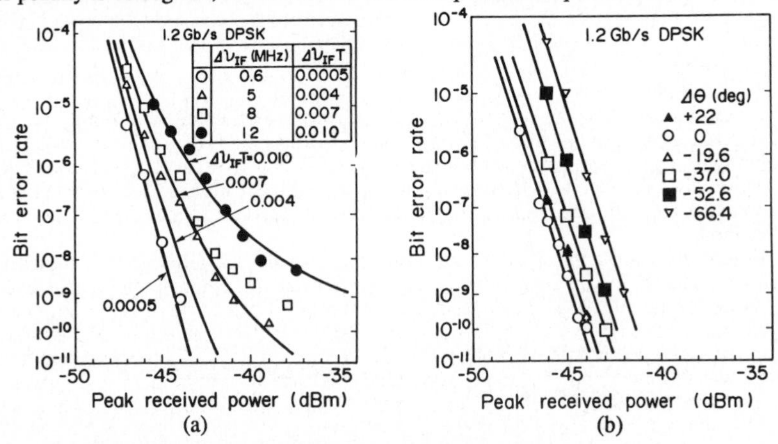

Fig.10 Measured and theoretical bit error rate dependence on (a) the laser phase noise and (b) deviation of phase modulation from 180 degrees .

E. 1.2 Gb/s DPSK Transmission Using Mach-Zehnder Modulator

As described in section 3.1, a Mach-Zehnder modulator was used as a phase modulator and DPSK transmission characteristics investigated. The experimental setup was almost the same as shown in Fig.7, except for the Mach-Zehnder modulator. The MZ modulator has a insertion loss of 1.9 dB, a bandwidth of 7 GHz and a halfwave voltage of 10.2V. Figure 11 shows the measured BER. The receiver sensitivity was -40.8 dBm (closed circles). The open circles in Fig.11 show the case for which a straight line phase modulator was used. The 4.4dB difference in sensitivity was mainly due to the imperfect phase modulation caused by an asymmetric phase change in the two waveguides of the MZ phase modulator. Our MZ modulator has an asymmetric electrode pattern, and investigation of the applied field in the waveguide shows difference of three times the modulation efficiency. The phase change is, therefore, estimated to be 90 degrees, which causes a power penalty of about 4.5 dB. This value is anticipated from the investigations described above. By preparing a MZ modulator with a more symmetric structure, we can expect a higher performance.

4.2 Polarization Diversity Reception

In a coherent lightwave transmission system, overcoming signal fading caused by

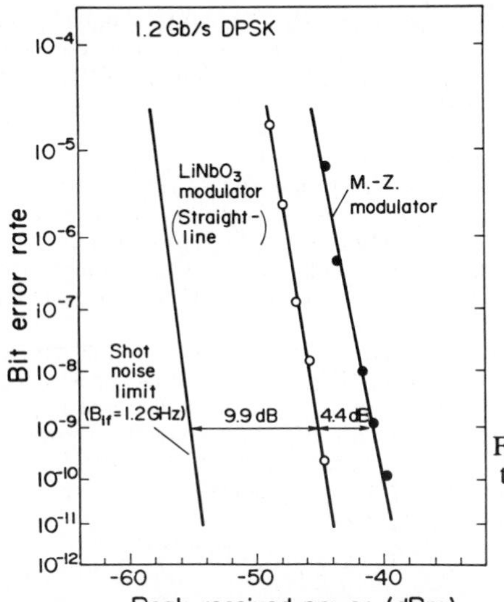

Fig.11 Measured BERs in 1.2 Gb/s DPSK transmission using Mach-Zehnder modulator.

polarization fluctuation in the transmitting fiber is crucial for a practical system. Among the several possible solutions, polarization diversity reception [18], is the most attractive, considering its high-speed response and feasibility for OFDM. We developed a dual-balanced polarization-diversity receiver (DPR) using two balanced receivers [19] and demonstrated 1.2 Gb/s optical DPSK transmission [20]. The IF was monitored and stabilized using a novel AFC method.

Figure 12 shows the system configuration for the 1.2 Gb/s DPSK transmission experiment. The DPR is enclosed by dashed lines. The received signal and local laser output were divided separately by polarization beam splitters (PBSs) into two orthogonally polarized components. The components of the signal and local laser with the same polarization were then mixed in the polarization-maintaining fiber couplers (PMCs). The output pair from the couplers was detected by a pair of DBORs. Each PBS has an extinction ratio of greater than 20 dB and an insertion loss of about 2 dB. Separately demodulated signals were summed and passed though an LPF to regenerate the baseband data signal.

In case of baseband recombining it is not easy to extract the IF signal stably for AFC. If we simply combine the IF signals extracted from the two branches of the diversity receiver, the combined signal for AFC disappears when the extracted two IF signals are equal in amplitude and of opposite phase. To overcome this problem, two IF signals were extracted separately, then one of these signals was modulated at a low frequency (20 kHz) before combining. With this scheme the combined IF signal does not disappear even when both IF signals have the phase difference π, because the phase of the combined signal changes periodically with the frequency of the extra modulation, which is sufficiently higher than the

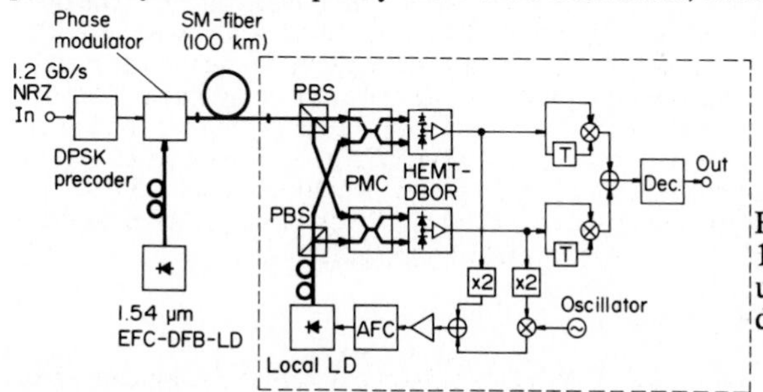

Fig.12 Configuration of 1.2Gb/s DPSK transmission using dual-balanced polarization diversity receiver.

response of AFC loop. With this scheme the IF center frequency deviation was stabilized within 6 MHz.

The DPSK transmission was performed at 1.2Gb/s using 100km of conventional fiber. BERs were measured for polarization states corresponding to: (i) all signal light incident on one balanced receiver, (ii) all incident on the other receiver, and (iii) equally divided between them. The best receiver sensitivity was -44.2 dBm at 10^{-9} BER for (i). Receiver sensitivity variation by changing polarization was measured to 1.4 dB, which is due to a sensitivity imbalance between the two branches (0.7 dB) and imperfect multiplexing caused by deviation from square-law characteristics (0.7 dB). The difference between the best measurement -46.9 dBm, obtained without polarization diversity as shown in Fig.9, was 2.7 dB. This can be atrributed to the thermal noise (1.1 dB), signal recombination at the baseband stage (0.4 dB), and other factors (1.2 dB).

5. SYNCHRONOUS DETECTION USING CARRIER RECOVERY

5.1. Transmission Experiment

Figure 13 shows the system configuration for the 560 Mb/s carrier-recovered PSK (CR-PSK) transmission experiment [21]. The CR-PSK demodulator described in section 3.2 is enclosed by dashed lines. The output from the EFC-DFB-LD modules was modulated by a straight line travelling-wave type $Ti:LiNbO_3$ phase modulator with a 560 Mb/s NRZ (PN: 2^{15}-1) signal. The signal was transmitted through a 232 km of conventional single mode fiber (0.21 dB/km) and detected by the DBOR with a HEMT amplifier. The local laser output power was +6.6 dBm and $\Delta\nu_{if}$ was about 2 MHz. The f_{if} was set to 2.24 GHz.

Figure 14 shows the IF beat spectrum without phase modulation and the carrier spectrum recovered from the phase modulated signal. The two spectrums are frequency shifted by about 15 MHz for easier observation. A stable recovered carrier of nearly the same spectral linewidth as that of the IF beat spectrum was obtained.

The BER characteristics are shown in Fig. 15. The receiver sensitivity at a BER of 10^{-9} was -51.6 dBm or 96 photons/bit for CR-PSK, and -50.7 dBm for DPSK using the same front-end and local laser. No sensitivity degradation for CR-PSK was observed after transmission through 232 km of single mode fiber. The 0.9 dB difference in sensitivity was mainly due to the suppression of the laser phase noise in CR-PSK and the difference in the demodulated waveform between the two schemes. The demodulated eye pattern for the CR-PSK receiver, shown in Fig. 16, has good symmetry. The power penalty from the shot noise limit for CR-PSK is 6.8 dB, which is attributed to the quantum efficiency of the photodetector (1.6 dB), the thermal noise (0.7 dB), filter and amplifier imperfections (2.9 dB), the decrease in the CNR in the carrier recovery route (0.8 dB), and the phase noise (0.4 dB). The abrupt increase of the BER at an input signal power below -52 dBm was caused by phase instabilities of the recovered carrier due to the decreased CNR, and was mainly affected by the frequency halver used in our experiment.

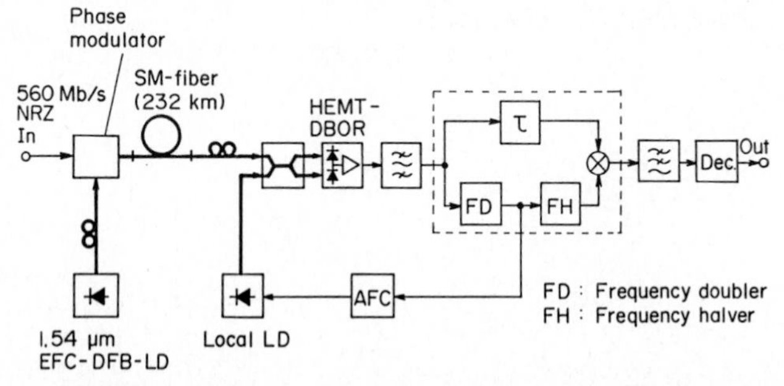

Fig.13 Configuration of carrier recovery PSK transmission.

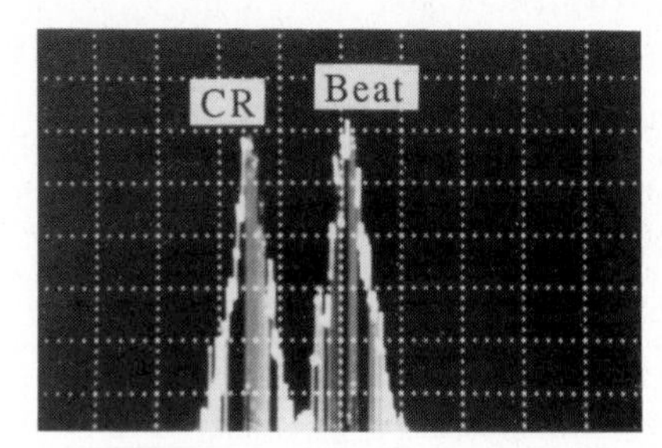

Fig.14 IF beat spectrum without phase modulation and recovered carrier spectrum from phase modulated signal.

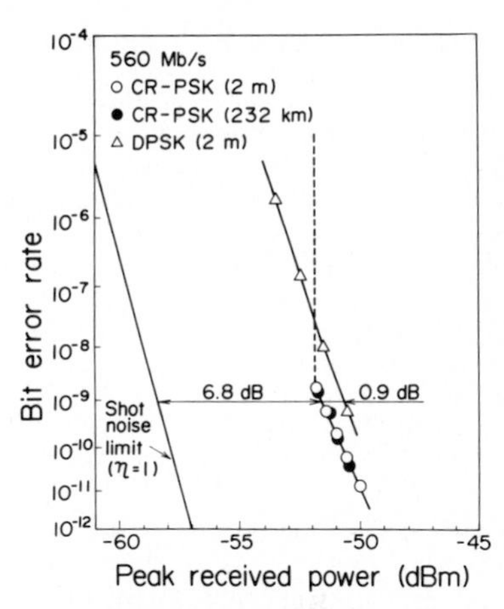

Fig.15 Measured BERs in 560Mb/s CR-PSK transmission.

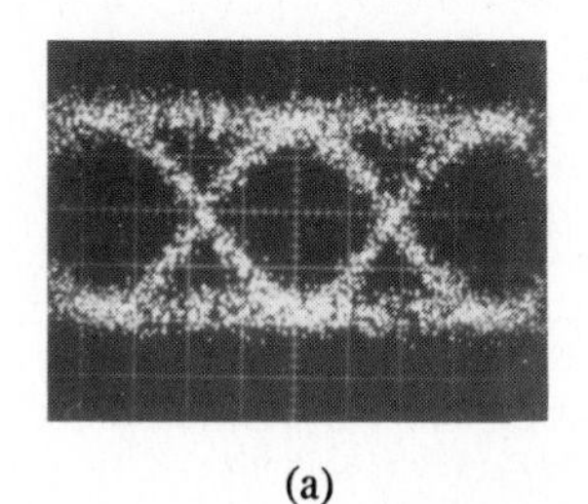

(a)

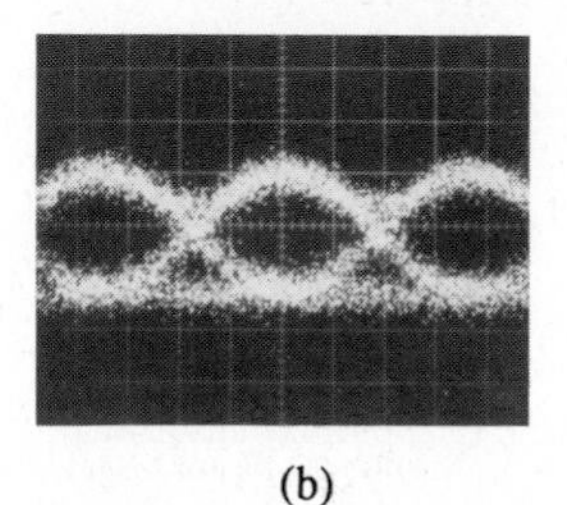

(b)

Fig.16 Demodulated waveforms in 560 Mb/s (a) CR-PSK, and (b) DPSK.

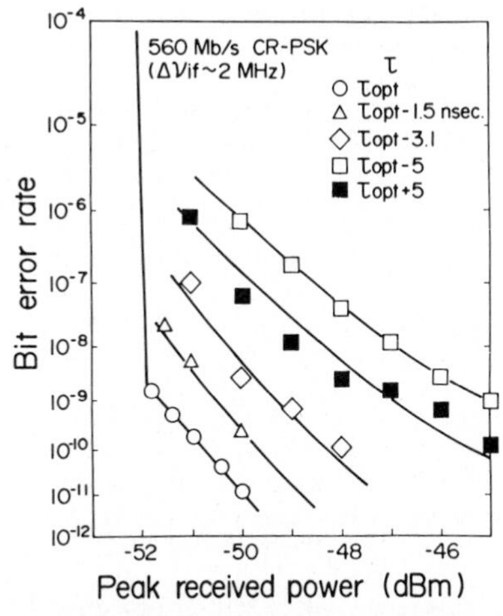

Fig.17 Measured BERs in 560 Mb/s CR-PSK by varying the delay time.

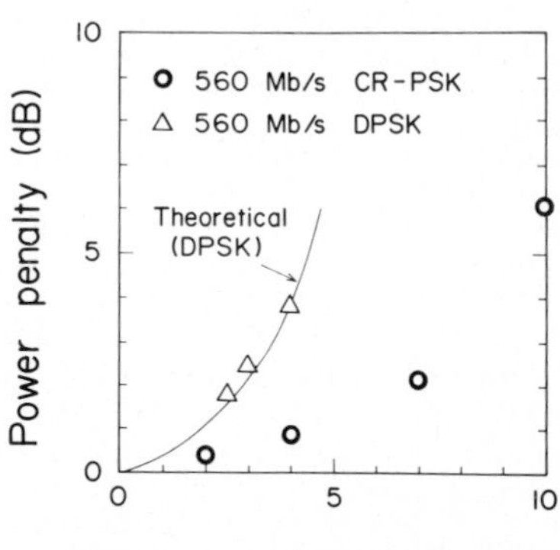

Fig.18 Power penalty dependence on beat linewidth in CR-PSK and DPSK.

To confirm the phase noise suppression effect in the CR-PSK scheme mentioned in section 3.2, we measured the BER for five values of τ by varying the length of the IF signal route (Fig.17). The power penalty increases as τ deviates from the optimum point, where the phase noise is minimum. The power penalty was also measured while varying the linewidth of the beat spectrum, and shown in Fig. 18. The power penalty caused by the laser phase noise in CR-PSK was suppressed to less than 1/3 of that of DPSK, and we can use lasers with more than twice the linewidth in CR-PSK systems.

5.2. Application to Multi-value PSK Demodulator

Figure 19 shows the demodulator for a quarternary PSK scheme using carrier recovery. The signal and recovered carrier are divided and multiplied in each branch having a phase difference of 90 degrees. Carrier recovery can similarly be applied to multi-value transmission. Phase noise suppression mentioned in binary PSK also takes effect by controlling the differences of delay time in the two demodulators τ_1, τ_2 between each IF signal and the carrier recovery route.

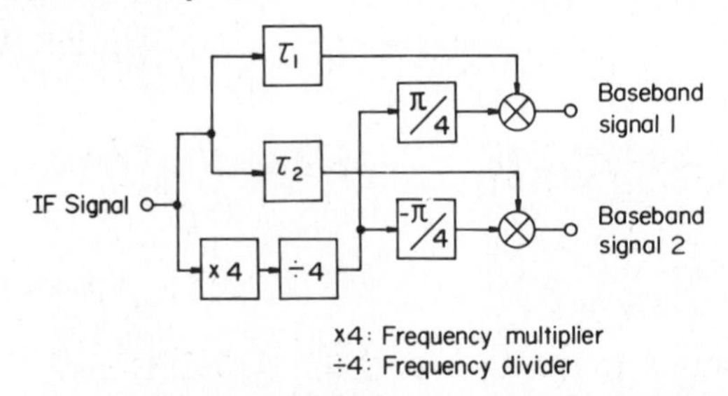

Fig.19 Demodulator configuration of quarternary PSK using carrier recovery synchronous demodulation.

6. MULTI-CHANNEL TRANSMISSION

One major advantage of coherent lightwave transmission is its greatly enhanced frequency selectivity. By using OFDM, it can be a promising system for future broadband ISDN service applications. We investigated DPSK multi-channel transmission and demonstrated high-definition TV (HDTV) signal transmission. An image rejection reception technique enabling extremely dense channel spacing was also investigated.

6.1. Multi-channel HDTV Signal Transmission Experiment

Using above mentioned technology, we developed transmitter and receiver equipment and demonstrated 2-channel HDTV signal transmission by OFDM. Figure 20 shows the configuration of our demonstration system. On the transmitter side, four channel transmitters were prepared using EFC DFB-LD modules. Two HDTV analog signals from a video tape recorder (VTR) and TV-camera, which have a bandwidth of about 30 MHz, are converted directly to digital signals of 560 Mb/s and fed to two of these four channels. The optical signals with frequencies f_1 and f_2, are modulated by 560 Mb/s binary DPSK signals. They are mixed in a 4x4 star coupler and transmitted through 50 km conventional single mode fiber. The signals were received by a heterodyne polarization diversity receiver which selects channels by local laser tuning. The resultant error-free non-bandcompressed HDTV signals were clearly observed on a HDTV monitor.

Figure 21 is a photograph of our system. We demonstrated stable performance of this coherent HDTV transmission at the Fujitsu Technology'88 exhibition and the International Optoelectronics Exhibition'88, held in Tokyo in May 1988 and July 1988, respectively.

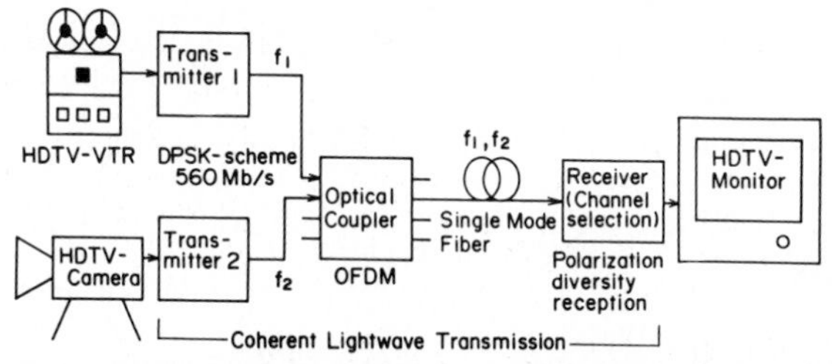

Fig.20 Configuration of multi-channel coherent lightwave HDTV transmission.

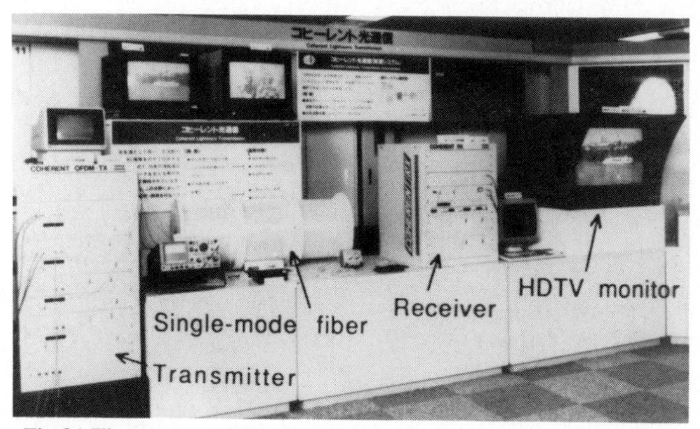

Fig.21 The system exhibited at Fujitsu Technology '88 (May 1988).

6.2. Image Rejection Reception

An image rejection receiver (IRR) is attractive because it enables extremely dense channel spacing in a coherent OFDM transmission system, which is important in the limited tuning range of the local laser. We developed an IRR and measured the crosstalk penalties in a 2-channel 560 Mb/s DPSK transmission system [22].

Figure 22 shows the configuration of the experiment using the IRR. The EFC-DFB-LD modules were used as 2-channel signal and local oscillator light sources. Two-channel signal lightwaves were phase modulated individually with a 560 Mb/s DPSK encoded NRZ pseudo

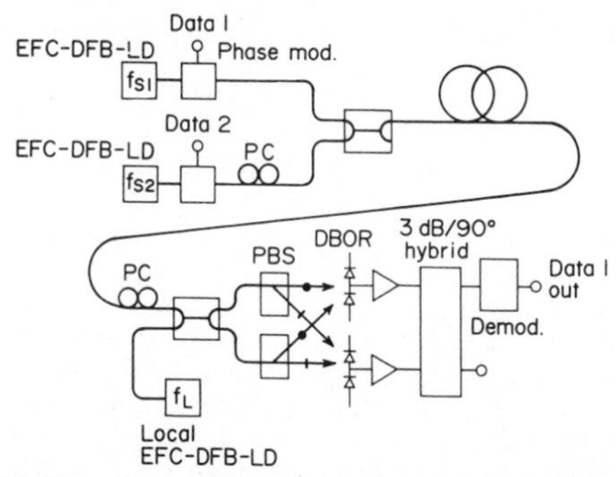

Fig.22 Experimental set-up of a two channel 560 Mb/s DPSK heterodyne transmission experiment using the IRR.

random (2^{15}-1) data stream. They were combined in a 3 dB fiber coupler, then were transmitted through conventional single mode fiber. The IRR consists of 90-deg. optical hybrid circuits, two DBORs, and a 90-deg. microwave hybrid coupler. In the 90-deg. optical hybrid circuits, the signal lightwaves (frequencies f_{S1} and f_{S2}) and the local lightwave (frequency f_L) are first combined in a 3 dB coupler, then divided by two PBSs. The states of polarization of the 2-channel signal lightwaves were adjusted for circular polarization, while that of the local lightwave was set for linear polarization. One output from the IRR provides the real IF band signal ($f_{S1} > f_L$), while the other output provides the image IF band signal ($f_{S2} < f_L$). The real IF signal is demodulated by a 1-bit delay demodulator.

The IRR developed provides image suppression of more than 18 dB over a 1.5-3.0 GHz IF region. Receiver sensitivities (at 10^{-9} BER) are -44.6 dBm without the image band signal and -43.5 dBm with the image band signal. The power of image band signal was equal to that of the real band signal and $f_{S1} - f_L = f_L - f_{S2}$. The power penalty was about 1.1 dB with the image band signal. This corresponds exactly to the case of 20 dB suppression of the IRR and a 14.9 dB CNR (at 10^{-9} BER), obtained considering the influence of the linewidth of the lasers.

Figure 23 shows crosstalk penalties as a function of the channel allocation. Theoretical calculations are shown by solid lines and experimantal data are shown by open circles (without IRR) and closed circles (with IRR). When the image band signal overlaps the real band signal in the IF domain, the signal can not be received without the IRR, while IRR can receive signals with small power penalty. If this penalty is permitted, the channel spacing can be decreased significantly (Fig.24). This IRR is thus advantageous for high density OFDM.

One problem in the IRRs is their severe sensitivity to the polarization fluctuation of the transmitted signals. A polarization-insensitive IRR is required in practice, and a diversity scheme and an active control scheme are proposed to overcome this problem [23].

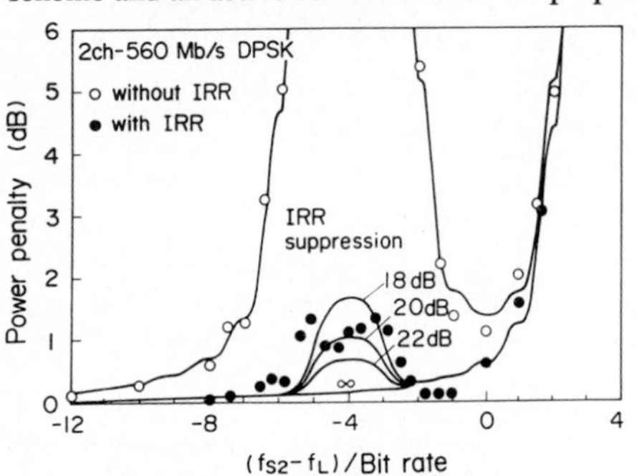

Fig.23 Crosstalk penalties as a function of channel allocation.

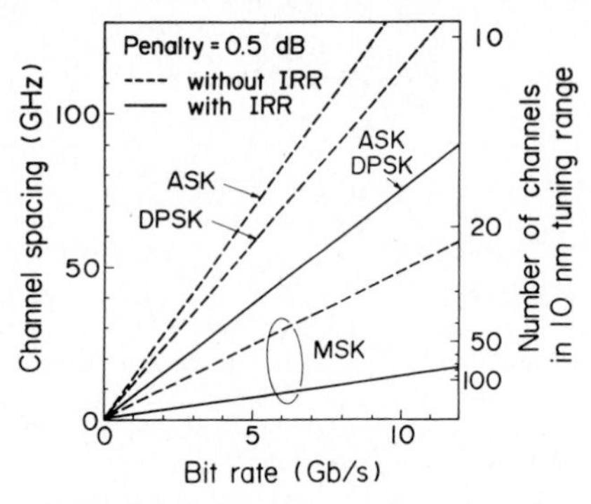

Fig.24 Channel spacing and number of channels accommodated in 10 nm tuning range.

7. DEVICE TECHNOLOGIES FOR CP-FSK MODULATION

CP-FSK is another attractive modulation scheme because of its simple modulation structure, relatively narrow spectrum, and high sensitivity. It is suited not only to trunk line systems but for subscriber systems. For demodulation of CP-FSK, delayed demodulation can be used as DPSK scheme. CP-FSK scheme, however, depends largely on the device characteristics of the semiconductor lasers. The conventional DFB laser has a dip in the FM response curve in the several MHz region, caused by the effect of thermal response and carrier fluctuation. This dip is harmful in FSK modulation to obtain a clear eye opening when modulated by pseudo random signals.

To overcome this problem, a monolithic multi-electrode DFB-LD having a narrow linewidth was developed [24]. Figure 25 is a schematic of the device structure as an example.

The $\lambda/4$ shift in the corrugation was at the center of the 1200-μm-long cavity. The p-side electrode was divided into three sections with lengths of 300, 600, and 300 μm. The center current Ic is applied to the center electrode and the side current Is is applied to the two side electrodes which are connected to each other. In the long cavity DFB laser the field intensity is strong at the center of the laser cavity, causing spatial hole burning. By varying the carrier density around the center of the cavity, the spatial hole burning can be enhanced or suppressed. Non-uniform injection along the cavity enables a change in the lasing condition.

The lasing wavelength and the spectral linewidth are shown in Fig. 26 as functions of the center current. The lasing wavelength was varied continuously towards longer wavelengths with an increase in the center current. A wide tuning range of 1.5 nm was obtained without mode jumping, while keeping the linewidth less than 1MHz. An output power of more than 20 mW was obtained. In Fig. 27 the FM response obtained for our laser is shown by the solid line, and the results for a conventional DFB laser is shown by the dashed line. The modulation current was superposed on Ic. The dip was not observed in our new laser and sufficiently flat response up to 10 GHz was obtained. This FM response was measured using a birefringent fiber interferometer. A NRZ pseudo random pulse was applied at bitrates of 155 Mb/s, 622 Mb/s, 2.5 Gb/s, and 10 Gb/s. At each bitrate a clear eye opening was observed, and the case for 10 Gb/s is shown in Fig. 28.

This type of long cavity multi-electrode DFB laser exhibits a narrow linewidth, a wide tuning range, and a flat frequency response. This laser is quite suitable for use in CP-FSK, as well as in wide frequency deviation asynchronous demodulation schemes. The impact of easy direct modulation will be great in the practical implementation of coherent lightwave technology.

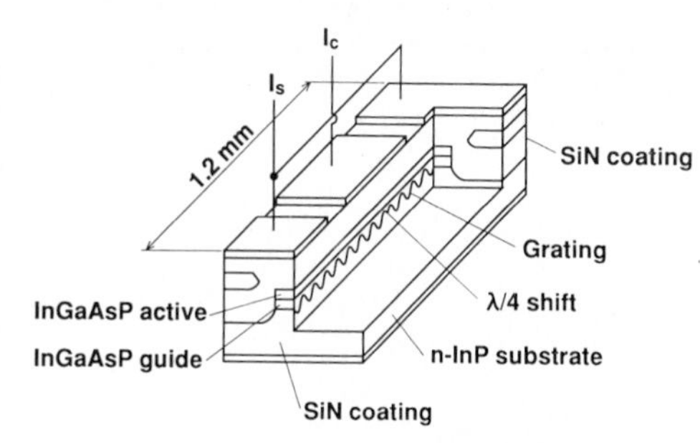

Fig.25 Cross section of $\lambda/4$ shifted multi-electrode DFB laser.

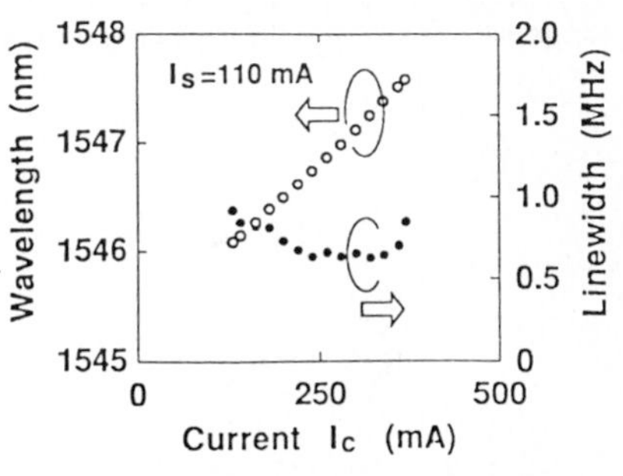

Fig.26 Wavelength tuning and linewidth characteristics.

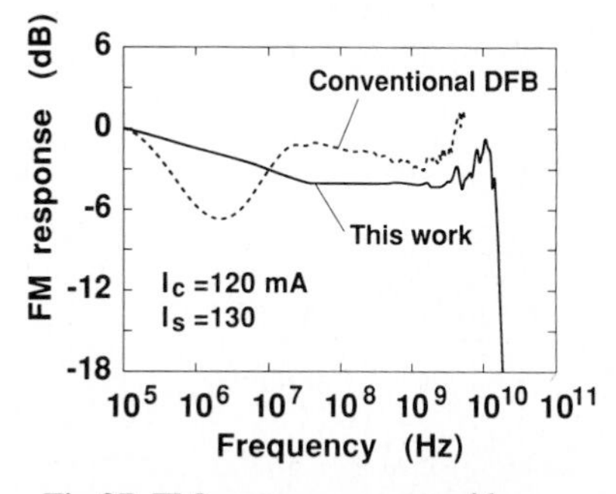

Fig.27 FM response measured by a birefringent fiber interferometer.

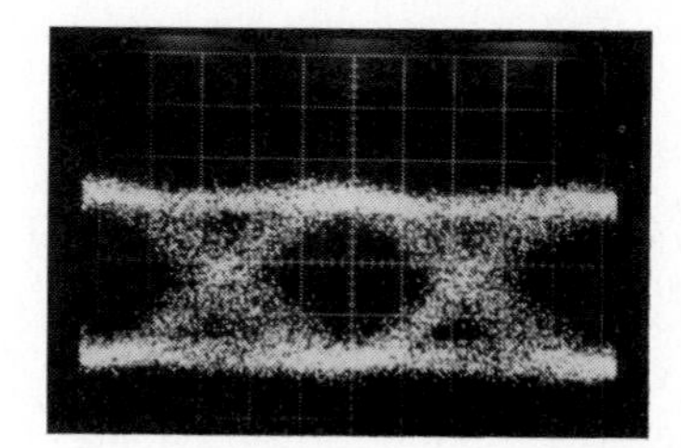

Fig.28 Eye diagram under 10 Gb/s NRZ pseudo random pulse FSK modulation.

8. CONCLUSIONS

Among the various modulation and demodulation schemes used in coherent lightwave transmission, PSK/DPSK and CP-FSK are the most practical, especially in trunk line systems, and have been investigated intensively. Our recent investigations of PSK/DPSK modulation/demodulation, transmission experiments, and recent device technologies for CP-FSK were described.

PSK/DPSK modulation using a straight line modulator, a Mach-Zehnder modulator, direct modulation, and demodulation using delayed demodulation and carrier recovery synchronous demodulation were compared, and the advantages and disadvantages discussed. DPSK transmission at 1.2 Gb/s and 560 Mb/s were performed with high sensitivity using narrow linewidth light sources and wideband balanced optical receivers. Polarization-insensitive transmission was performed by developing polarization diversity receivers. To overcome the phase noise affects we proposed carrier recovery PSK demodulation, which will also useful in multi-value coherent lightwave transmission. Multi-channel transmitters and tunable receivers were described, including demonstrations of 2-channel HDTV transmission. Image rejection reception was performed and its effectiveness to realize dense OFDM heterodyne transmission was described. Recent progress in semiconductor lasers having narrow linewidth and flat FM response was shown for FSK direct modulation. These are substantial steps toward the development of coherent lightwave transmission.

Acknowledgements

We are grateful to T. Naito for the transmission experiment, H. Miyata for providing the external fiber cavity laser modules, T. Kiyonaga and Y. Onoda for providing the DBOR, M. Seino for providing the phase modulator, and Y.Kotaki and S.Ogita for developing the multi-electrode lasers. We thank H. Takanashi, T. Nakagami and T. Touge for their guidance and encouragement.

References

[1] R.C.Steel et al., ECOC'88, Tech. Digest, PD61-64, Brighton, UK, 1988.
[2] S.Ryu et al., Electron. Lett., vol. 24, pp.399-400, 1988.
[3] T.Chikama et al, ECOC'87, pp.349-352, 1987.
[4] K.Iwashita et al., Electron. Lett., vol.24, pp.759-760, 1988.
[5] Y.Ichihashi et al., IOOC'89, Paper 20PDA-12, Kobe, Japan, 1989.
[6] T.Okoshi et al., J. Opt. Commun., vol. 2, pp.89-96, 1981.
[7] A.E.Elrefaie et al., Electron.Lett., vol.23, pp.756-758,1987.
[8] Y.Kotaki et al., IOOC'89, Paper 19A2-4, Kobe, Japan, 1989.
[9] T.Chikama et al., Electron., Lett., vol. 24, pp.636-637, 1988.
[10] M.Shirasaki et al., Electron Lett., vol.24, pp.486-488,1988.
[11] R.S.Vodhanel, CLEO'89, Baltimore, FG4, 1989.
[12] S.Watanabe et al., Electron. Lett., vol. 25, pp.588-590, 1989.
[13] K.Kikuchi et al., Electron. Lett., vol. 19, pp.417-418, 1983.
[14] H.Onaka et al., CLEO'88, Paper TuC4, 1988.
[15] M.Makiuchi et al., ECOC'89, ThA19-2, Gothenburg, Sweden, 1989.
[16] G.Nicholson, Electron. Lett., vol. 20, pp.1005-1006, 1984.
[17] N.M.Blachman, IEEE Trans Comm., vol. COM-29, pp.364-365, 1981.
[18] T.Okoshi et al., IOOC'83, Paper 30C3-2, 1983.
[19] H.Kuwahara et al., ECOC'86, Tech. digest, pp.407-410, Barcelona, Spain,1986.
[20] S.Watanabe et al., ECOC'88, Tech. digest I, pp.90-93, Brighton, UK, 1988.
[21] S.Watanabe et al., ECOC'89, ThA21-4, Gothenburg, Sweden, 1989.
[22] T.Naito et al., Electron. Lett., vol. 25, pp.895-896, 1989.
[23] T.Chikama et al., IOOC'89, Paper 19C2-5, Kobe, Japan, 1989.
[24] S.Ogita et al., ECOC'89, TuA7-6, Gothenburg, Sweden, 1989.

CONSIDERATIONS FOR FIELD DEPLOYMENT OF COHERENT SYSTEMS

M. C. BRAIN

British Telecom Research Laboratories
Martlesham Heath, Ipswich, IP5 7RE, UK

Coherent optical transmission techniques, used in combination with optical amplifier repeaters, offer the prospect of long-distance communication channels which are transparent to both bit-rate and wavelength. For the customer, this means greater versatility in the use of the communication medium; for the network operator, the availability of many wavelengths provides additional capacity from the same fibre, a new mechanism for routing calls with unspecified format, and improved reliability with the elimination of much electronics from the transmission path.

By using existing networks of single-mode fibre, the transition from direct to coherent detection systems can be seen as one further step in the continuing evolution of optical communications. The step is a significant one, however, since progress is required in a range of device and control technologies to provide terminal equipment suitable for field deployment.

The world's first demonstration of these techniques in the operational environment featured the transmission of DPSK-modulated data at 565 Mbit/s over 176 km of fibre, already installed in the BT network. This paper describes the miniaturised external cavity lasers and the automated endless polarisation control techniques which formed the basis for this demonstration. Subsequent developments have lead to improved sensitivity at the receiver, and demonstrations of transmission via optical amplifier repeaters. The viability and merits of important alternative system configurations are also considered, and the opportunities they may create for new network applications are discussed.

1 INTRODUCTION

Worldwide, there have been many laboratory demonstrations of coherent optical transmission systems, spanning a wide range of modulation formats, system bit-rates, and device technologies. For field deployment, however, the present challenges still relate directly to the fundamental differences between direct and coherent detection, namely how to provide a stable, wavelength-tunable, spectrally-pure optical carrier, how to modulate it, and how to match the states of polarisation (SOPs) of the weak signal and the strong optical local oscillator (LO) at the receiver, using technology appropriate to the operational environment.

This paper therefore begins by considering the possible options for the transmitter and receiver for the various available modulation formats, and their merits and limitations for field use.

Since field deployment presumes a practical application, the later sections of the paper concentrate on the particular advantages of coherent transmission, and relate new networking opportunities to the earlier discussion of technology options. We conclude that systems based both on phase and frequency modulation offer opportunities to improve greatly the transparency, versatility, and reliability of existing fibre networks.

2 COHERENT TRANSMITTERS

DPSK and FSK modulation systems have received much attention: both offer comparable power budgets [1], and are compatible with optically-amplified multi-wavelength transmission with relative freedom from crosstalk. In contrast, ASK modulation may involve a less effective use of transmitter power, an unfavourable quantum-limited receiver sensitivity, and crosstalk between different wavelength-channels in semiconductor laser amplifiers.

DPSK and FSK systems have been the subject of experiments at BTRL, though progress has been most rapid with DPSK modulation and heterodyne detection (which formed the basis for the field demonstration system [2]). The technology associated with each format produces solutions with complementary merits, however, and both remain important for potential field applications.

2.1 DPSK Modulation with External Cavity Lasers

A DPSK transmitter typically comprises a narrow-linewidth laser source followed by an isolator and an external modulator, usually a $LiNbO_3$ guided-wave device. An optical power amplifier may then be used to provide a high level of launched power into single-mode fibre. Where simple polarisation-sensitive isolators, modulators, and amplifiers are used, they would ideally be interconnected by polarisation-maintaining (PM) fibre, though more complicated arrangements can be configured which operate in parallel on separated orthogonal components of polarisation, to produce devices which are thus insensitive to the input SOP.

The critical element in the transmitter is the laser. Phase modulation places the greatest demands on source spectral linewidth, which must be less than 0.3% of the bit-rate for DPSK heterodyne detection (eg ~2 MHz for 600 Mbit/s) [3]. DFB lasers have been under development as single-mode sources for direct detection systems for many years, but commercially-available devices have linewidths at least ten times greater than the required value, and only very recently have DFB devices been reported with linewidths well within the theoretical bounds [4]-[6].

Experimental work has therefore proceeded with long external cavity (LEC) lasers [7]. Such devices are formed by anti-reflection (AR) coating one facet of a conventional Fabry-Perot semiconductor laser, and restoring laser action by collimating the output from that facet onto a reflector orientated normal to the optical axis through the active region. This has the effect of narrowing the linewidth of each mode, but also greatly increases the spectral density of the modes. The addition of a wavelength-selective element into the cavity, such as a diffraction grating in place of the reflector, is therefore needed to provide single-longitudinal mode (SLM) operation.

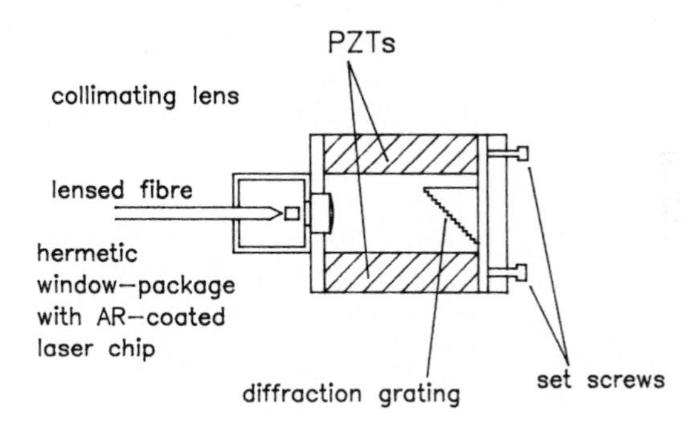

Fig 1. Schematic of miniaturised external cavity laser.

Since the product of mode-spacing and cavity length is 15 GHz.cm at 1.5 μm wavelength, and a mode-spacing corresponds to a change in cavity length of only half a wavelength, it follows that these assemblies are extremely sensitive to environmental influences such as temperature and vibration. In laboratory-based assemblies, this can cause the laser to hop between different cavity modes, resulting in hops in the optical carrier frequency. Nevertheless, a miniaturised LEC laser has been developed for field use [8], and is illustrated in Fig 1. The output from the AR-coated semiconductor laser chip is collimated by a 0.45 NA plano-convex GRIN lens onto a 1200 grooves/mm diffraction grating, which is blazed for 1.5 μm wavelength. The cavity length is ~ 2 cm, and produces a linewidth of < 50 kHz. Coarse alignment by three adjustment screws provides a wavelength-setting range of typically 50 nm (~1.51 - 1.56 μm). Fine-tuning of the carrier frequency by up to 50 GHz is provided by piezo-electric transducers between the laser and grating mount, which permit both the grating tilt and cavity length to be varied simultaneously to maintain continuous SLM operation. Peltier controllers maintain a temperature stability of 0.1°C: cycling the ambient temperature between 5 and 40°C produces a change in the optical carrier frequency of only ~ 8 GHz [9].

These devices were used for the transmitter and LO sources in the field demonstration experiment, and produced error-free operation routinely during many measurement periods of several hours, without special provision for vibration isolation. These devices therefore show great potential for field application, though additional controls would ideally be provided to ensure that mode-hopping never happened in an operational system. This is a subject of continuing study.

A further simplification to the transmitter may be afforded by applying phase-modulation via a semiconductor optical amplifier at the transmitter output stage [10]-[11]. Early experiments suggest that most of the 7.5 dB additional penalty measured in a 565 Mbit/s system was due to the limited modulation bandwidth of the laser amplifier. The idea nevertheless has practical potential, since the bandwidth may be improved to perhaps several GHz by attention to parasitic impedances in both the laser structure and its surrounding package.

2.2 FSK Modulation with Distributed Feedback (DFB) Lasers

FSK modulation (combined with non-synchronous second detection) is more tolerant of laser linewidth, which may be as much as 10% of the bit-rate without significant penalty [12]. Moreover, it is well-known from direct-detection systems that changes in bias current produce "chirping" of the optical carrier frequency: this can be used as an effective means of directly FSK-modulating the laser for coherent transmission. In principle, therefore, the DFB laser is a strong contender as a coherent optical transmitter source, particularly since laser action is sustained by distributed feedback rather than Fabry-Perot resonance, and hence should be inherently free from mode-hopping.

There are a number of further practical considerations, however. Firstly, in order to maximise the available power budget, the energy in <u>both</u> binary signal states should be detected. This could be achieved in dual-filter or frequency-discriminator receivers; but if the source linewidth is excessive, the bandwidth demanded from the receiver may lead to thermal noise penalties. Some improvement in linewidth and launched power is therefore desirable over that available in present commercial devices.

Again recognising that commercial DFB lasers are intended for direct detection systems, it must be borne in mind that their use in coherent detection systems renders them far more susceptible to spurious optical reflections [13], and hence they require optical isolation within the transmitter laser package. However, this also becomes a requirement for higher bit-rate direct detection systems, and appropriately-packaged devices are now becoming more readily available.

When these "simple" DFB devices are directly modulated, it emerges that the FM response is dependent on the bit pattern [14]. Modulating the bias current produces two opposing effects on the refractive index of the active layer: a fast effect due to changes in carrier concentration, and a slow effect due to heating (which therefore has a greater effect when there are long sequences of like pulses). The combination of these effects produces a non-uniformity in the FM index at low frequencies (typically in the MHz range) which cannot be completely compensated within the laser bias circuit, though it could be avoided by encoding the data to remove spectral content from that frequency band [15]-[16] (eg using AMI and Manchester code). A more attractive solution may be provided by multi-terminal lasers, however, in which the effect is far less pronounced [17].

A final practical difficulty lies in the limited tuning range of DFB lasers (~1 nm). This has caused problems in obtaining pairs of wavelength-matched devices for transmitter and LO in single-wavelength systems, and would require that devices be fabricated with different gratings to accommodate wavelength-division multiplexing (WDM). A solution may again lie in a more complicated device structure, namely the Distributed Bragg Reflector (DBR) laser, whose optical guide has distinct active and passive sections. The wavelength-tuning range is increased by locating the grating in a non-active section of the guiding structure, where greater variations in carrier concentration are therefore possible [18].

2.3 Discussion

It can be seen that the differing requirements for source linewidth for DPSK and FSK transmitters lead to coherent systems solutions with complementary advantages and disadvantages. Both have comparable power budgets, but the LEC laser needed for DPSK modulation uses a relatively simple semiconductor laser chip, offers a wide wavelength-tuning range, is relatively insensitive to reflections, but does not yet guarantee continuous SLM operation indefinitely. The DFB laser is already a more complicated device, whose packaging demands the incorporation of optical isolation, and whose structure ideally would further incorporate a multi-electrode pattern for good FM performance, and multiple phase-shifts in the grating for line-narrowing. Its great advantage is its inherent SLM operation (ie freedom from mode-hopping), but combined with only a very limited wavelength-tuning range. It remains to be seen whether both features co-exist in the DBR laser.

Thus none of the existing solutions meets all the foreseeable demands of coherent transmission, and there are opportunities for both approaches. The LEC laser has allowed coherent transmission to be demonstrated in the field, but the DFB or DBR approach may have the greatest potential to be made relatively inexpensively for longer-term mass application. Much practical investigation and development remains to be done.

3 COHERENT RECEIVERS

At the receiver of a coherent system, the weak transmitted signal is combined with a strong optical local oscillator signal, and the two are mixed in a square-law photodetection process to produce a beat-frequency component of photocurrent from which the original message is derived. In order to minimise penalties and avoid signal fading, the states of polarisation (SOPs) of the signal and LO must be matched before detection. However, conventional single-mode fibre exhibits stress-birefringence, and hence the SOP of the transmitted signal is unpredictable and may vary over a virtually limitless range.

There are three principle strategies available for ensuring efficient signal reception, namely polarisation control, diversity reception, and scrambling. These are summarised below, followed by a brief description of the design of the receiver itself.

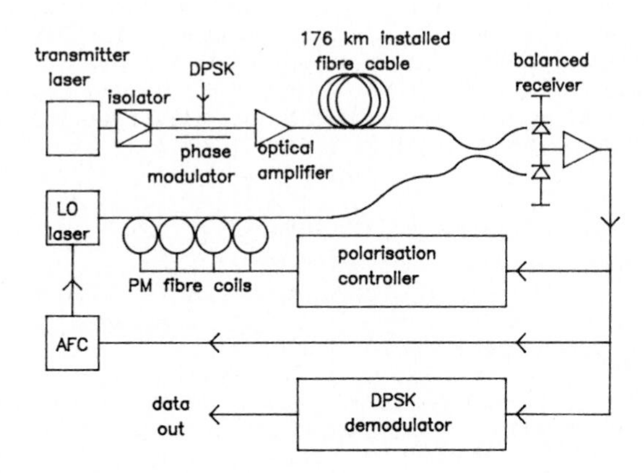

Fig 2. Field demonstration system schematic.

3.1 Polarisation Control

An example of the application of this strategy is given in the system schematic for the field demonstration shown in Fig 2. In broad terms, a system of transducers is placed at the output of the LO source, and a small "dither" is applied to the transducer bias to induce a small variation in the LO SOP. The effect on the strength of the recovered signal at the receiver output can then be used as an indication of the matching between the LO and the transmitted signal: if the dither has a discernible effect, the SOPs are not matched, and the control algorithm will vary the transducer bias to direct it to produce a maximum signal, by continuously assessing the presence of the dither in the IF signal until it is eventually minimised. The challenge is to produce a system of transducers which can provide unlimited range of adjustment ("endless" control) which are suitable for field use, and to provide a reliable algorithm with adequate speed of response.

Firstly, it is possible to provide a limitless range of adjustment using a small number of finite range transducers [19]. This has been demonstrated using four piezo-electric transducers (PZTs) [20] to induce stress birefringence by squeezing the fibre at the LO output. Understanding how this achieved is greatly helped by reference to the Poincare sphere representation for SOPs, shown in Fig 3. Briefly, each point on the surface of the sphere uniquely represents a single SOP: linear polarisation states are represented by points on the equator; left and right circular states correspond to the two poles; and the intermediate points represent the continuum of elliptical states. If light linearly polarised at 45° is launched into a fibre which is squeezed in the vertical direction, then as the applied stress is increased, the phase delay will increase between the horizontal and vertical components of the propagating electric field, and the output SOP will become elliptical. On the Poincare sphere, the output SOP will start at the point P (45°) on the Poincare sphere, progress through one pole to the point Q (-45°), then through the second pole to return to P, as the phase delay increases from 0 to 2π. The SOP thus describes a circle about the axis HV, which corresponds to the direction of the applied stress. The general result is that when a fibre is stressed, the SOP of the propagating optical signal is transformed about an arc of a circle on the surface of the sphere, about an axis through the equator which corresponds to the direction of the applied stress.

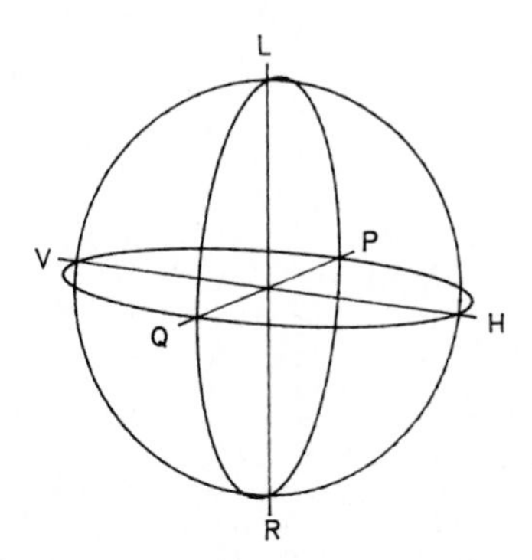

Fig 3. The Poincare sphere representation for states of polarisation.

By positioning transducers in series orientated at 45° to each other, transformations of the SOP can be assembled using arcs of circles about axes at 90° to each other. It can be seen that only three such transducers are needed to transform any state to any other state, but four are needed to allow continuous transformation between arbitrarily-varying states.

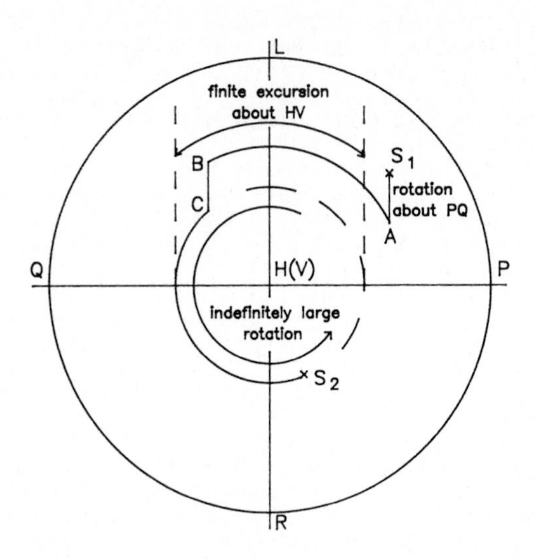

Fig 4. Endless range of SOP transformation on the Poincare sphere, using finite-range transducers.

The provision of endlessness is illustrated using Fig 4. Suppose S_1 is the LO SOP, and S_1ABCS_2 is the transformation needed to match the LO to the transmitted signal state S_2. If the transmission fibre is subjected to increasing stress in the vertical direction, S_2 will move continuously about a circle around the axis HV. Eventually the squeezer providing the transformation CS_2 will reach a range limit. However, the movement of S_2 can continue to be tracked by keeping the arc-length CS_2 constant and instead varying the intermediate states B and C. Now, the arc CS_2 will move about HV, but the point B need only move back and forth over a finite arc-length on a larger circle about HV. Thus unlimited movement of S_2 can be accommodated by having two finite range transducers in series, with an intermediate transducer at 45°. Similarly unlimited movement about PQ can be accommodated by adding a fourth transducer at 45°, leading to the general conclusion that arbitrary continuous transformations may be provided over unlimited range using only four finite-range transducers.

A second approach has been demonstrated using an X-cut Z-propagating $LiNbO_3$ guided-wave device [21]. By providing one electrode over the guide and one either side of the guide, it is possible to vary both the strength and direction of the electric field in the vicinity of the guide. This is equivalent to varying the arc-length <u>and</u> the direction of the axis through the equator of the Poincare sphere about which the SOP is transformed. In principle, only one transducer is needed for any given transformation; in practice, two sets of electrodes were provided to allow the algorithm to be simplified.

Devices which were produced for practical evaluation performed according to expectation, and provided endless polarisation control in laboratory measurements. Use in the field requires hermetic packaging, however, due to the high fields generated between the electrodes.

A third controller has therefore been devised, based on polarisation- maintaining fibre [22]. PM fibre maintains polarisation if linearly polarised light is launched parallel to one of the principle axes of the fibre core; otherwise, the SOP varies rapidly with distance along the fibre, and is greatly affected by small additional stresses. A transducer can thus be formed by winding PM fibre onto a piezo-electric cylinder, such that the SOP can be varied according to the applied voltage bias. The operation of these transducers is analogous to the fibre squeezers described above, hence four transducers are required for endless control, with the principle axes of adjacent transducers orientated at 45° to each other. However, the necessary changes in SOP require strain levels of only ~10^{-4}, which is well within the proof strain of the fibre.

Compared with the $LiNbO_3$ device, this type of controller has a low insertion loss, a high return loss, and has no need of hermetic packaging. Since four transducers are needed, however, the algorithm is more complicated and may have a more limited speed of response. The choice of one or other may therefore be determined by experience in the field - does the algorithm react fast enough to track all practical variations in SOP in the transmission fibre? - or by long-term reliability.

3.2 Polarisation Diversity Receivers

The principle uncertainty associated with polarisation control concerns speed of response: although variations in SOP in fibre would only be expected to occur at acoustic rates (originating in mechanical or thermal disturbances), much calculation has to be performed for each iteration of the dither procedure, which may restrict the speed of the algorithm. Also, longer-term applications of coherent detection based on Wavelength-Division Multiple Access (WDMA) would require a receiver which could respond almost instantly to an incoming signal regardless of its SOP: the time for a controller to acquire polarisation-matching may be unacceptably long.

In these circumstances, a polarisation-diversity receiver would be attractive [23]. Many variations are possible, but the basic idea is illustrated in Fig 5. The output from the LO laser is split into two equal orthogonal components of polarisation before being combined with the incoming signal.

The latter is similarly split, though the ratio of the split will depend on the SOP of the signal and hence will be uncertain. The components of like polarisation are combined in PM couplers feeding into separate optically-balanced receivers, whose outputs are added. Thus, regardless of the SOP of the incoming signal or the rate at which it varies, there will inevitably be a component of LO power with which the signal can beat to allow the original message to be recovered.

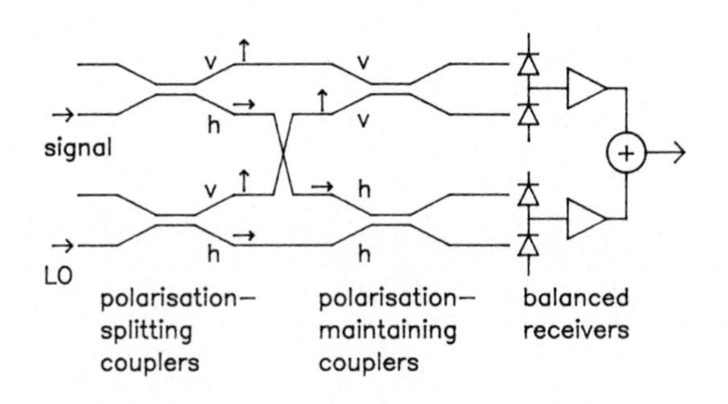

Fig 5. Schematic of polarisation diversity receiver.

This advantage is obtained at a price. Firstly, LO power is not used to maximum advantage; and secondly, the additional receiver preamplifier will add to the thermal noise of the receiver. Both effects will tend to increase the sensitivity penalty from thermal noise, unless high levels of LO power are available.

From the viewpoint of field deployment, however, the greatest concern is the availability of the components. Ideally the coupling optics would be in guided-wave form for ruggedness, but suitable wavelength-insensitive devices with acceptable performance are not yet available. Also, the opto-electronic receivers need to be accurately matched for optimum performance, which becomes more difficult with increasing bit-rate.

3.3 Polarisation Scrambling

The objective of scrambling is the same as with diversity reception, namely to ensure that reception of the signal is guaranteed regardless of SOP, but the diversity is incorporated at the transmitter by varying the signal SOP between orthogonal states many times within a bit-period [24]. The disadvantage is that signal power is now wasted, and use is restricted to lower bit-rates by the need to scramble the SOP within a bit-period. Again, specialised optics would be needed for field use, which are not yet available.

3.4 Discussion

All of the above techniques have been demonstrated in the laboratory and represent complementary solutions to the problem of polarisation-matching. Polarisation control is more dependent on the precise nature of the operational environment, but several versions of appropriate technology have already been demonstrated. Diversity reception and scrambling should both be independent of rate of change of SOP, but require technology for which solutions can be envisaged, but which are not yet available. Both of the latter could be of longer-term interest, however: for WDMA-based networks in the case of diversity reception; and for low-bit-rate distribution networks in the case of scrambling, where the cost and complexity of the receiver is a prime consideration.

3.5 Receiver Opto-electronics

As Fig 2 shows, the signal and LO output are combined in a passive single-mode fibre coupler, whose two outputs are connected to separate PIN photodiodes feeding a single preamplifier. This balanced optical input [25] has the advantage of minimising the sensitivity penalty due to thermal noise in the receiver, by using all the available signal and LO power. That penalty can still be significant, however, and it is therefore advantageous to design the preamplifier for low noise.

The receiver used in the field demonstration was assembled from discrete components, and yielded combined penalties from quantum efficiency and thermal noise of around 5 dB for a LO power level of 0 dBm. This was reduced to ~ 1 dB by adopting a GaAs IC preamplifier hybrid integrated with two PIN photodiode chips, giving a sensitivity at 565 Mbit/s of -52 dBm [26]. Producing the preamplifiers in this way also simplifies the assembly of the receiver and makes them potentially more reproducible, and hence is relevent to diversity reception.

4 APPLICATIONS

Coherent transmission has a number of advantages which combine to suggest new ways of exploiting the enormous unused spectral capacity of optical fibre to great advantage.

Firstly, coherent transmission results in low levels of material dispersion in silica fibre, since the spread of frequencies transmitted corresponds only to the modulation bandwidth on the optical carrier, compared with broad source linewidth or chirp in present day systems. Hence for systems spanning hundreds of kilometres or more, the need for digital regeneration rather than amplification is determined by repeater noise rather than channel bandwidth. Optical amplifiers based on semiconductor and fibre lasers can provide simple low-noise multi-wavelength repeaters for WDM applications; the former is the more mature technology, though the latter has potential advantages of insensitivity to input SOP, low coupling losses and reflectivities, and higher gain. This therefore suggests the use of coherent transmission with existing fibre and optical amplifier repeaters, to provide channels between terminals spanning hundreds of kilometres, which are transparent to bit-rate and wavelength [27].

Secondly, coherent receivers are inherently wavelength-selective. Effectively, filtering of the optical signal is provided in the electrical domain, since the coherent receiver responds only to optical signals near the LO carrier frequency, such that the difference frequency lies within the electrical pass-band of the receiver. This provides effective filtering of accumulated spontaneous emission noise from optical amplifier repeaters, and allows many wavelength channels to be closely-spaced to provide effective use of spectral capacity of fibre and amplifiers.

Thus coherent transmission offers high-performance all-optical links over long distances. This in turn suggests novel uses for optics in future networks, such as using wavelength as a means of routing calls using either passive optical filters [28] or broadcasting techniques [29]. Such novel uses will tend to demand a greater power budget, either to accommodate the receiver noise from accumulated spontaneous emission from amplifiers, or to accommodate broadcasting losses through power-sharing between terminals. Broadly speaking, every additional 3 dB could double the number of terminals served by the network, or the number of amplifiers (and hence system span) between terminals. Coherent receivers provide the additional sensitivity which allows the overall power budget to be improved.

This suggests future long-haul transmission networks in which the transmission rate (and indeed the reliability) are no longer restricted by the electronic regenerators and switches in the transmission path. As the technology matures, it would be expected to penetrate into short-haul applications, such as providing local wideband services.

5 FIELD DEMONSTRATION SYSTEM

The world's first demonstration of a coherent optical transmission system in the operational environment was performed in the BT network, featuring DPSK transmission at 565 Mbit/s, and heterodyne detection [2]. Repeaterless links with lengths up to 176 km were configured using conventional single-mode fibre in an 18-fibre cable installed between Cambridge and Bedford. A schematic of the system is given in Fig 2.

The DPSK transmitter comprised a miniaturised LEC laser, a commercial 30-dB isolator in a fibre-fibre package, a lithium niobate phase modulator, and a semiconductor laser power amplifier giving 11 dB gain. The power launched into the fibre was +1 dBm at 1.534 μm wavelength.

At the receiver, a second LEC laser formed the LO, whose output passed via a PM-fibre-based polarisation controller into a conventional single-mode fibre fused-taper coupler. The two coupler output ports were in turn coupled to two PIN photodiodes in a balanced optical receiver, whose sensitivity at 10^{-9} BER was -47.6 dBm. The 176 km route had an insertion loss of 49 dB, resulting in a measured BER of 5×10^{-9}. Transmission over shorter loop lengths provided sufficient power margin to obtain error-free operation routinely for several hours at a time, corresponding to a BER better than 10^{-13}.

The receiver sensitivity has since been improved by adopting a GaAs IC preamplifier hybrid-integrated with two PIN photodiodes, yielding a receiver sensitivity of -52 dBm at 565 Mbit/s (~80 photons/bit). Moreover, transmission through a cascade of 5 semiconductor amplifiers, with a sensitivity penalty of only 6.5 dB, has demonstrated the scope to extend the loss budget [30]. Transmission through a diode-pumped erbium-doped fibre amplifier also confirmed the polarisation-insensitive and low-noise properties of these devices [31].

6 DISCUSSION AND CONCLUSIONS

The above sections are by no means a comprehensive review of coherent systems development to date, but the series of experiments described in Section 5 serves to illustrate the technological considerations described in Sections 2 and 3 in transferring coherent systems from the laboratory into the operational environment, and to indicate progress towards the applications considered in Section 4.

In demonstrating coherent transmission on an operational link in the BT network, we have shown firstly that conventional single-mode fibre can be used for this purpose, secondly that control transducers and algorithms can be used to provide endless polarisation-matching at the receiver, and thirdly that a narrow-linewidth tunable SLM source can be operated without the need for anti-vibration precautions.

The choice of technology was strongly influenced by what was readily available. DFB lasers designed for FSK modulation and packaged with optical isolation were not (and still are not) readily available, though they may ultimately provide an extremely attractive option when they do emerge, both for guaranteed freedom from mode-hopping and potential volume applications. Meanwhile, the miniaturised LEC laser has provided excellent performance in both laboratory and field environments: featuring wide ranges of wavelength adjustment and SLM tuning, narrow spectral linewidth, and low susceptibility to optical reflections, it has much potential for practical systems use.

Similarly, polarisation control presented a quicker solution for field use than diversity reception, with negligible penalties, and with excellent performance. Diversity receivers still await suitable optical devices for maintaining/splitting the signal polarisation state at the optical input; their electronics will benefit from the development of GaAs IC receivers to simplify assembly and provide the reproducibility needed for good matching, though their suitability for high-bit-rate operation remains to be seen.

In the broader sense, it is interesting to note that both coherent and high-bit-rate direct detection systems are beginning to make similar demands of technology. Multi-contact lasers for coherent FSK sources could also provide low-chirp devices for direct detection (as could LEC lasers); such sources also begin to need optical isolation and external modulation at high bit-rates, introducing a dependence on polarisation previously restricted to coherent systems. From the applications viewpoint, interest in using direct detection for WDM through amplifiers suggests the use of narrow-band optical filters, such as DFB amplifiers which again use technology common to both direct and coherent detection sources.

We may thus conclude that future networks, whether based on coherent or direct detection, will require progress in the same device technologies (eg semiconductors, fibre-based devices, lithium niobate, etc). However, field use also requires that the substantial contribution made by assembly and packaging should not be under-estimated: hermeticity, optical feed-throughs with good coupling to guided-wave devices (via isolators in many cases), electrical feedthroughs for high-speed or high-voltage connections, and good mechanical and thermal properties are just some of the many features demanded by these new devices. However, coherent systems inherently offer the low dispersion, wavelength selectivity, and large power budget, required to exploit existing fibre networks to provide greater transparency, versatility, and reliability.

ACKNOWLEDGEMENTS

The author thanks the Research and Technology Board of British Telecom for permission to present this paper.

REFERENCES

[1] M C Brain, D W Smith, Proc SPIE, Vol 716, pp 153-161, 23-24 Sept 1986.

[2] M J Creaner, R C Steele, I Marshall, G R Walker, N G Walker, J Mellis, S Al Chalabi, I Sturgess, M Rutherford, J Davidson, M Brain, Electronics Letters, Vol 24, No 22, pp 1354-1356, 27 October 1988.

[3] G Nicholson, Electronics Letters, Vol 20, pp 1005-1007, 1984.

[4] Y Kondo, K Sato, M Nakao, M Fukuda, K Oe, Electronics Letters, Vol 25, No 3, pp 175-177, 1989.

[5] S Takano, T Sasaki, H Yamada, M Kitamura, I Mito, Electronics Letters, Vol 25, No 5, pp 356-357, 1989.

[6] S Ogita, Y Kotaki, M Matsuda, Y Kuwahara, H Ishikawa, Electronics Letters, Vol 25, No 10, pp 629-630, 1989.

[7] M W Fleming, A Mooradian, IEEE J Quantum Electronics, Vol QE-17, No 1, pp 44-59, January 1981.

[8] J Mellis, S Al Chalabi, K H Cameron, R Wyatt, J C Regnault, W J Devlin, M C Brain, Electronics Letters, Vol 24, No 16, pp 988-989, 4th August 1988.

[9] S Al-Chalabi, to be published.

[10]J Mellis, Electronics Letters, Vol 25, No 10, pp 679-680, 11 May 1989

[11]J Mellis and M J Creaner, Electronics Letters, Vol 25, No 10, pp 680-681, 11 May 1989.

[12]I Garrett, G Jacobsen, J Lightwave Technology, Vol LT-4, No 3, pp 323-334, March 1986.

[13]R W Tkach, A R Chraplyvy, J Lightwave Technology, Vol LT-4, No 11, pp 1655-1661, Nov 1986.

[14]K Emura, M Shikada, S Fujita, I Mito, H Honmou, K Minemura, Electronics Letters, Vol 20, No 24, pp 1022-1023, 22 November 1984.

[15]R Noe, M W Maeda, S G Menocal, C E Zah, Proc 14th European Conference on Optical Communication (ECOC 88), Brighton, UK, Sept 1988, p 175

[16]R C Steele, M Creaner, Electronics Letters, Vol 25, No 11, pp 732-734, 25 May 1989.

[17]S Yamazaki, K Emura, M Shikada, M Yamaguchi, I Mito, Electronics Letters, Vol 21, No 7, pp 283-285, 28 March 1985.

[18]T L Koch, U Koren, R P Gnall, C A Burrus, B I Miller, Electronics Letters, Vol 24, pp 1431, 1988.

[19]R Noe, Electronics Letters, Vol 22, No 15, pp 772-773, 1986.

[20]N G Walker, G R Walker, Electronics Letters, Vol 23, No 6, pp 290-292, 12 March 1987.

[21]N G Walker, G R Walker, J Davidson, Electronics Letters, Vol 24, No 5, pp 266-268, 3 March 1988.

[22]G R Walker, N G Walker, Electronics Letters, Vol 24, No 22, pp 1353-1354, 27 October 1988.

[23]T Okoshi, S Ryu, K Kikuchi, Proc IOOC '83, Tokyo, Japan, pp 386-387, 1983.

[24]T G Hodgkinson, R A Harmon, D W Smith, Electronics Letters, Vol 23, No 10, pp 513-514, 7 May 1987.

[25]G Abbas, V W S Chan, T K Yee, Optics Letters, Vol 8, No 8, pp 419-421, August 1983.

[26]P Smyth, A Sayles, N Back, A McDonna, M Creaner, ECOC '89, Goteborg, Sweden.

[27]Y Yamamoto, T Kimura, IEEE J Quantum Electronics, Vol QE-17, No 6, pp 919-935, June 1981.

[28]G R Hill, Br Telecom Technol J, Vol 6 No 3, pp 24-31, July 1988.

[29]M C Brain, P Cochrane, IEEE International Conference on Communications (ICC '88), Philadelphia, USA, June 12-15, 1988.

[30]D J Malyon, R C Steele, M J Creaner, M C Brain, W A Stallard, Electronics Letters, Vol 25, No 5, pp 354-356, 2 March 1989.

[31]M J Creaner, T J Whitley, R C Steele, G Garnham, C A Millar, M C Brain, ECOC '89, Goteborg, Sweden.

RECENT TRENDS IN COHERENT OPTICAL FIBER COMMUNICATIONS RESEARCH WITH EMPHASIS ON DIVERSITY-TYPE RECEIVING TECHNIQUES

TAKANORI OKOSHI

Research Center for Advanced Science and Technology (RCAST)
University of Tokyo
4-6-1 Komaba, Meguro-ku, Tokyo 153, Japan

Recent trends in coherent optical fiber communications research is reviewed, with special emphasis on various diversity techniques. The diversity techniques, i.e., the polarization diversity and phase diversity, have become in the past several years the key technologies in coherent optical fiber communications. The technical and historical backgrounds are reviewed first. In the latter half of the paper, a new diversity scheme called the double-stage phase-diversity (DSPD) is described and compared with the conventional schemes. Results of experiments are also shown.

1. INTRODUCTION

The research and development of coherent fiber optical communications have now a history of more than ten years. At present, the most important technical task is probably to establish countermeasures against the fluctuation phenomena in the communications systems, i,e.,

(1) the fluctuation of state-of-polarization (hereafter SOP) of received signal, and
(2) that of the phase of received signal and/or local oscillator light.

The latter is particularly important when homodyne detection is used at the frontend of the receiver (1).

A remarkable trend in the past several years is that various diversity techniques (use of multiport receivers) have increasingly become common as the countermeasures. To be more specific, the research along this direction can be classified into

(1) Polarization-diversity techniques,
(2) Phase-diversity techniques, and
(3) Polarization/phase double diversity techniques.

In this paper, the technical and historical backgrounds of these diversity techniques are reviewed first. In the latter half, a new diversity scheme called the double-stage phase-diversity (DSPD) is described; this features phase-diversity-type optical homodyne detection, and the subsequent phase-diversity-type frequency up-conversion which produces a heterodyne-like second IF signal. It has the advantages of both heterodyning and homodyning.

2. POLARIZATION DIVERSITY

When the SOP of the received signal light fluctuates and disturbs the match with the SOP of the local oscillator light, the receiver sensitivity is degraded. So far three countermeasures have been investigated to prevent this degradation:

(1) Use of a single polarization fiber as the lightguide,
(2) Use of an automatic SOP control scheme at the frontend of the receiver, and
(3) Polarization diversity.

The polarization-diversity scheme was proposed and experimented with first in 1983 (2). However, in this early proposal the so-called IF-combining scheme was employed, which is not widely used today (IF: intermediate frequency). Instead, the baseband-combining scheme proposed in 1986-1987 by three groups independently (3)(4)(5) is widely used today in coherent communications system experiments.

Figures 1 and 2 shows the examples of the IF-combining (2) and baseband-combining (4) receivers, respectively.

3. PHASE DIVERSITY

The receivers used in coherent optical fiber communications are classified basically into heterodyne and homodyne types (1).

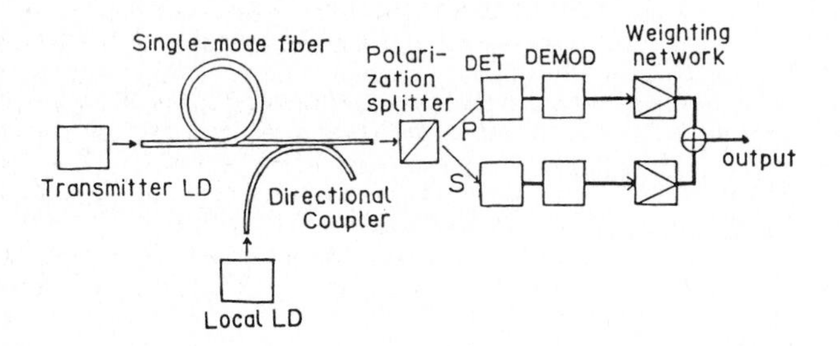

FIGURE 1
An experimental arrangement for the IF-combining polarization-diversity scheme (after Okoshi et al. (2)).

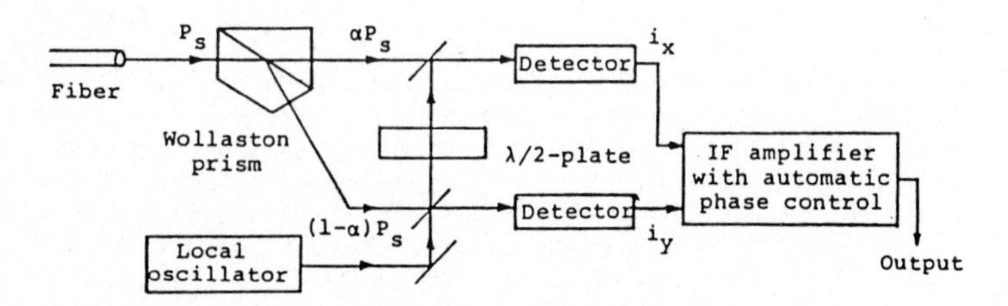

FIGURE 2
An experimental arrangement for the baseband-combining polarization-diversity scheme (after Imai et al (4)).

When the signal bit-rate is very high (several Gb/s), the higher edge of the intermediate frequency (IF) band in a heterodyne receiver becomes as high as 10-20GHz, resulting in the deterioration of receiver performance because of the microwave technology limitation. On the other hand, the heterodyne scheme has an advantage in that the fiber-delay equalization (6) is possible at the IF stage.

In contrast to the heterodyne scheme, the homodyne scheme enables us to realize a receiver having highest frequency equal to the signal baseband. However, this has two drawbacks: the difficulty in realizing the frequency/phase match between the signal and local oscillator (LO), and that the fiber-delay equalization is not possible because the optical signal spectrum is "folded" in the baseband.

To overcome the first difficulty, phase-diversity homodyne scheme has been proposed (7)(8) and is investigated intensively. However, the second drawback mentioned above remains in this scheme. Figures 3 and 4 show the constructions of the three-phase diversity (7) and the two-phase (in-phase and quadrature) diversity (8) receivers, respectively.

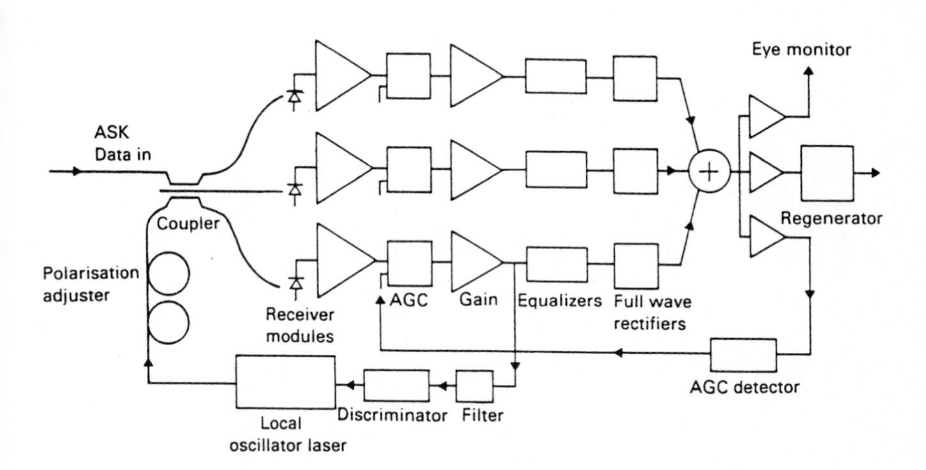

FIGURE 3
An experimental arrangement for the three-phase phase-diversity scheme (after Davis and Wright(7)).

4. DOUBLE-STAGE PHASE DIVERSITY

The double-stage phase-diversity (DSPD) scheme proposed recently by the author (9) is a novel scheme having the advantages of both heterodyning and homodyning. It features the phase-diversity homodyne detection at the frontend which generates a baseband signal, and heterodyne-like second IF stage where the delay equalization as well as all sorts of signal demodulations are possible.

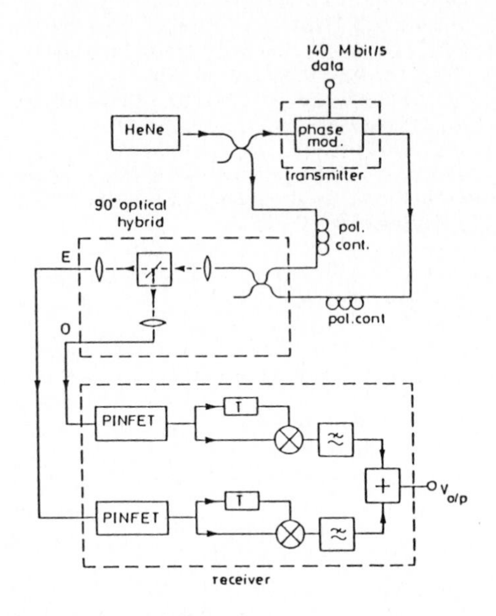

FIGURE 4
An experimental arrangement for the in-phase and quadrature phase-diversity scheme (after Hodgkinson et al. (8)).

The DPSD scheme is applicable, not only to optical communications, but also to all the electrical and/or radio communications ranging from VLF, LF, MF, HF, VHF, UHF, to microwave/millimeter-wave communications, and the associated measurements as well.

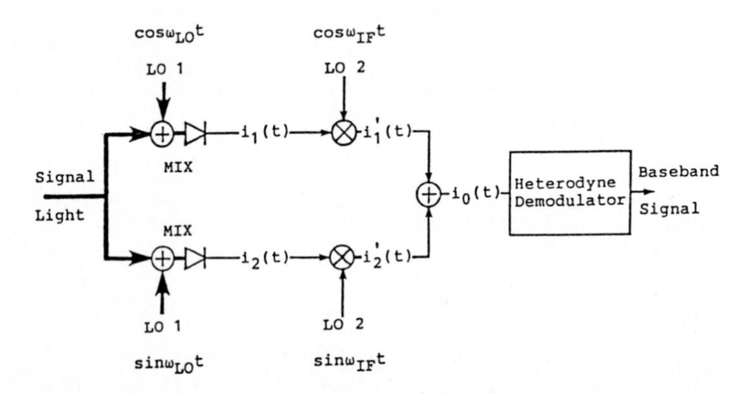

FIGURE 5
Basic construction of a double-stage phase-diversity receiver (after Okoshi et al. (9)).

A. Principle

Figure 5 shows the basic construction of a DSPD receiver. The received signal is converted into two baseband signals by using two LOs having mutually orthogonal phases (1st stage phase-diversity detection). The two baseband signals are amplified and then up-converted to IF signals by using two LOs again having mutually orthogonal phases (2nd stage phase-diversity frequency up-coversion). Finally, the two IF signals are added. It is easily proved by algebra that the added signal is equal to what we obtain in an ordinary heterodyne receiver.

Figure 6 illustrates how a heterodyne-like IF signal can be obtained in a DSPD receiver. It is seen that the signal spectrum, which is once "folded" at the first baseband stage, is "unfolded" at the second IF stage, and finally converted to a heterodyne-like signal regardless the offset frequency in the first stage phase diversity.

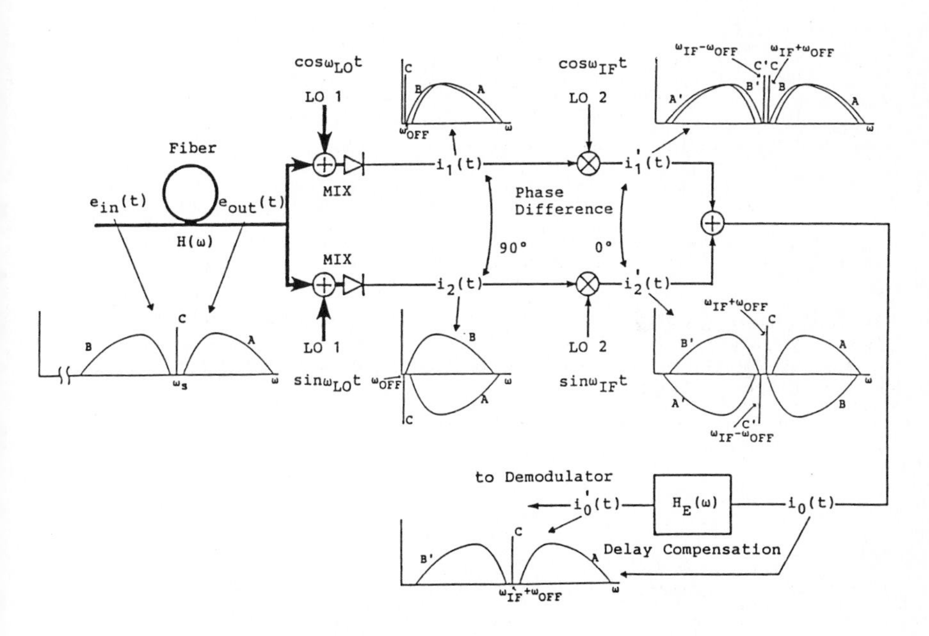

FIGURE 6
Illustration of spectrum-folding and-unfolding processes (after Okoshi et al.(9)).

B. Advantages and Disadvantages

Table I compares the advantages and disadvantages of ordinary direct detection, heterodyne, homodyne, phase diversity, and the DSPD optical receivers. It is seen that the DSPD scheme has excellent features in most of the items compared in this table. The fiber-delay equalization is possible as in an ordinary heterodyne receiver.

Item / Scheme	Heterodyne	Homodyne	Phase diversity	Double-stage phase diversity
Receiver sensitivity	10-25dB better than direct detection	3dB better than heterodyne (the highest)	Equal to or slightly worse than heterodyne	Equal to heterodyne
Required detector bandwidth	3 to 5 times of the BR	Half of the BR	Half of the BR	Half of the BR
Optical phase stability	(1) Synchronous demodulation: needed but the requirements are less stringent than in homodyne (2) Otherwise: Not strictly needed	Very strictly needed	Equal to the case (2) in heterodyne	Equal to heterodyne
Possible modulation formats	All sorts of modulations	All sorts of modulations except FSK	All sorts of modulations	All sorts of modulations
Possible demodulation schemes	Synchronous or asynchronous demodulations are possible for all sorts of modulations	Baseband signal obtained directly	Asynchronous demodulations only (See the text for details)	Synchronous or asynchronous demodulations are possible for all sorts of modulations
Frequency control	AFC signal from the IF stage	Optical PLL is needed	Offset frequency is needed for AFC	AFC signal from the 2nd IF stage
Delay equalization	Possible	Impossible	Impossible	Possible
Channel spacing in FDM systems	Broader than in homodyne	Narrow	Narrow	Narrow

BR: bit rate

TABLE 1
Comparison of various receiving schemes.

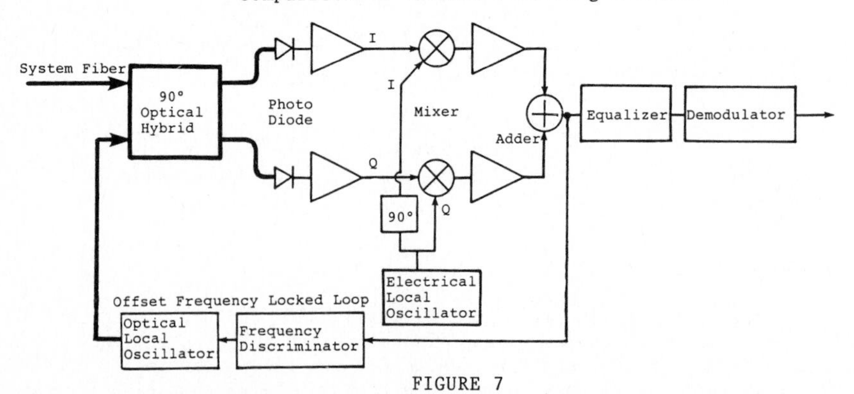

FIGURE 7
Construction of a DSPD optical communication receiver (after Okoshi et al.(9)).

C. DSPD Optical Communications Receivers

Figure 7 shows the construction of a DSPD optical communications receiver. Generally, a k-port DSPD receiver is possible (9); the constructions shown in Figs. 5, 6, and 7 are examples corresponding to k=2 (i.e., a two-port DSPD receiver).

Figure 8 shows a combination of a two-port DSPD receiver and the polarization-diversity scheme (3)(10)(11).

D. Bit-Error Rate (BER) Measurement

An experimental 100Mbit/s FSK coherent optical communication system em-

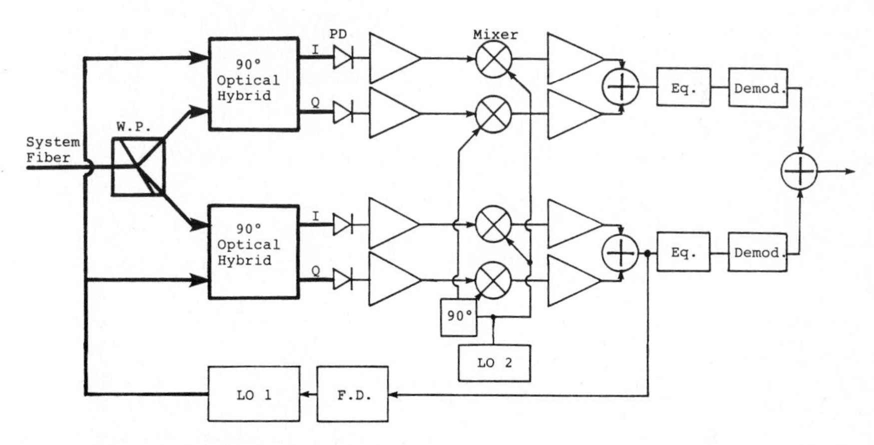

FIGURE 8
Construction of a polarization/DSPD double diversity receiver.

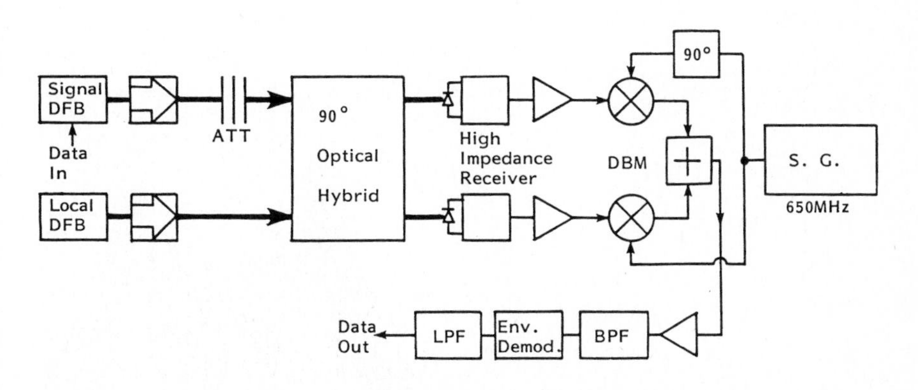

FIGURE 9
Experimental setup of DSPD system

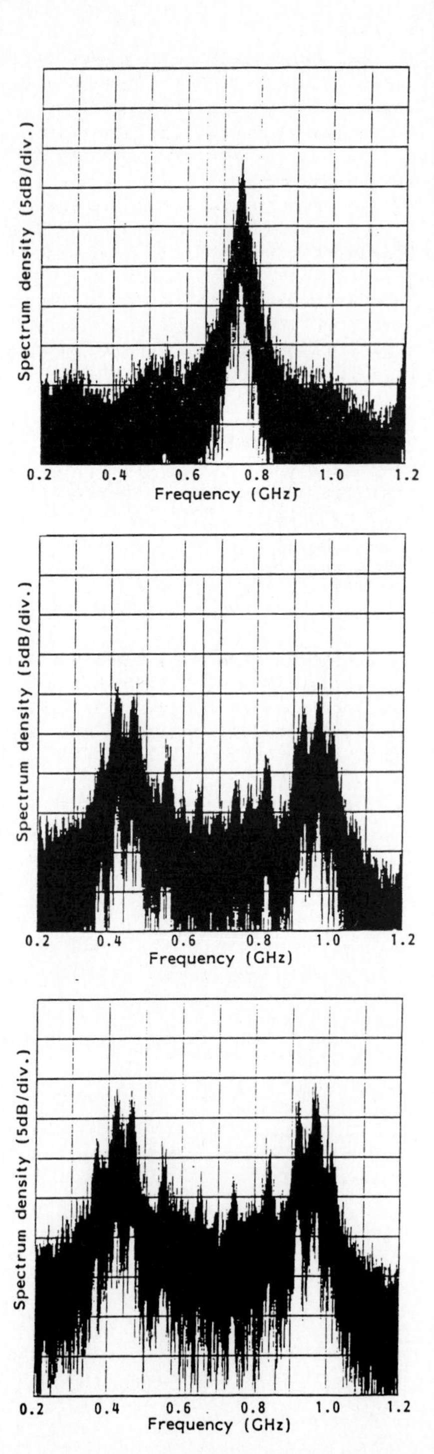

FIGURE 10(a)
Spectrum of added 2nd IF signal (unmodulated)

FIGURE 10(b)
Spectrum of the added 2nd IF signal (modulated)

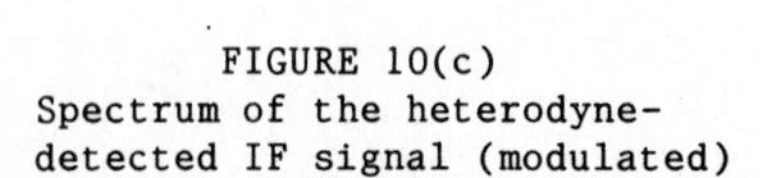

FIGURE 10(c)
Spectrum of the heterodyne-detected IF signal (modulated)

ploying the double-stage phase-diversity (DSPD) scheme was constructed, and the bit-error rate (BER) was measured. A heterodyne-like second IF signal could be obtained as expected. A sensitivity 0.6dB worse than that of heterodyne was achieved.

The experimental setup is shown in Fig. 9. Two 1.3 um DFB semiconductor lasers are temperature-controlled within 0.01K, and used as the transmitter and local oscillator (LO). The beat linewidth is about 30MHz. The transmitter is FSK-modulated via injection current with frequency deviation of about 600MHz. The data signal is a 100Mbit/s one-zero pattern. The LO frequency is adjusted at the center of the signal FSK spectrum. Isolators are inserted in front of the two lasers. An ND filter is inserted in the signal path to simulate a fiber. The 90 degree optical hybrid consists of a $\lambda/4$-plate and polarization beam splitter (PBS) (12).

The frontend is a phase-diversity receiver using InGaAs PIN photodetectors and high-impedance-type baseband amplifiers. The LO power at the photodetector surface is -3dBm. In the 2nd stage phase-diversity frequency up-conversion, the baseband signals are up-converted to 650MHz (central frequency) using two double balanced mixers (DBMs). The two up-converted IF signals are added by a resistive combiner, and fed to a conventional FSK heterodyne signal-filter demodulator consisting of a band-pass filter (center:1GHz, bandwidth:400MHz), an envelope demodulator, and a low-pass filter (bandwidth:50MHz).

Figure 10(a) shows the spectrum of the added 2nd IF signal (unmodulated). The offset frequency (the difference between the optical signal and LO frequencies) is 100MHz. The unwanted signal, which would appear at 550MHz if the cancellation is not complete, is found to be suppressed at least 20dB below the signal at 750MHz. This result demonstrates that the DSPD

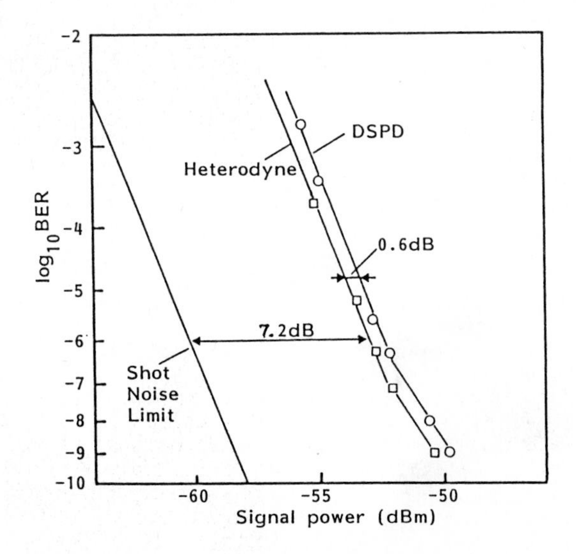

FIGURE 11
Bit-error rate as a function of received signal power

scheme is functioning as expected. The pure line spectrum at 650 MHz shows the spuriously coupled LO output; this can eventually be eliminated afterwards because it is outside the bandwidth of the BPF. (To make it negligible, the detected signal must be amplified up to a level comparable to the 2nd LO, i.e., a few dBm (13).) Figure 10(b) shows the 2nd-IF spectrum obtained when the signal is modulated. This spectrum is found to be identical to that in a conventional single-port heterodyne receiver using the same system, which is shown in Fig. 10(c).

The measured bit-error rate (BER) is shown in Fig. 11 as a function of the sum of the received powers detected in the two branches. The BER of the heterodyne receiver is also measured and shown for comparison. Theoretically, the sensitivities of DSPD and heterodyne schames are equal, whereas 0.6dB degradation is observed in Fig. 11. This degradation is most probably due to imperfect phase and amplitude match between the two branches.

5. CONCLUSIONS

Because of the increase of the signal bit rate, the importance of the phase-diversity technique is increasing. On the other hand, the progress in optoelectronic IC technology will give an advantage to the polarization-diversity technique as compared with other alternatives. Thus, the diversity techniques in general are becoming the key issues in coherent optical communications. The DSPD scheme described in Section 4 of this article will also become an important concept in various areas in rf and optical communications and measurements.

REFERENCES

(1) Okoshi, T. and Kikuchi, K., Coherent Optical Fiber Communications (KTK/Kluwer, Tokyo/Dordrecth, 1988).

(2) Okoshi, T., Ryu, S. and Kikuchi, K., IOOC'83 Tokyo (1983) Tech. Digest pp.386-387.

(3) Okoshi, T., Cheng, Y.H., and Kikuchi, K., Electron. Lett. 23, 8 (1987) pp. 377-378.

(4) Imai, T., Matsumoto, T., and Iwasaki, K., OFS'86 Tokyo (1986) Tech. Digest PP.283-286.

(5) Glance, B., IEEE/OSA Jour. Lightwave Technol., LT-5, 2 (1987) pp.274-276.

(6) Takachio, N. and Iwashita, K., Electron. Lett. 24, 2 (1988) pp.108-109.

(7) Hodgkinson, T.G., Harman, R.A., and Smith, D.W., Electron. Lett. 21, 19 (1985) pp.867-868.

(9) Okoshi, T. and Yamashita, S., Paper of Tech. Group (in Japanese), IEICE Japan, No. CS88-78/IE88-82(1988).

(10) Okoshi, T. and Yamashita, S., OFC'89 Houston (1989), Paper No. PD-19.

(11) Okoshi, T. and Yamashita, S., IEEE/OSA Jour. Lightwave Technol., in print.

(12) Leef, W.R., Arch. Electron. und Ubertragungstech. 37 (1983) pp.203-206.

(13) Emura, K. and Vodhanel, R.S., ECOC'88 Brighton (1988), Tech. Digest pp.147-150

(14) Okoshi, T. and Yamashita, S., Electron. Lett., being submitted.

LASERS FOR COHERENT SYSTEMS

D. Campi, G. Destefanis, M. Meliga

CSELT - Centro Studi E Laboratori Telecomunicazioni S.p.A.
Via G. Reiss Romoli, 274
I - 10148 TORINO (Italy)

The potential of coherent optical systems will be effectively exploited when narrow linewidth, high power and tunable single frequency lasers will be available. Here the present research on single frequency semiconductor lasers is discussed, with emphasis on multi-section and multi-quantum- well devices. In particular the tuning model of a three section distributed Bragg reflector source and the linewidth enhancement factor of multi-quantum-well lasers, calculated from the spectral gain, will be discussed, to demonstrate the possibilities of these structures.

1. INTRODUCTION

Optical coherent transmission systems around the 1.55 μm wavelength are a promising alternative to meet the requirements of the future broadband communication network.

The high sensitivity of a coherent system allows the distribution of broadband services to a high number of customers. Moreover in a coherent multi-channel (CMC) a great number of optical frequencies, closely packed in a narrow wavelength range, allow the multiplexing of hundreds of channels [1] among which a subscriber makes his selection by tuning the optical local oscillator (LO) contained in an heterodyne receiver.

The transmitter and LO sources are required to have a narrow linewidth, to limit the receiver bandwidth . Even in the less demanding case of a FSK modulation, a linewidth in the range 0.1-0.2 times the bit rate is required, so that in practical applications a linewidth lower than a few tens of MHz is necessary. Commercial DFB lasers, with linewidths ranging from 10 to 100 MHz, are then bounded to FSK applications, while in a modulation scheme like PSK, allowing a better system sensitivity, sub-megahertz linewidths are necessary. To reduce the linewidth the DFB laser can be mounted with an external cavity mirror, with drawbacks like mechanical stability problems, large external dimensions and high cost.

Thus the introduction of the CMC in the integrated broadband communication network is conditioned by the availability of low cost and reliable narrow linewidth single frequency semiconductor lasers, for which a target performance of 25 mW power and sub-MHz linewidth can be envisaged in a five year perspective.

In FSK systems the information is encoded as a frequency shift of the DFB laser transmitter, produced by modulating the drive current. This electronic change of index is fast, but, at low modulation frequencies, the temperature sweep due to the current produces a refractive index change that is opposite to the carrier effect; then the practical DFB show a dip in its FM characteristic (between 0.1 and 1 MHz) that limits the system performance. A solution to overcome this effect is the use of a dedicated equalizing network in the transmitter circuit, but this is not suitable for a widespread field application. Then lasers with a flat FM response up to 600 Mb/s must be studied.

In the CMC approach the frequency tunable LO laser should sweep across the multi-gigahertz band allocated to some hundred of optically multiplexed channels. Then a tunable semiconductor laser is necessary, with an high power, to enhance the heterodyne receiver sensitivity; likely targets are 5-10 nm tunability with 1 MHz linewidth across the whole frequency range.

To discuss the linewidth $\Delta\nu$ of a laser the expression [3]

$$\Delta\nu \sim \frac{R}{4\pi I}(1+\alpha^2)\,\chi^2 \qquad (1.1)$$

is used, where R is the spontaneous emission rate, I is the intensity of the lasing mode, χ is the fraction of optical length occupied by the active layer and α the linewidth enhancement factor [2]

$$\alpha = \frac{4\pi}{\lambda}\frac{\partial n/\partial N}{\partial g/\partial N} \qquad (1.2)$$

where λ is the radiation wavelength, n the refractive index, N the carrier density in the active layer, g the optical gain (the denominator of (1.2) is often quoted as differential gain). A detuning between the emission wavelength and the peak of the gain curve of the active layer has been successfully used in DFB lasers to increase the differential gain and reduce α.

At present new investigations span over multi-section devices, where the active layer and the grating may either be uniform or limited to portions along the cavity, and the current injection in two or three separate sections can be tailored according to the parameter to be optimized.

With a carrier distribution overcoming the longitudinal hole burning a narrower linewidth emission is obtained. A further improvement is achieved with a monolithic integration of a long passive cavity to the device, because of a reduction of χ.

Frequency tuning is obtained with the injection of different currents in the grating and in the passive phase control section, while flat FM characteristics are obtained with a non-uniform current injection.

In recent times quantum-well (QW) layers in the active region of the single frequency devices has been adopted to enhance the differential gain, allowing linewidth reduction and better bandwidth, efficiency and power.

The aim of this paper is to discuss the properties of multi-section devices and the effect of multi-QW layers on the laser performance, in the frame of the coherent systems. In particular a 3-section tunable DBR laser with a passive integrated cavity will be discussed as a tunable local oscillator and the linewidth enhancement factor will be analyzed for MQW lasers and compared with bulk active layer devices.

2. MULTI-SECTION SINGLE FREQUENCY LASERS

2.1 Multi-section DFB and DBR lasers

The three section DBR laser (Fig. 1) has been proposed [4] as a frequency tunable emitter. It consists of an active section (A), similar to a conventional heterostructure laser, that is monolithically integrated with an extended cavity, formed with a phase control (PC) section and a grating (G) section, that is a wavelength selective distributed Bragg reflector. An important prerequisite is that the device sections are electrically independent, since an electrical cross talk impairs the control of the carriers injected in each section, and also because an increased linewidth has been observed [5], due to current leakage from the active to the passive region, causing an increased optical loss. The principle of the device has been theoretically discussed by several authors [6,7]. The current I_G injected in the grating causes a refractive index decrease, through free-carrier plasma and bandfilling effect, producing a shift of the Bragg wavelength and of the longitudinal modes. The injection of a current I_P in the phase control region modifies similarly the refractive index and allows to re-phase the radiation reflected by the grating towards the active region and to keep the oscillating mode in the center of the filter band.

Among the experimental results on three section tunable lasers, the tuning range of 6.2 nm obtained by Kotaki et al. [8] would be adequate for CMC applications, but the linewidths reaches more than 20 MHz. The structure has been optimized, keeping the linewidth in a range 7-17 MHz, but in a tuning range limited to 3.4 nm [12]. A significant improvement in the linewidth of a tunable laser has been recently obtained with a phase-shifted three-section DFB (fig. 2) by Kotaki [9], that reports a linewidth 1-3 MHz in a continuous tuning range of 2 nm and sub-MHz linewidths in 0.7-0.8 nm intervals In this case the side currents were kept constant, while the central current was used to tune the cavity.

As for three-section DFB coherence, it results that increasing the current in the central section to overcome the longitudinal spatial hole burning [10], a minimum linewidth of 0.79 MHz has been obtained with a device length of 1200 μm.

A sub-MHz linewidth (0.58 MHz @ 2 mW) at a fixed frequency has been obtained with a DBR structure [11]. The device, shown in fig. 3, consists of an active and a passive phase shift region integrated longitudinally with two Bragg reflectors. The length of the cavity and an highly selective reflectivity of the gratings produce a linewidth-power product as low as 1.3 MHz.mW with an output power limited to 2-3 mW; a power of 5-6 mW is obtained with the deposition of an anti-reflective coating onto one facet of the device, then the linewidth increases to 0.75-1.1 MHz Another composite cavity structure that has been studied [13] consists in the integration of a DFB laser with a long (2000μm) passive and a short active cavity (fig. 4) whose behavior is a combination of DFB and DBR. When both DFB and active regions are driven with a proper current, in order to adjust the frequency of the gaining mode on a resonant mode of the passive cavity, a minimum linewidth of 1 MHz can be attained.

As for frequency modulation characteristics, a three electrode DFB laser (fig. 2), allowing an efficient control on the longitudinal carrier distribution, produced a flat FM response about 0.7-0.8 GHz/mA up to 550 MHz with a linewidth 10-20 MHz[14]. More recently a DFB-passive-active device showed a flat 0.7 GHz/mA response with a linewidth of 1 MHz [13].

2.2 Tuning characteristics of three-section DBR laser diodes

The principle of the device, shown in fig 1, has been outlined above: the active section (A) comprises an InGaAsP active layer, emitting at 1.55 μm, that is grown on the InP substrate in a first epitaxial step. After an etching, the A region is defined and, in a second epitaxy, the InGaAsP waveguide, with a composition corresponding to a gap wavelength of ≈1.3 μm, and a cladding InP layer are grown. The lateral confinement of carriers and radiation can be obtained with a buried heterostructure, assuring a strong index guiding to the structure. The InGaAsP waveguide, is almost transparent to the 1.55 radiation and forms a passive integrated cavity. The grating can be realized either on the top of the waveguide or on the substrate, but in this model we consider it merely as a periodic perturbation to the refractive index of the waveguide. Three separate metal contacts are provided to inject independent currents in the sections.

To model the tuning of the device, the carrier densities in the three sections are calculated from the currents, taking into account the recombination mechanisms in the active layer (section A) and in the waveguide (section P and G): non radiative recombination associated with defects and interfaces, bimolecular recombination and non-radiative Auger recombination [7].

For a wide wavelength tuning, current densities in the order of tens of kA/cm^2 are necessary in the passive sections; then the temperature in each section has been evaluated [7] calculating the heat produced at the junctions and also by the currents crossing the different layers of the structure. So the influence of temperature on the refractive indices and gain or absorption of the active and waveguide layers and on the carrier leakage above the heterobarriers can be easily accounted for in the calculations. The temperature dependence of the Auger non-radiative recombination coefficient must be considered too.

The laser operation of the device is described through a condition for longitudinal resonance [6], imposing that the complex reflectivities ($r_{L,R}$) for left and right hand propagating radiation satisfy the relationship

$$r_L(\lambda, N_A)r_R(\lambda, N_P, N_G) = 1 \qquad (2.1)$$

at the interface between the A and P sections, for the wavelength λ. Both reflectivities contain the radiation complex wavenumber and the cavity length (thus the modal indices, with their depression due to injected carrier densities $N_{A,P,G}$, gain and loss are represented). The rightward reflectivity contains also the power transmission at the interface between active and phase control section, and the selective reflectivity of the grating, in terms of its coupling coefficient K and spatial period Λ. The interface between A and P represents a discontinuity in the optical waveguide, where the transmission of radiation is affected by the modal index mismatch of the two sections and by the scattering, due to composition and thickness discontinuity, and to localized defects produced in the epitaxial regrowth.

Equation 2.1 is solved by simultaneously satisfying two conditions describing the gain and the phase of the radiation. The first one gives the carrier density in the active area N_A, as a function of the optical frequency, with N_P and N_G as parameters. The threshold carrier density, calculated for a typical device is reported in fig. 5 as a function of the photon wavelength, referred to the Bragg wavelength of the grating

$$\lambda_o = m \frac{\mu_G \Lambda}{2} \qquad (2.2)$$

that is a function of the current I_G through the modal index μ_G, being m the mode order. The phase condition is set in the form

$$\frac{h(\lambda)}{2\pi} = \text{integer} \qquad (2.3)$$

A typical behavior of h vs. wavelength, is reported in fig 6. The only lasing mode, among the wavelengths giving an integer $h/(2\pi)$, corresponds to the specific wavelength having the lowest carrier density, obtained from fig. 5.
When the current in the P section is kept constant and the current in the grating is increased, the Bragg wavelength shift corresponds to a displacement of the working point on the the curves of fig. 5-6. The wavelength shift is continuous until the work point runs through the minimum of the carrier density curve, but a wavelength jump appears when an adjacent mode reaches a conditions for a lower carrier density. This discontinuous wavelength tuning is represented in fig. 7. The trend of the wavelength shift, corresponding to the decrease of the Bragg wavelength (2.2) due to carrier injection, shows a saturation at high current. This is explained in terms of carrier density saturation , that is substantially determined by the temperature increase and by Auger recombinations.

When the current in the grating is kept constant and the current in the phase control section is varied, the Bragg wavelength remains fixed, while the h curves of fig. 6 shift with respect to the N curve. Then the work wavelength can be swept continuously across the minimum of the N curve, in a range equal to a mode spacing. To keep constant the emitted power during the tuning, the active section current must be increased, to overcome the increasing losses in the P or G region.

To obtain a continuous tuning across the whole available Bragg wavelength shift it is necessary to change the current in the phase control section, in order to shift the wavelength inside the mode spacing, and, in the same time, the grating current, to follow the mode with the selective mirror. The mathematical model makes possible to produce tuning maps (fig. 8) where each line represents the combination of values (I_P, I_G) corresponding to a mode jump, namely the condition for an equal threshold carrier density N_A for two contiguous modes. Then the regions of the plane between the curves correspond to areas of continuous tuning of a definite mode of the device. This results of the model enables the choice of the most effective tuning path to obtain the widest wavelength span without mode jumps. A closer analysis of the device shows that a ripple appears on the mode-jump curves of fig. 8 and that it increases with the reflectivity at the interface between active and passive cavity, then, in particular cases, the areas of continuous tuning in the map may reduce to series of adjacent small cells and the continuous tuning is limited.

The theoretical analysis, in accordance with experimental results, show that the tuning range of the three section DBR laser is mostly determined by the maximum refractive index change in the P section, with a saturation at a current density of about 20 kA/cm^2. As for the device geometry, an high confinement factor and a long P section are favorable for a wide wavelength range, but a long P cavity produces closely spaced modes, reducing the extension of the continuous tuning.

3. MULTI-QUANTUM-WELL SINGLE FREQUENCY LASERS

3.1 Device characteristics

Multi-quantum-well active layer lasers show an higher differential gain in comparison with conventional bulk active layer lasers, then the potential of high speed, high efficiency, narrow linewidth.

Using a multi-quantum-well active layer in a three section tunable DBR laser the linewidth was kept between 5 and 25 MHz [15] for a 1.4 nm tuning. More recently [16] a two section DFB-MQW laser was continuously tuned in a 1.8 nm range, with a linewidth lower than 5 MHz at a constant power of 3 mW. The same device shows a FM efficiency higher than 1.7 GHz/mA; according to the current distribution in the two sections, the dip in the FM characteristics can be kept under at 10 kHz. allowing a 2.4 Gb/s CPFSK heterodyne system.

To show the potential of MQW materials as an active medium in a semiconductor laser, the gain spectra of a InGaAs/AlInGaAs MQW structure will be studied in the following , to show the improvement of differential gain and linewidth enhancement factor

3.1 Optical gain and linewidth in MQW materials

In a quantum-well structure a series of discrete levels are formed due to the quantization in the direction perpendicular to the growth surface. The joint density of states in such confined system exhibits a step-like behavior, with very large values in the vicinity of the threshold transition: this narrows the gain spectrum compared to that of the bulk material, thereby leading to an increase of the differential gain. Here we report a calculation of the dispersion of dn/dN, dg/dN, and α in a test structure comprising is a 5 well InGaAs undoped active layer, lattice matched to InP, with $In_{.53}Ga_{.24}Al_{.23}As$ barriers with a well width of 7 nm. A comparison of the obtained results with those on a DH structure with bulk active layer demonstrates a reduction of α.

When the recombination is dominated by radiative processes, the linear bulk gain can be derived using a model which includes direct band to band transitions with momentum conservation and lifetime broadening effects [17]. The radiative processes, not including lifetime effects for the moment, are described by the following equation

$$G(E)=\frac{1}{nEL}\frac{e^2}{2\pi ch\varepsilon_0}\frac{1}{m_0}\Sigma_{ij}\ |<\Psi_i|\Psi_j>|^2\ |P_{ij}|^2\ \rho_{ij}\ (F_i - F_j) \qquad (3.1)$$

Here ρ_{ij} represents the joint density of states of the j-th heavy or ligth-hole subband and the i-th electron sub-band, $|<\Psi_i|\Psi_j>|$ is the overlap integral, and F_i and F_j are the occupation factors. P_{ij} is the momentum matrix element, which, in general, is energy-dependent and behaves differently for the TM and TE modes and for ligth and heavy holes transitions [17]. The constants n, L, c, h, ε_0, m_0 and e are the refractive index, the total length of the structure, the velocity of light, Planck's constant, the vacuum permittivity, the bare electron mass and the electronic charge, respectively. The lifetime broadening due to intra-band relaxation, is accounted for by the following convolution integral

$$g(E)=\frac{1}{2\pi}\int_0^\infty dE'\ G(E')\ \frac{\Gamma}{(\Gamma^2/4)+(E-E')^2} \qquad (3.2)$$

where Γ is the broadening parameter, whose assumed value was 20 meV. It is worth noting that this value is not critical for the conclusion on the linewidth enhancement that are derived from this calculation. Equation 3.2 is plotted in fig. 9 for different carrier concentrations. The band gap shrinkage was taken into account by shifting the curves toward the low energy side by a quantity assumed to be proportional to $(N_{2D})^{1/3}$, being N_{2D} the two-dimension carrier density [19]. From eq. 3.2 the differential gain can readily be obtained. The calculation of α can be performed using the Kramers-Kronig transform giving the real part of an optical function in terms of the imaginary part. Differentiation of this relation with respect to carrier density N at constant photon energy gives.

$$\frac{dn(E)}{dN}=\frac{hc}{2\pi^2}\ P\int_0^\infty dE'\ \frac{1}{E^2-E'^2}\ \frac{dg(E')}{dN} \qquad (3.3)$$

where P denotes the prinicpal part of the integral.Then α is readily calculated from eq. 1.2. From the simple considerations about the density of states in the above, an increase of the denominator of (1.2) can be expected with the use of QW active layers. The numerator, however, is also expected to enhance through eq. 3.3. It is therefore difficult to predict the behavior of α without explicit calculations. Figure 10 displays the results of the (exact) numerical calculation of α based on eq. 1.2 and eq. 3.3 for different carrier concentrations. This results is comparatively smaller than the value $|\alpha| = 4\text{-}5$ measured on lasers with bulk InGaAsP active layer [22]. It is also clear that α depends strongly on N, its absolute value increasing for larger N. Owing to this strong dependence, it is very important that the number of quantum-wells is optimized so that the threshold current is minimum. It should be stressed that this could be done only if the total loss is known [21].

Moreover, some additional considerations can be made basing on the rough approximation that most of the thermalized carriers lie within the lowest sub-band. In this limit, the quasi-Fermi level is roughly proportional to the product of the width of the structure, multiplied by the injected carrier density. In fact, α crosses zero for a photon energy roughly proportional to the quasi-Fermi level; thus, from the consideration above, we can see that the magnitude of α at a given energy increases with increasing well width.

The calculation presented here demonstrates the effectiveness of using a quantum-well active layer in reducing the linewidth-enhancement factor, though this is dependent on using optimized structural parameters, together with appropriate wavelength selection through a DFB grating. Considering as an example, a region of energies above the maximum gain, where both g and dg/dN can be kept high, MQW have $1+\alpha^2 \approx 1.25\text{-}2$, whereas a value of 17-26 results for bulk material [22], then a linewidth reduction of more then 10 times can be expected with MQWs.

4. CONCLUSIONS

The research in single frequency lasers with narrow linewidth and tunable frequency for coherent transmission applications is presently devoted to the exploitation of multi-cavity or multi section DFB and DBR devices and of MQW active layers.
The former allows to tailor the current distribution in the device, yielding sub-MHz linewidths, wavelength tunability and flat FM response. The modelling of a three section DBR laser predicts the tunability range and shows clearly the saturation effect due to thermal effects and Auger recombination at high current injection.
The MQW active layer results in a higher differential gain producing narrow linewidth and high efficiency devices. The calculation of the linewidth enhancement factor for a InGaAs/AlInGaAs MQW material predicts in principle a linewidth reduction of more than 10 times, if compared with bulk active layer lasers.

ACKNOWLEDGEMENTS

The activity on tunable lasers is partially supported by Progetto Finalizzato Telecomunicazioni of the Consiglio Nazionale delle Ricerche. The activity on MQW lasers is partially supported by EEC under Project R1057 "AQUA".

REFERENCES

[1] Bachus, E.-J et al., Electron. Lett. 22 (1986) 1002
[2] Henry, C. H., IEEE J. Quant. Electron., QE-18 (1982) 259
[3] Henry, C. H. J. Lightwave Technol., LT-4 (1986) 288
[4] Coldren, L.A.; Corzine, S.W., IEEE J. Quantum Electron. QE-23 (1987) 903
[5] Kotaki, Y., Ishikawa, H., Proc. of the 11th IEEE International semiconductor laser conference, Boston, Aug. 29- Sept. 1, 1988, 128-129
[6] Pan, X. et al, IEEE J. Quantum Electron. 24 (1988) 2423
[7] Caponio, N.P. et al., to be published in IEEE J. Select. Areas in Comm.
[8] Kotaki, Y. et al., Electron. Lett. 24 (1988) 503
[9] Kotaki, Y. et al., Proc. of the Seventh Intern. Conf. on Integrated Optics and Optical Fiber Communications, Kobe (J), July 18-21, 1989, paper 19A2-4
[10] Soda, H. et al, Electron. Lett. 22 (1986) 1047
[11] Tohmori, Y et al., Proc. of the Seventh Intern. Conf. on Integrated Optics and Optical Fiber Communications, Kobe (J), July 18-21, 1989, paper 19A3-3
[12] Murata, S. et al., Proc. of the 11th IEEE International semiconductor laser conference, Boston, Aug. 29- Sept. 1, 1988, 122-123
[13] Liou, K.Y et al., Proc. of the Seventh Intern. Conf. on Integrated Optics and Optical Fiber Communications, Kobe (J), July 18-21, 1989, paper 21D3-2
[14] Leclerc, D. et al., Electron. Lett. 25 (1989) 45
[15] Koch, T.L. et al., Electron. Lett. 24 (1988) 1431
[16] Kitamura, M. et al., Proc. of the Seventh Intern. Conf. on Integrated Optics and Optical Fiber Communications, Kobe (J), July 18-21, 1989, paper 19C1-5
[17] Zielinski, E. et al., IEEE J. Quantum Electron., QE23 (1987) 969
[18] Asada, M. et al., IEEE J. Quantum Electron., QE20 (1984)745
[19] Kleinman, D.A., and Miller, R.C., Phys. Rev. B, 32 (1985) 2266
[20] Westbrook, L.D., and Adams, M.J., IEE Proceedings J, 135 (1988) 223
[21] Arakawa, Y. et al., IEEE J. Quantum Electron., QE-21 (1985) 1666
[22] Westbrook, L.D, Electron. Lett. 21 (1985) 1018

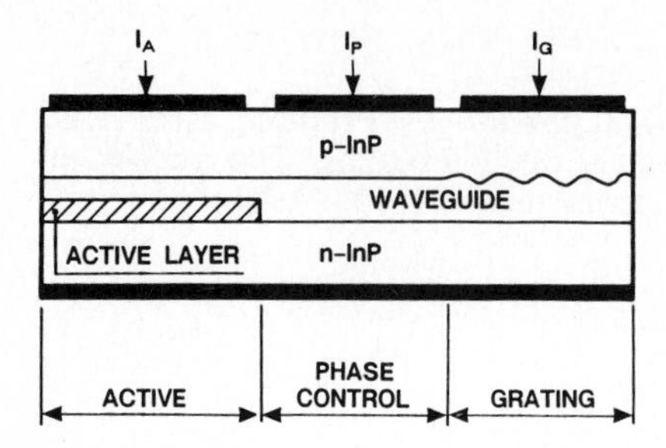

Fig. 1. Schematic structure of a three-section tunable DBR laser.

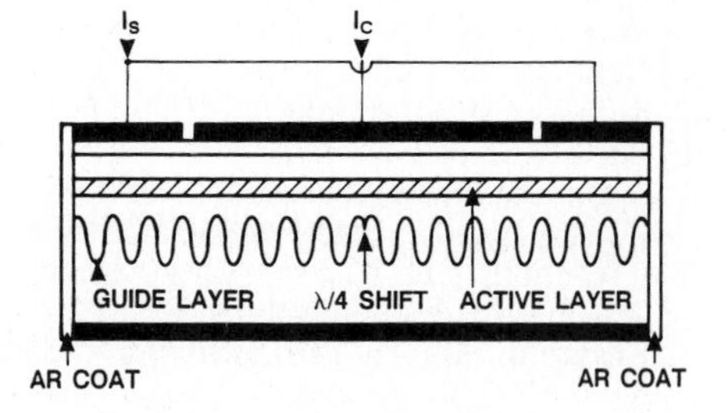

Fig. 2. Schematic structure of a three-section multi-electrode DFB laser, after Kotaki et al. [9].

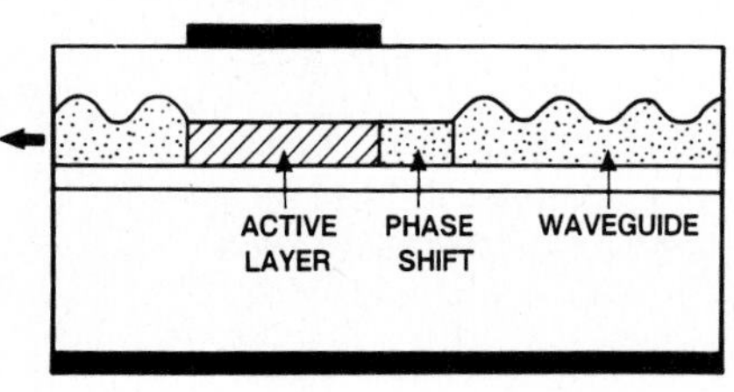

Fig. 3. Schematic structure of a butt-jointed DBR laser, after Tohmori et al. [11].

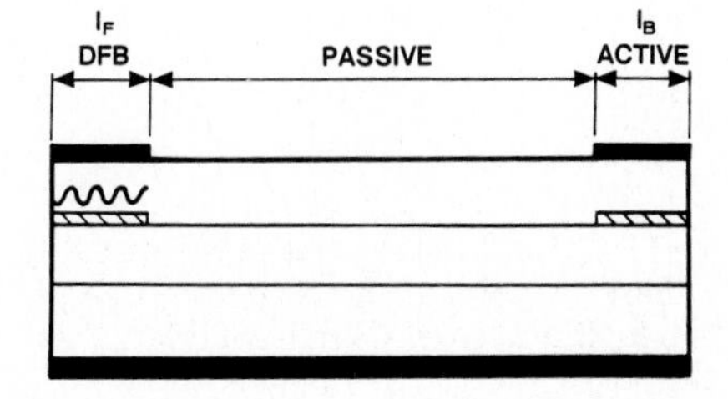

Fig. 4. Schematic structure of a DFB-passive-active laser, after Liou et al. [13].

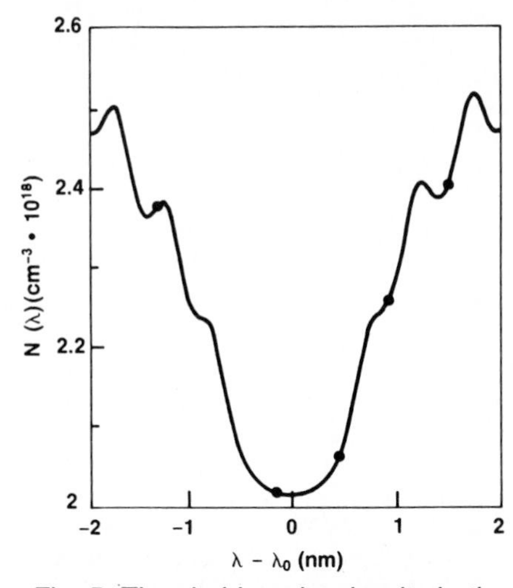

Fig. 5. Threshold carrier density in the active layer of a three-section tunable DBR laser (Fig. 1), as a function of the wavelength deviation from Bragg wavelength λ_0 [7]. The dots mark the longitudinal modes of the complex cavity, satisfying eq. (2.3).

Fig. 7. Discontinuous tuning of the device of Fig. 1 by means of the current I_G injected in the grating section.

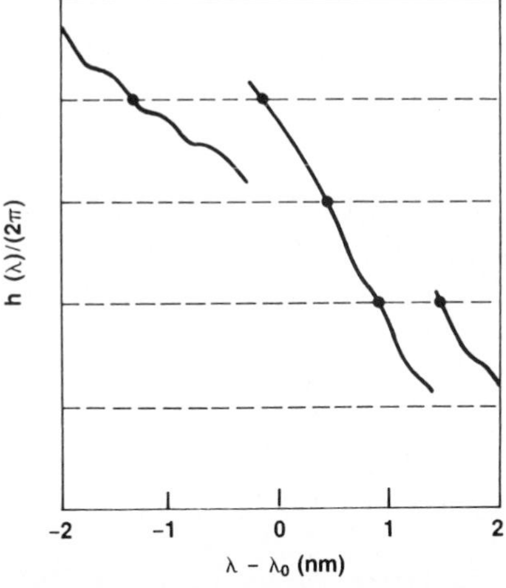

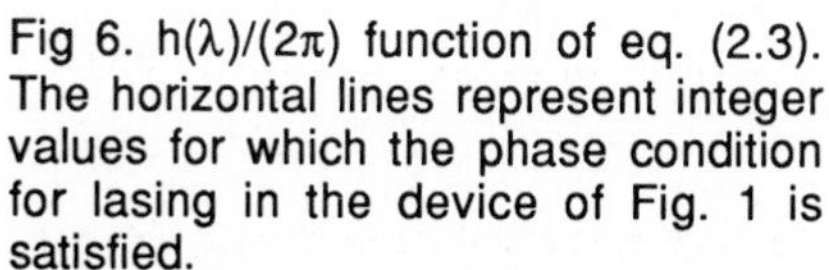

Fig 6. $h(\lambda)/(2\pi)$ function of eq. (2.3). The horizontal lines represent integer values for which the phase condition for lasing in the device of Fig. 1 is satisfied.

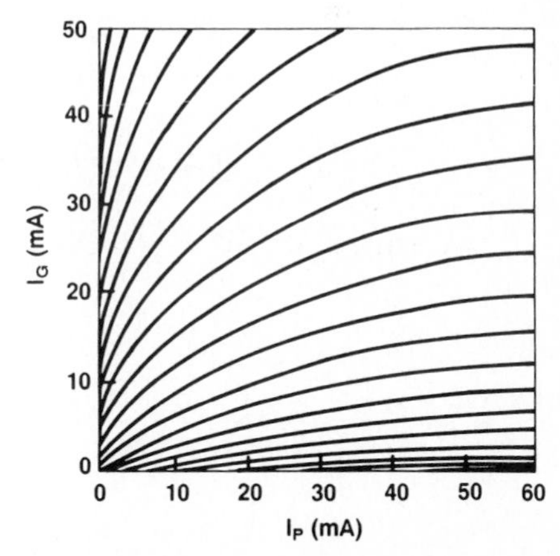

Fig. 8. Tuning map of the device of Fig. 1 , when both I_G and I_B are varied. Each curve represents a mode-jump of the cavity.

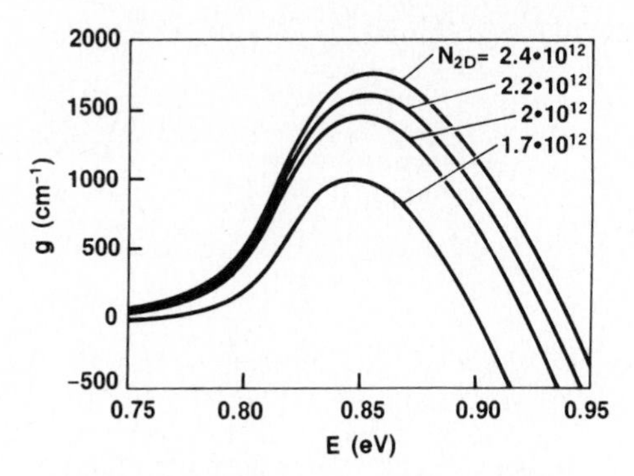

Fig. 9. The gain as a function of the photon energy E, calculated for various carrier densities, for the MQW active layer described in the text.

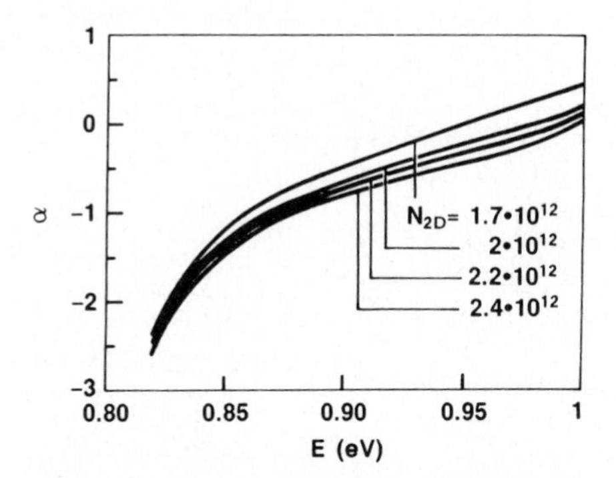

Fig. 10. Dispersion of the linewidth-enhancement factor a, calculated for different carrier densities, for the MQW active layer described in the text.

PROGRESS IN OPTICAL AMPLIFIERS

M.J.O'Mahony

British Telecom Research Laboratories
Martlesham Heath, Ipswich, IP5 7RE, UK

Interest in all-optical amplifiers has increased significantly in recent years because of their potential as direct light amplifiers in optical systems. Optical amplifiers can be based either on semiconductor laser structures or on optical fibre. Currently semiconductor amplifiers are the most developed, but erbium fibre amplifiers are now being demonstrated with very promising results. Non-linear devices based on semiconductor laser amplifiers, for example wavelength convertors, are also very much of interest for future networks.

1. INTRODUCTION

Optical amplifiers will play an important role in future optical systems as they enable the direct amplification of light with a minimum of electronics and thus eliminate the electronic bottleneck associated with many current systems. They can be used in both linear and non-linear modes of operation, but the initial applications will be as linear amplifiers in both long haul and local optical networks, using either coherent or direct detection. The linear amplifier has a very wide bandwidth, in the order of THz, and its incorporation into an optical system or network will allow future upgrading of capacity by an increase in the number of wavelengths (direct detection WDM or coherent FDM) or by an increase in the bit rate. In an optical system using linear amplifiers this upgrading can be achieved without modification of the hardware, ie the system is transparent to changes in signal format. For local networks amplifiers may be used to overcome the splitting losses associated with passive networks and increase fan-out capability. Indeed for many of the future networks and systems being proposed, optical amplifiers are essential components. This paper, therefore, reviews the characteristics of semiconductor and fibre amplifiers and comments on recent experiments and their results.

2. OPTICAL AMPLIFIERS

Optical amplifiers are based on either semiconductor laser structures or specially doped optical fibre. Semiconductor amplifiers are more developed at present, as they are based on existing laser structures with anti-reflection coatings applied to the facets. Recent developments have resulted in devices exhibiting high gain and low polarisation sensitivity combined with high saturation power. The progress in fibre amplifier technology has also been very rapid in the last year and semiconductor pumped erbium amplifiers have been demonstrated in many system experiments. Fibre amplifiers have the particular advantage that they can be spliced into the system fibre with very low loss and this is in contrast to the large coupling losses associated with semiconductor amplifier fibre coupling. At present therefore fibre amplifiers exhibit higher net gains than semiconductor amplifiers.

2.1 Semiconductor Laser Amplifiers

The semiconductor laser amplifier is based on the normal semiconductor laser structure [1,2,3]. Gain occurs in the active region and the device length is normally between 250-500μm. In a standard laser the end facets have effective reflectivities of approximately 30 % due to the difference in refractive index between the active material and air. As the bias current is increased the optical gain in the active region increases and the facet reflections provide the necessary optical feedback to stimulate oscillation when the device is used as a lasing source. In a semiconductor laser amplifier, operation is at a bias current less than the lasing threshold current, ie the device is operating as an optical amplifier rather than an oscillator. Light entering one facet appears amplified at the other facet. As with any amplifier, noise is also present.

The facet-facet gain of a semiconductor amplifier has the form [1]:

$$G = \frac{(1-R_1)(1-R_2)G_s}{(1-\sqrt{R_1R_2}\,G_s)^2 + 4\sqrt{R_1R_2}\,G_s\sin^2\phi}$$

where R_1,R_2 are the input and output facet reflectivities respectively, G_s the single pass gain through the device and f the phase shift ($=2\pi NL/\lambda$, where N is the refractive index and λ the wavelength). An amplifier with significant optical feedback ($G_s\sqrt{R_1R_2}$ large) is generally referred to as a Fabry Perot amplifier, as the passband is dominated by the Fabry Perot resonances associated with the optical cavity. Such amplifiers are characterised by narrow bandwidths (in the order of GHz) and a high sensitivity to temperature, bias current and signal polarisation fluctuations. For most applications however, wide bandwidths are essential and it is necessary to reduce the feedback. This is normally achieved by applying anti-reflection (a.r.) coatings to the facets. Thus wideband semiconductor amplifiers are based on laser structures with additional a.r. facet coatings

Perfect anti-reflection coatings would result in a travelling wave amplifier (TWA) with no internal reflections and a bandwidth determined by the active material. In this case the single pass gain through the amplifier is given by G_s=EXP(gL), where g is the nett gain per unit length (which is a function of bias current) and L is the amplifier length. In practice there is always some residual reflectivity and the intrinsic gain and bandwidth characteristics of the material are modified by the optical cavity, ie resonance effects are introduced. This is observed as a gain ripple across the amplifier passband. The ripple period is a function of the amplifier length and its amplitude is determined by the residual facet reflectivity and the single pass gain of the amplifier. An interesting rule of thumb is that if one assumes a ratio of peak gain to ripple amplitude of 3 dB, the cavity gain equation above can be shown to reduce to the approximation:

$$G \approx 0.25/\sqrt{R_1R_2}$$

Thus with a residual facet reflectivity of 0.1 % a peak gain of 24 dB is possible with a gain ripple of 3 dB. This equation illustrates the importance of reducing the facet reflectivity. State of the art devices have residual reflectivities better than 0.1%, allowing a facet-facet gain of 24 dB with a bandwidth of approximately 50 nm (7 THz). This performance is unequalled by any electrical component.

Fig.1 shows the gain versus current characteristic of a state of the art low polarisation semiconductor amplifier measured for orthogonal signal polarisations [4]; the residual facet reflectivity was 0.03%.

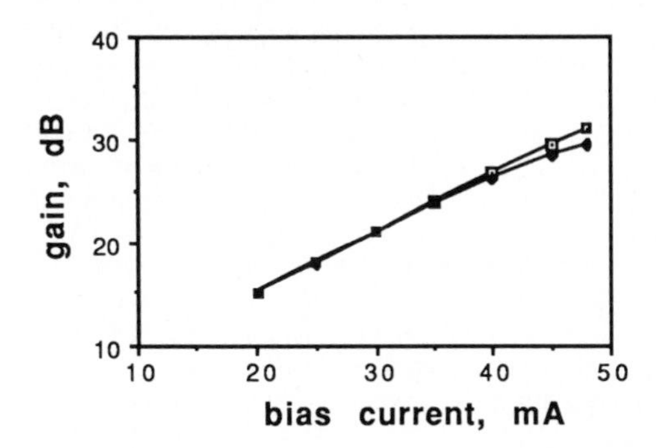

Fig.1: Amplifier gain vs bias current
TE:upper curve, TM: lower curve

The main features of the results in Fig.1 are that (a) gains in excess of 30 dB are possible when the amplifier is biased close to threshold (approximately 50 mA) and (b) the gain variation with signal polarisation is less than 1 dB at a gain of 25 dB.

An important feature of all amplifiers is the variation of gain with input signal power and Fig.2 shows the measured characteristic of a recent buried heterostructure device.

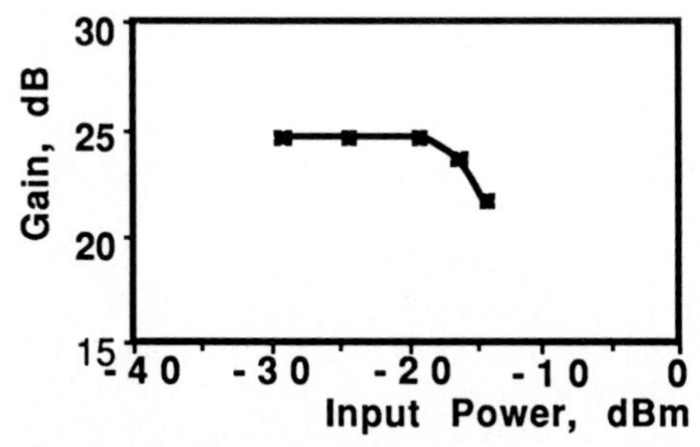

Fig.2: Gain vs input power

The curve shows that the gain has fallen by 3 dB when the input power has increased to -14 dBm. This corresponds to an output saturation power of +7 dBm.

The gains measured above were facet to facet gains. In a real system it is necessary to couple the light to and from the amplifier and with a semiconductor amplifier there are associated coupling losses. Typically these losses are about 3.5 dB/facet and so the fibre to fibre gain of the amplifier will be 7 dB lower than the facet-facet gain. For this reason 20 dB is likely to be the maximum realistic gain of a wideband semiconductor amplifier.

Semiconductor amplifiers were the first optical amplifiers to be used in system orientated experiments. From these experiments a number of problems were identified, although these are becoming less significant with the progress in device development and coating technology. The major drawbacks are: the loss of gain associated with fibre coupling, the backward travelling waves associated with the finite facet reflectivities, and low saturation power. Saturation power in particular is a most important parameter for future system applications. For example, it determines whether amplifiers can be used as power boosters or to increase fan-out capability. Also in many of the future networks being proposed a large number of channels may be required to pass through the amplifier and this demands a high power handling capability. For these reasons, amongst others, attention was paid to amplifiers based on glass fibres.

2.2 Fibre amplifiers

There are a number of varieties of fibre amplifier, but their physical forms are similar. Fig.3 is the schematic diagram of a fibre amplifier. The amplifier comprises a length of specially doped optical fibre together with an associated optical pump (generally a high power laser) to achieve population inversion (the optical pump performs the same function as the bias current in the semiconductor laser amplifier). In operation pump and signal are combined in the amplifying fibre.

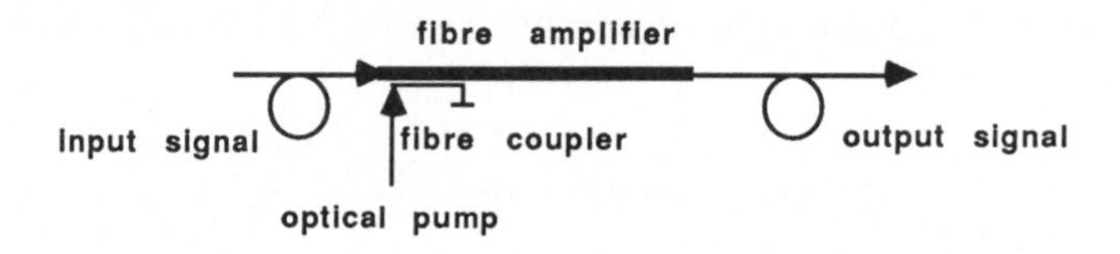

Fig.3: Fibre amplifier schematic

In a *Raman fibre amplifier* [5] the amplification mechanism is Raman scattering in the fibre which causes an energy transfer from the pump to the signal. The pump wavelength is lower than the signal, for example, amplification of a 1.55 μm signal requires a pump wavelength of 1.45 μm. To enhance the gain, the spot size of the fibre must be reduced in order to increase the pump intensity within the fibre. This is frequently achieved by germania doping of the silica fibre to increase the refractive index difference between the core and cladding. Even with the best designs, however, pump powers in the order of 300 mW in the fibre are necessary to provide gains in the order of 15 dB. A key feature of Raman amplifiers is that output signal powers in the order of 200 mW can be obtained, which is an order of magnitude increase on the semiconductor device. The gain bandwidth is similar to that of a semiconductor amplifier ie 40 -50 nm. The major problems are the length of the amplifier, which is generally greater than 1 km, and the requirement for a high power semiconductor pump.

In *rare earth fibre amplifiers* [6-8] the pump energy is absorbed at specific wavelengths and amplification is at a wavelength.determined by the rare earth dopant. For example, with erbium doping amplification occurs in a band centred on 1.53 μm and the pump wavelengths currently being considered are 980 nm and 1470-1500 nm. A typical state of the art result, for an erbium fibre amplifier, would be a signal gain of 25 dB with a 35 nm bandwidth for a pump power of 50 mW at 1.49 μm [9]. Maximum signal output saturation powers to date are in the order of 15 mW, which are similar to the best semiconductor devices.

In addition to the low coupling loss associated with erbium amplifiers there are other particular advantages, namely, the amplifier is inherently polarisation insensitive, the lifetimes associated with the process are such that that crosstalk in the presence of a number of wavelengths is significantly reduced and the amplifier is an almost true travelling wave device, which should eliminate the problems of backward propagating signals. Erbium amplifiers also have a lower noise figure than semiconductor amplifiers, partly because of the intrinsic mechanisms, but mainly because of the reduced coupling loss at the input.

The limits to saturation power in erbium amplifiers have not yet been fully explored, but it would seem that saturation powers much greater than that available with semiconductor devices can be achieved provided sufficient pump power is available.

The key feature in the deployment of fibre amplifiers is the development of high power semiconductor pumps and there has been significant progress in this area in the past year. Pumps with a wavelength in the region of 1470-1500 nm are suitable for both Raman and erbium amplifiers and devices with powers of 170 mW (facet output) have now been demonstrated [10].

2.3 Noise in Amplifiers

The performance of optical amplifiers in systems depends upon a number of factors, such as gain, bandwidth, linearity, stability with respect to temperature and bias current fluctuations etc. Particularly important, however, is the noise performance. In optical amplifiers noise is introduced by the background spontaneous emission of photons and in both semiconductor and fibre amplifiers the output noise comprises a number of components which are functions of the amplified signal power and the spontaneous noise power, and their crossproducts [11]. Assuming the broadband spontaneous noise is removed by filtering, the amplifier noise performance may be characterised by its noise figure. This has the form $2\gamma/k$, where γ is the population inversion parameter and k the input coupling efficiency. The best noise figure, therefore, is 3 dB. For a semiconductor amplifier $\gamma \approx 1.5$ and $k \approx 0.5$, yielding a noise figure in the order of 8 dB. Fibre amplifiers with a higher coupling efficiency and better population inversion have noise figures closer to the ideal. A recent erbium amplifier experiment, for example, yielded a noise figure of 4.8 dB [12].

A significant feature of fibre amplifiers, therefore, is the improvement in the noise figure due to the lower input coupling loss.

3. APPLICATIONS AND EXPERIMENTS

The number of possible applications of optical amplifiers to future systems is large and ever increasing. Some of the main applications are briefly discussed below.

3.1 Linear Repeater

In transmission systems using single longitudinal mode lasers the effects of fibre dispersion may be small and the main limitation on regenerator spacing is the signal attenuation due to fibre loss. Such systems do not require a complete regeneration of the signal at each repeater and linear amplification is sufficient. Optical linear amplifiers can therefore be used as repeaters for intensity modulated or coherent systems. Indeed optical linear amplifiers are currently the only viable form of coherent repeater Some particular advantages of a linear optical amplifier repeater are:

(a) *reduced complexity and size* in comparison with conventional electro-optic repeaters and reduced power requirements, the latter is of particular importance in undersea systems.

(b) *transparency.* Because of their very wide bandwidth, linear amplifier systems are transparent to bit rate and system capacity may be upgraded by increasing the bit rate or using additional wavelengths

(c) *bidirectional operation* . Transmission in either direction is possible and indeed simultaneous transmission in both directions has been demonstrated. This considerably simplifies the hardware needed in a real system.

There are also a number of disadvantages, namely:

(a) *The effective repeater gain is less* than that available with a conventional opto-electronic repeater and this implies a closer repeater spacing for multi amplifier systems. The fibre section loss with a conventional 565 Mbit/s repeater system is approximately 35 dB, for example, but with a semiconductor amplifier this is unlikely to exceed 20 dB.

(b) *A linear repeater system is an analogue system* hence noise and jitter accumulate along the chain of repeaters limiting the number that can be cascaded before full regeneration is necessary.

(c) *In a semiconductor repeater* an amplified signal is propagated in the backward as well as the forward direction. The backward wave is caused by the finite refectivity of the amplifier facets. Unless this reflectivity is sufficiently reduced, a multi amplifier system can become unstable.

In recent years a number of system experiments using optical amplifier repeaters have been reported. The majority of these have used semiconductor amplifiers

[13,14], but in the past year there have been an increasing number of erbium fibre amplifier experiments. A particular requirement of amplifier repeaters is the need for low polarisation sensitivity and this has led to the recent development of low polarisation semiconductor amplifiers. NEC, for example, have developed a window facet structure amplifier in which the active region is terminated in a low refractive index window region [15]. At 1.564 μm a gain of 16 dB (fibre to fibre) was achieved with an effective facet reflectivity of 0.03% and a saturation power of +6.5 dBm. The maximum TE-TM gain difference was 1.3 dB. British Telecom have also developed a buried heterostructure type device with a modified structure. In this case residual facet reflectivities of approximately 0.03% have been measured with a TE-TM gain difference of less than 1 dB and an output saturation power of +7dBm [4].

System experiments have moved forward on a number of fronts, in particular, demonstrations of an increased number of cascaded amplifiers, multi-channel operation and the performance with picosecond pulses. Recently KDD reported a 516 km direct detection 2.4 Gbit/s/s system using ten amplifiers in cascade [16] and a 4 Gbit/s system using three packaged amplifiers over a distance of 188 km. The simultaneous amplification of a ten channel coherent FDM system has also been demonstrated by NEC.

The feasibility of using erbium amplifiers in systems has been demonstrated in a number of recent experiments. KDD, for example, have reported a 267 km, 1.2 Gbit/s system using two fibre amplifier repeaters [17]. NTT have similarly demonstrated a 210 km 1,8 Gbit/s system using a diode pumped erbium repeater with a 20 dB gain [18].

An important area of current research concerns the simultaneous transmission of signals of different wavelengths though optical amplifier systems; for example a recent experiment has demonstrated the transmission of twenty channels through a semiconductor amplifier [19]. The main area of investigation is the level of crosstalk in such systems [20]. Crosstalk is to be expected for a number of reasons. In an intensity modulated system if one channel is operating in the non-linear region, ie with gain saturation, the amplifier gain will be modulated causing crosstalk in the other channel. The degree and nature of this crosstalk differs for semiconductor and fibre amplifiers because of the different time constants associated with each device. For standard optical systems fibre amplifiers should be much better in this respect. Four wave mixing between signals will also generate new components which may fall within the passband of other signals [21].

3.2 Receiver Preamplifier

Optical amplifiers can be used as preamplifiers to increase the sensitivity of optical receivers. By providing optical gain prior to the photodiode, the signal and the associated amplifier noise are increased above the receiver noise level. Optical preamplifiers, therefore, offer the advantage of reducing or eliminating the effects of receiver thermal noise, a similar function to that provided by coherent detection. The improvement can be particularly marked for bit rates in excess of 1 Gbit/s and enables the development of sensitive wide-band receivers, which are necessary for future high capacity systems.

An early study of semiconductor laser preamplifiers was reported by Mukai et al [11], which covered much of the theoretical analysis and also gave the results of experiments at 850 nm. These showed an improvement in sensitivity of approximately 7 dB by the use of a preamplifier in front of a low noise silicon APD receiver operating at 100 Mbit/s. In 1986 the first long wavelength results were reported at 500 Mbit/s and 2 Gbit/s [22,23] respectively, showing that preamplifiers could exceed and match the performance of the best conventional optical receivers.

More recently preamplifier experiments using erbium fibre amplifiers have been reported and these results have demonstrated the low noise figure associated with fibre amplifiers. For example a sensitivity of -40 dBm was reported by ATT at a bit rate of 1.8 Gbit/s [12].

3.3 High Speed Pulse Amplifiers

The dynamic bandwidth of semiconductor and fibre amplifiers has been demonstrated in a number of experiments using high speed pulses. Systems operating at 5 Gbit/s through semiconductor amplifiers have been reported. More recently the demonstration of a 12 Gbit/s experiment using erbium fibre amplifiers has been reported by Fujitsu [24]. In the picosecond pulse regime optical amplifiers can amplify pulses to high peak powers. Erbium fibre amplifiers are particularly attractive in this respect because of the longer associated time constants. Soliton amplification and transmission using erbium amplifiers has recently been demonstrated. For example NTT recently reported a 20 Gbit/s soliton experiment [25] using an erbium amplifier for amplification and transmission.

4. NON-LINEAR AMPLIFIER DEVICES

Semiconductor laser amplifiers can also be operated in a *non-linear mode* [26]. In Fabry-Perot amplifiers, the change in refractive index with input signal is sufficient to observe non-linear behaviour. This can be used, for example, to provide optical pulse shaping,and is part of the requirement for an all optical regenerative repeater. Pulse shaping and thresholding functions have also been demonstrated using amplifier absorber devices. Work on dispersive non-linear devices has shown that the narrow bandwidths involved (in the order of 5 GHz) would make it extremely difficult for the devices to be used in real systems. Absorptive non-linear and bistable devices should in principle have wider bandwidths as these devices can operate with anti reflection coated facets. Switching threshold as low as -50 dBm have been demonstrated. Maximum repetition rates are limited by carrier recombination times to 1 GHz, but may be increased by reverse biasing the absorber.

Four wave mixing in wideband semiconductor amplifiers is also of current interest. With two signals present at the amplifier input, at appropriate levels, additional frequency components can be observed at the output This phenomenon can be used to demonstrate frequency translation, an important function for future multi wavelengths systems. A recent experiment at Heinrich Hertz Institute, for example, demonstrated the translation of a 140 Mbit/s DPSK signal by 1500 GHz with an efficiency of -10 dB [27].

A more recent example of the use of a semiconductor amplifier is as a phase modulator in a coherent system [28]. Modulation of the injection current varies the refractive index of the active region and provides the required modulation. This type of modulator provides gain in contrast with the loss associated with the standard Lithium Niobate component and has lower modulation voltages. System experiments have demonstrated a 16 mA p-p current requirement for a π phase shift.

5. SUMMARY

Progress in the development of optical amplifiers has made rapid strides in the past few years as the potential advantages have become more obvious and the technology has improved. The particular advantage that appeals to system planners is the promise of transparent systems using linear optical repeaters. Such systems can be upgraded in capacity at some future date without any modification of the repeaters.

Currently semiconductor laser amplifiers are the most developed and have already been demonstrated in installed optical systems. Development programs have in hand to optimise laser structures for amplifier applications and the initial results have been reported in the past year. Thus low polarisation high power amplifiers are now becoming available. These devices will have fibre to fibre gains in the order of 15-20 dB and 10-20mW saturated output power. Polarisation sensitivities in the order of 1 dB have been demonstrated.

In the past year the main thrust has been in the use of rare earth doped fibre amplifiers with reported gains of 30 dB and bandwidths in the THz region. The availability of semiconductor laser pumps has triggered a great number of system experiments demonstrating the suitability of fibre amplifiers as repeaters, preamplifiers and power boosters.

Non-linear applications such as frequency translation using four wave mixing and optical pulse shaping are still in the early stages of research, but are still very much of interest for future systems.

ACKNOWLEDGEMENTS

The author wishes to thank the Directors of British Telecom Research Laboratories for permission to publish this paper and colleagues in both the systems and devices divisions for their contribution to the work.

REFERENCES

[1] Yamamoto,Y "Noise and error rate performance of semiconductor laser amplifiers in PCM-IM optical transmission systems." IEEE J. Quantum Electron., Vol. QE-16, No.10, Oct. 1980,.pp.1073-1081

[2] Simon,J.C., "Semiconductor laser amplifier for single-mode optical fiber communications," Journal of Optical Communications., Vol. 4, No. 2, 1983, pp. 51-62

[3] O'Mahony, M.J. "Semiconductor Laser Amplifiers for Future Fibre Systems", J.Lightwave Technology,Vol.6, No.4, April 1988, pp.531-544

[4] Devlin,W.G., Elton,D.G., Reid, G.,Sherlock,A.D.,Ellis,A., Cooper,D., Isaac,J., Spurdens,P., Cockburn,A., Stallard,W."Polarisation Insensitive High Output Power 1.3 μm and 1.5 μm Optical Amplifiers made by MOVPE", IOOC'89, Kobe Japan, 1989, Paper 20C2-1.

[5] Stolen, R.H. "Non-linearity in Fiber Transmission", Proc. IEEE, Vol. 68, No.10, October 1980, pp.1232-1237

[6] Urquhart, P. "Review of rare earth doped fibre lasers and amplifiers", IEE Proceedings, Vol. 135 Pt. J, No. 6, December 1988, pp. 385-407

[7] Desurvire,E, Simpson,J.R., Becker,P.C., "High gain erbium doped travelling wave fiber amplifier", Optics letters, Vol.12, No.11, Nov. 1987, pp.888-890.

[8] Laming, R.I.,Reekie,L., Payne,D.N. "Optimal pumping of Erbium Doped Fibre Optical Amplifiers", ECOC 88, Brighton, Postdeadline Vol.,pp.25-27.

[9] Atkins,C.,Massicott,J.,Armitage,J.,Wyatt,R.,Ainslie,B.J., Craig-Ryan,S. "A High Gain, Broad Spectral Bandwidth Erbium Doped Fibre Amplifier Pumped near 1.5 μm",Electronics Letters, Vol.25, No.14, July 1989, pp.910-911

[10] Mito,I., Yamazaki,H., Yamada,H., Sasaki,T., Takano,S., Aoki,Y., Kitamura,M. "170 mW High Power CW Operation in 1.48-1.51μm InGaAs MQW-DC-PBH LD", IOOC'89, Kobe, Japan, 1989, Paper 20PDB-13.

[11] T.Mukai, Y.Yamamoto, and T.Kimura, "S/N and error rate performance in AlGAs semiconductor laser preamplifier and linear repeater systems, "IEEE Trans Microwave Theory Tech., Vol. MTT-30, No. 10, 1982, pp.1548-1556.

[12] Giles,C.R., Desurvire,J.L., Zyskind, J.L., Simpson, J.R. "Near Quantum Limited Erbium doped Fiber Preamplifier with 215 Photons/bit Sensitivity at 1.8 Gbit/s", IOOC'89, Kobe, Japan, 1989, Paper 20PDA-5.

[13] Malyon,D.,Stallard,W., "565Mbit/s FSK Direct Detection System Operating With Four Cascaded Photonic Amplifiers", Electronics Letters, Vol.25, No.8, April 1989,pp.495-496

[14] Olsson, N. A., Oberg, M. G., Koszi, L. A. and Przbylek, G. J. "400 Mbit/s, 372 km coherent transmission experiment using in line optical amplifiers", Electron. Lett. Vol, 24, No. 1, (1988),p36.

[15] Cha,I.,Kitamura,M., Honmou,H., Mito,I.,"1.5μm Band Travelling Wave Semiconductor Optical Amplifiers with Window Fact Structure",IOOC'89, Kobe, Japan, 1989, Paper 20C2-2

[16] Yamamoto,S.,Taga,H.,Edagawa,N.,Mochizuki,K.,Wakabayashi,H. "516km,2.4 Gbit/s Optical Fiber Transmission Experiment using 10 Semiconductor Laser Amplifiers and Measurement of Jitter Accumulation", IOOC'89, Kobe, Japan, July 1989, Paper 20PDA-9

[17] Edagwawa,N., Suzuli,M., Mochizuki,K.,"267km, 1.2 Gbit/s Optical Transmission Experiment Using Two In -Line LD-Pumped Er Doped Optical Fibre Amplifiers and an Electroabsorption Modulator", IOOC'89, Kobe, Japan, 1989, Paper 21B4-1

[18] Takada,A., Hagimoto,K., Iwatsuki,K., Aida,K.,Nakagawa,K., Shimizu,M. "1.8 Gbit/s Transmission over 210 km Using an Erbium Doped Fiber Laser Amplifier with 20 dB Repeater Gain in a Direct Detection System", IOOC'89, Kobe, Japan, 1989, Paper 21B3-3

[19] Coquin,G, Kobrinski,H, Zah,C.E, Shokoohi,F,Caneau,C. Menocal,S. "Simultaneous amplification of 20-channels centred at 1.54 um in a multiwavelength distribution system", ECOC 88, Brighton, Postdeadline Vol., p.41.

[20] G.Grosskopf, R.Ludwig, and H.G.Weber, "Crosstalk in optical amplifiers for two channel transmission," Electron. Lett., Vol. 22, No. 17, 1986, p. 900

[21] Inoue, K. "Observation of crosstalk due to four wave mixing in a laser amplifier for FDM transmission", Electron. Lett., Vol. 23, No, 24, 1987, pl293

[22] Olsson, N.A. and Garbinski, P. "High sensitivity direct detection receiver with a 1.5 μm optical preamplifier", Electron. Lett., Vol. 22, No. 21, 1986, p1114

[23] Marshall, I.W. and O'Mahony, M.J. "10 GHz optical receiver using a travelling wave semiconductor laser preamplifier", Electron. Lett., Vol, 23, No. 20, 1987, p.1052

[24] Nishimoto,H., Yokota,I., Suyama,M., Okiyama,T., Seino,M., Horimatsu,T., Kuwahara,H., Touge,T. "Transmission of 12 Gbit/s over 100 km Using an LD pumped Erbium doped Fiber Amplifier and a Ti:LiNbO3 Mach Zehnder Modulator", IOOC'89, Kobe, Japan, 1989, Paper 20PDA-8.

[25] Nakazawa,M., Suzuki,K., Kimura,Y. ,"20 Gbit/s soliton amplification and transmission with an Erbium doped fibre", IOOC'89, Kobe, Japan, 1989, Paper 20PDA-3

[26] M. J. Adams, H. J. Westlake, M. J. O'Mahony, and I. D. Henning, "A comparison of active and passive bistability in semiconductors," IEEE J. Quantum Electron., Vol. QE-21, No. 9 , Sept. 1985, pp.1498-1504

[27] Groskopf,G., Ludwig,R., Schnabel,R., Weber,H. "Frequency Conversion with Semiconductor Laser Amplifiers for Coherent Optical Frequency-Division Switching", IOOC'89, Kobe, Japan, 1989, Paper 19C4-4

[28] Mellis,J. "Optical Phase Modulation using Semiconductor Laser Amplifiers", IOOC'89, Kobe, Japan, 1989,Paper 21B4-3

RECENT DEVELOPMENTS IN COHERENT TRANSMISSION SYSTEMS FOR APPLICATION TO SUBMARINE CABLE SYSTEMS

Shiro RYU, Kiyofumi MOCHIZUKI,
and Hiroharu WAKABAYASHI

KDD Meguro R&D Laboratories
2-1-23, Nakameguro, Meguro-ku, Tokyo 153, Japan

Recent research and development works on coherent transmission systems, especially with emphasis on the application to submarine cable systems, are described. We are convinced that we can fully utilize the advantages of the coherent communication techniques in the all optical network with in-line optical amplifiers and frequency-division-multiplexed carriers.

1. INTRODUCTION

Research and development works on coherent optical communication systems are progressing rapidly and coherent technologies are steadily getting closer to their practical use. In realization of coherent optical communication systems, several problems to be solved confronted us, and some of them were already overcome by the efforts of many scientists.

We also made intensive works in order to solve these problems, from the view point of application of coherent technologies to submarine cable systems. One of the most severe problems is the *polarization fluctuation* of the signal light in the optical fiber submarine cable due to the external disturbances. For overcoming this problem, we have developed the *polarization diversity* optical receiving technique which is immune to polarization fluctuation, and demonstrated its feasibility by the *first sea trial* of FSK system.

After that, the improvement of the first system was implemented in terms of two important points. One is the realization of flat FM response in the conventional DFB laser diode by *a newly developed external-cavity laser module*, and the other is the improvement of the receiver sensitivity by using *a novel polarization coupling/dividing module* for realization of the balanced-receiver type polarization-diversity receiver.

Final application of coherent technologies in optical fiber submarine cable systems for the moment will be the all optical network which connects the different landing stations with *transparent* lines which consists of direct optical amplifiers, and frequency-division-multiplexed (FDM) signals are used for different destinations so that the desired signals can be tuned from them.

For the investigation of the feasibility of the all optical network, we examined the *nonlinear phenomena* in semiconductor laser amplifiers in FDM systems and also performed *546km transmission experiment using 10 cascaded semiconductor laser amplifiers.*

In the following sections, we describe the detailed results of our feasibility studies.

2. FSK HETERODYNE TRANSMISSION EXPERIMENT USING POLARIZATION DIVERSITY TECHNIQUE

2.1. Transmission Experiments in the Laboratory Environment [1]

In order to realize the coherent system, polarization matching between the signal and the local oscillator waves is essential. Various solutions against this problem have been proposed so far [2].

We chose the polarization diversity scheme [3] from among the several techniques for overcoming the polarization fluctuation problem, because this scheme is simple and achieves almost the same receiver sensitivity as the ideally polarization-matched coherent receiver, and maximum response speed against polarization fluctuation is fairly fast owing to the electrical signal processing.

In the polarization diversity optical receiving system, signal light with an arbitrary state of polarization (SOP) is split into two orthogonal polarization components, which are separately mixed with local oscillator light signals having constant SOP. The two demodulated IF signals are then combined again. The square-law combining method, in which each signal is squared, introduces only 0.4dB degradation [4] as compared with maximal-ratio combining [3], and we can use delay-line demodulators for the demodulation as well as the weighting elements in a polarization diversity receiver.

Fig.1 shows the experimental setup of the FSK polarization diversity transmission experiment. DFB laser modules operating at 1.56μm with an external grating was used both for the signal and the local oscillators. The SOP of the local oscillator light was adjusted by the polarization controller so that the equal amounts of local oscillator light (=-8.5dBm) were coupled into each PD. The beat signal linewidth was about 500kHz. Received IF signals were amplified, filtered, and then demodulated by delay-line demodulators, and the demodulated data streams were combined by an adder. The center frequency of the IF signal was stabilized at 980±2MHz by controlling the injection current of the local oscillator laser. A 560Mbit/s MSK (minimum shift keying) modulation with fixed (1,0) pattern was applied to the signal laser and the MSK signal was demodulated by one-bit delayed demodulation.

In Fig.2, the bit error rate (BER) characteristics were measured and plotted when the received power ratio at receivers 1 and 2 was set with a value of 1:0, 0.5:0.5 and 0:1, respectively by the manual SOP control of the signal light. The receiver sensitivities for the cases of 1:0 and 0:1 agree well, however, for the 0.5:0.5 case, it degrades by 0.5dB. This is probably due to the slight disagreement of the characteristics of the two delay-line demodulators. The receiver sensitivity penalty of 1.5dB for the diversity configuration as compared with the case of a single receiver is because of the square-law combination and the receiver excess noise due to the two optical receivers.

Then, we measured the variation of BER over the time axis when the

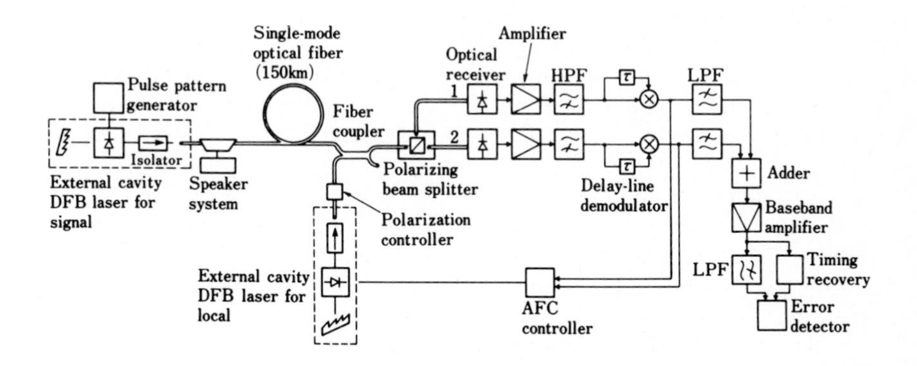

FIGURE 1
Experimental setup of the polarization diversity receiver.

periodic 30Hz polarization fluctuation was applied to the input part of the fiber (length = 3.5m), and the BER was measured with diversity on and off. Although the polarization fluctuation on account of the fiber vibration induced about 10dB peak level fluctuation of the beat signal spectrum at one receiver, we could measure the BER stably with diversity operation.

By this experiment, we confirmed the stable operation of the polarization diversity receiver even under the severe polarization fluctuation conditions.

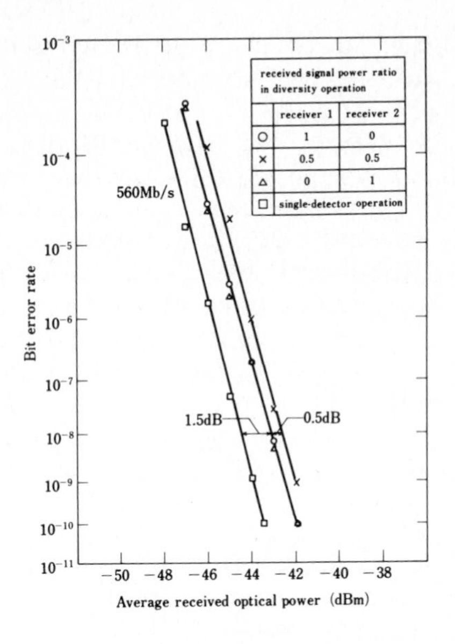

FIGURE 2
BER characteristics of the polarization diversity receiver.

2.2. First Sea Trial of Coherent FSK Transmission System [5]

Next, we performed the first sea trial of the coherent FSK system in

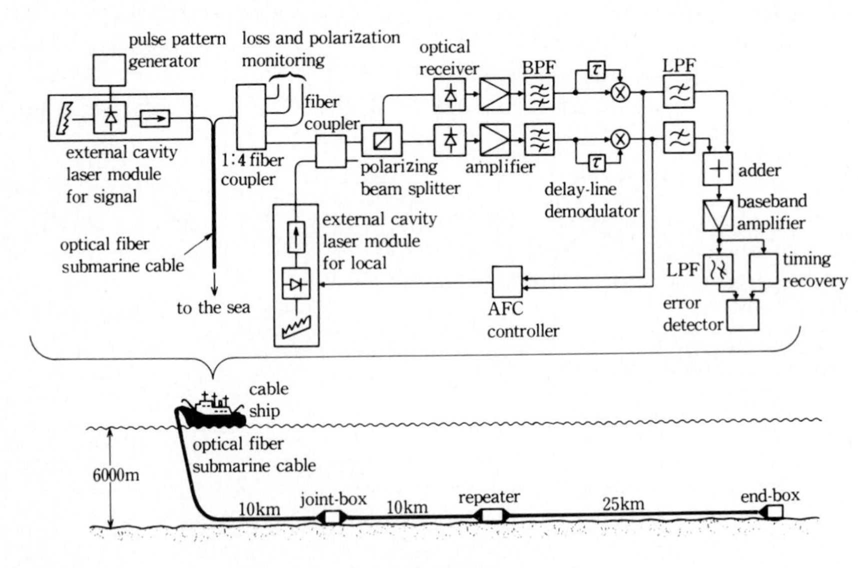

FIGURE 3
System configuration of the first sea trial.

order to show the feasibility of the polarization diversity system.

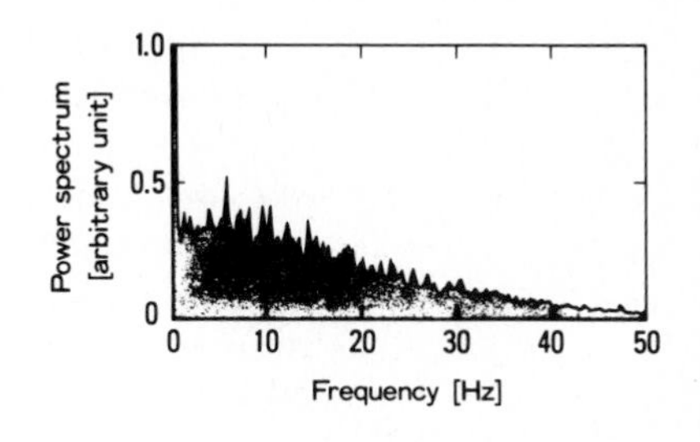

FIGURE 4
Power spectrum of polarization fluctuation in cable laying.

The experiment was carried out to the east coast of *Torishima* in the *Ogasawara* Island Group in Japan at a water depth of 6000m. In the experiment cable was laid and recovered in the 6000m deep sea. Fig.3 shows the configuration of the system. The length of the cable was 45km and it consisted of six single-mode fibers. One pair of fibers passed through the optical repeater and was loop-backed at the end-box, so that a 90km transmission line was obtained. The total loss of the transmission line including splices was about 20dB.

Here, we used essentially the same transmitter and receiver as the previous section. Light from the signal laser module, after propagating through the 90km optical fiber, was divided by a four port single-mode optical fiber coupler and one port was connected to the polarization diversity optical receiver. The regenerated data stream was detected at the error detector. The remainder of the ports were connected to the loss and SOP monitoring equipments. During the experiment, cable laying speed was changed from 0 to 3 knots. Maximum tension of the cable was about 4 tons in laying, 4 tons in holding, and 6tons in recovering processes, respectively.

For three cases, that is, cable laying, holding, and recovering processes, the polarization fluctuation was measured in terms of Stokes parameters and the power spectrum of the optical power after the polarizer. It was found that in every case the SOP of the received light varied rapidly and the maximum fluctuation frequency was 50 Hz. One example of the power spectrum of the polarization fluctuation in cable laying with the speed of 3knots is shown in Fig.4. In this condition it was completely impossible to receive the data at the error detector with single-detector (i.e. without polarization diversity) operation.

The operation of the system was then measured in terms of the BER variation with diversity on and off when the BER with diversity on was set at about 10^{-8}. In this experiment, the cable laying speed was 3 knots. Fig.5 shows the results. In diversity operation we could detect the data stream and the measured BER values distributed within some range. Without diversity operation, however, it was impossible to detect the data at all.

As a result, it was confirmed that the polarization diversity is a powerful technique for application of coherent systems to submarine cable systems.

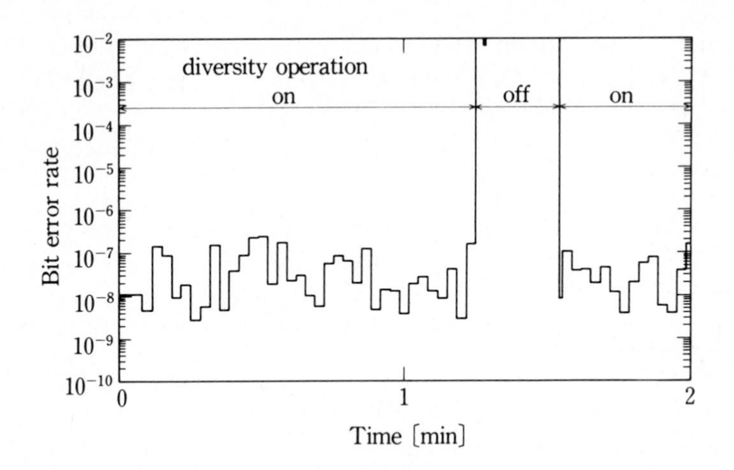

FIGURE 5
Variation of BER in cable laying with and without diversity operation.

2.3. Improvement of the system performance by some new technologies [6]

We have successfully demonstrated the first sea trial of the coherent FSK transmission system in the previous section. However, in the experiment we were confronted with two problems.

One is the nonuniform FM response of the conventional DFB laser diode, which caused, what is called, the pattern effect. Consequently we had receiver sensitivity degradation in the transmission of the pseudorandom bit sequence. The other one is the not so good receiver sensitivity because of the local oscillator intensity noise.

For the first problem, we propose a new FSK modulation method by using a laser module with a built-in $LiNbO_3$ waveguide phase modulator as an external cavity which realizes not only flat FM response but also narrow linewidth.

Against the second problem, balanced receiver configuration [7] is efficient because of the differential operation against the intensity noise of the local oscillator light. Here, we propose a novel polarization coupling/dividing module for this application.

By using these techniques we successfully carried out 280Mbit/s, 251km pattern independent FSK transmission experiment.

Fig.6 shows the experimental setup. The laser module used for the signal transmitter and the local oscillator has an external cavity made by $LiNbO_3$ waveguide phase modulator (Fig.7) with one facet HR coated (facet reflectivity≈99%) and the other facet AR coated. Phase modulator used here has an 8μm-wide, titanium-diffused waveguide and the π voltage at 1.55μm was 15V. Then, we can change the phase of the light in the external cavity continuously by changing the applied voltage to the phase modulator, so that we can continuously change the oscillation frequency of the output signal from a laser module. We observed flat FM response within a wide frequency range as compared with the case of the direct current modulation.

Actual module output wavelength was 1.555μm and the linewidth was maintained at about 700kHz for signal and 350kHz for local light with 36mm external cavity length while the optical frequency was changed about 1.3GHz by applying voltage from -20V to +20V.

Light output from the transmitter (+3.2dBm) was transmitted through

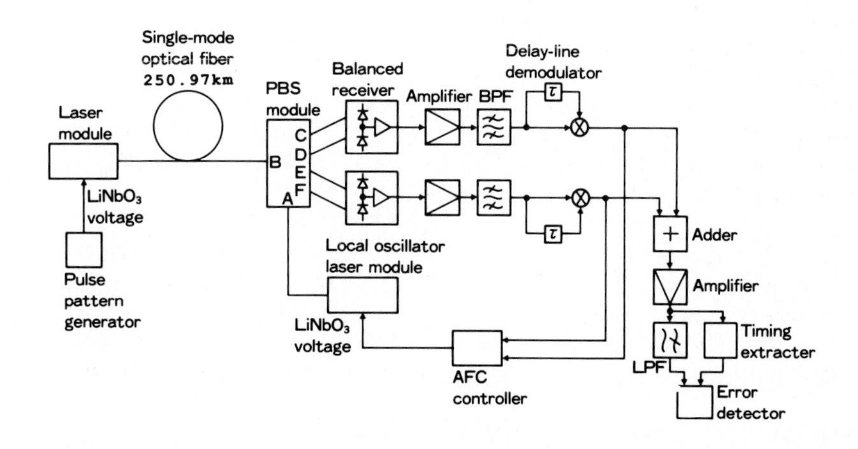

FIGURE 6
Experimental setup of the pattern independent FSK transmission.

250.97km single-mode fiber (total loss=45.2dB) and received in the polarization diversity receiver with balanced receiver configuration. In the receiver, we used a compact polarization coupling/dividing module as shown in Fig.8. In the figure, light from the local oscillator is propagated through the polarization maintaining optical fiber (PMF) and the axis of the PMF is oriented 45degrees against the axis of the polarizing beam splitter (PBS) of a port A, so that the equal local oscillator power is divided into each receiver. In this module, we used two elements. Each of them consists of one PBS and one half-mirror (HM), and we have the possibility of integrating these four optical elements into a single device. This module has fiber-to-fiber excess loss of 1dB and the local oscillator power at each port of the receiver was -3.2dBm. IF frequency stability of ±2MHz was achieved by the AFC circuit which controlled the external-cavity voltage of the local oscillator.

We performed 280Mbit/s FSK transmission experiment. Modulation index of 1.0 was obtained when the applied voltage to the modulator was $3V_{p-p}$. Then, the polarization dependence of the polarization diversity receiver was measured in case of $LiNbO_3$ voltage FSK modulation. It was found that this receiver has the small polarization sensitivity of 0.7dB. Fig.9 shows the BER characteristics when the signal power was equally divided into each receiver. From the figure, we can know that the perfect pattern independent transmission can

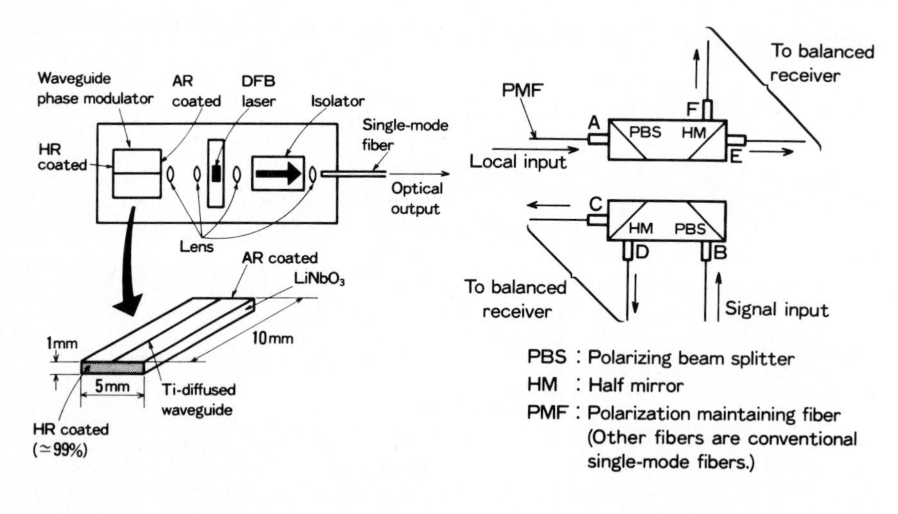

FIGURE 7
Newly developed laser module.

FIGURE 8
Compact polarization coupling/dividing module.

be performed by the $LiNbO_3$ voltage modulation, whereas large pattern dependence is observed in case of direct current modulation.

As a result, by using these techniques, we achieved much improvement in the essential points for realization of coherent transmission systems.

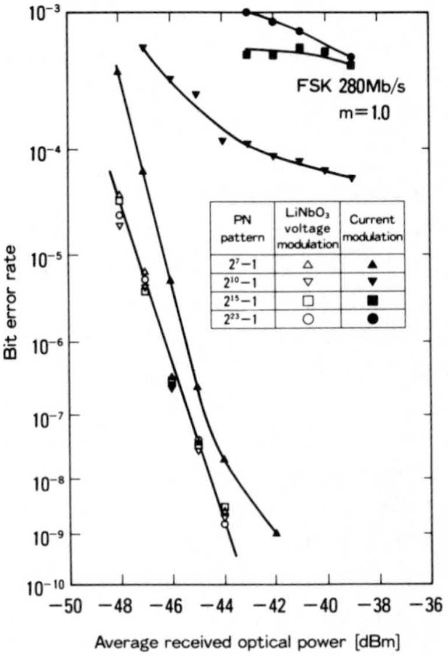

FIGURE 9
Comparison of BER characteristics in case of $LiNbO_3$ voltage and direct current modulation.

3. COHERENT SYSTEMS WITH IN-LINE OPTICAL AMPLIFIERS

3.1. All Optical Network

Future optical fiber submarine cable system will be the *all optical network* in which the different stations are connected by the *transparent* lines using in-line optical amplifiers. Signals transmitted through the all optical network are the FDM signals and each station selects the desired signals from the FDM signals like the satellite communication systems.

In all optical network with coherent techniques, several problems arise due to the inherent characteristics of the direct optical amplifiers, that is, the nonlinear phenomena as well as the accumulation of the spontaneous emission noise. In the following sections we investigate these problems in the experimental basis.

3.2. Influence of Nondegenerate Four-Wave Mixing on Coherent Systems with Semiconductor Laser Amplifiers

3.2.1. Basic Principle

When FDM signals are incident on the semiconductor laser amplifier, undesired optical signals are induced due to the carrier density modulation in the amplifier by the interference between the input carriers [8], [9], [10]. This effect is, so called, the nondegenerate four-wave mixing (NDFWM) in the semiconductor laser amplifier. These new lights fall on the frequency of neighboring carriers, so that the transmission quality of the signal carrier is degraded.

3.2.2. Effect of NDFWM on the Multistage Amplifier System [11]

In order to investigate the influence of NDFWM on multistage amplifier system, we measured the amount of the induced light in one, two, three, and four stage amplifier systems.

Fig.10 shows the experimental set-up. In this experiment, two light sources were used, and four travelling-wave (TW) laser amplifiers were cascaded. Gain of TW1, TW2, TW3, and TW4 were about -0.4dB, 2.5dB, 1.3dB, and -2.3dB, respectively, including fiber coupling loss. Input power of each carrier was adjusted to be same.

Measurement results are shown in Fig.11. From the figure, we can say that the induced light power at the previous stage amplifiers is added to the newly induced light power at the respective amplifiers. Difference of the separation of each line is due to the different induced light power at the individual amplifiers and the different span loss values. Anyway, in the design of the multistage amplifier systems, we should take account of the accumulated light power because of NDFWM.

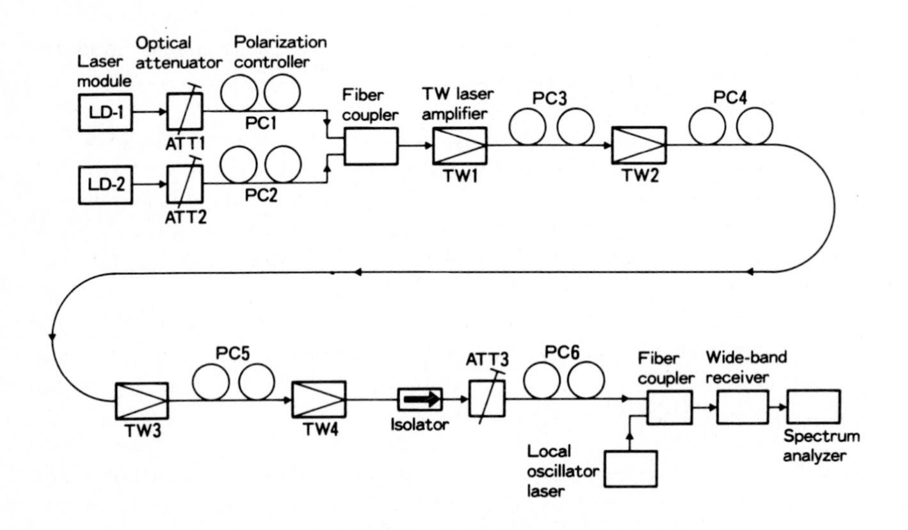

FIGURE 10
Experimental setup for investigating the accumulation of the induced power by NDFWM.

3.3. Coherent FSK Transmission Experiment Using Multistage Semiconductor Laser Amplifiers

3.3.1. System Performance of the Coherent System with Multistage Semiconductor Laser Amplifiers

In the all optical submarine cable network, repeaters made by optical amplifiers are cascaded to cover the whole range of the network. In such a system, spontaneous emission noise accumulates and causes the transmission quality degradation.

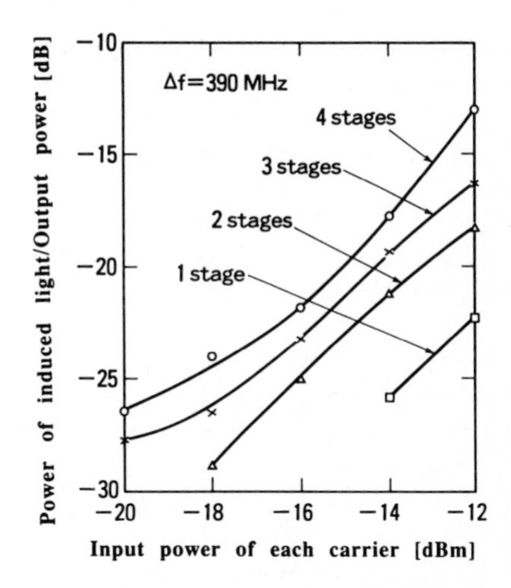

FIGURE 11
Power of induced light by NDFWM in multistage amplifier system.

For the first step of the experimental investigation of the system, we examined the transmission performance degradation and the system dynamic range in 10 cascaded semiconductor laser amplifier system.

Fig.12 shows the experimental setup. Semiconductor laser modules using $LiNbO_3$ waveguide external cavity same as those in Section 2.3 were used for both as signal and local oscillator lasers. 140Mbit/s FSK modulation with modulation index of 2.0 was applied to the signal laser. As for the polarization diversity receiver with PBS module, we also used the same configuration as that in Section 2.3. Travelling-wave semiconductor laser amplifiers used in this experiment have the facet reflectivity of below 0.05% and the maximum gain ripple of 0.5dB. Peak gain wavelength of these amplifiers was around 1.500μm, so that the signal operating wavelength (1.555μm) was about 55nm apart from the gain peak, which caused about 5~6dB gain reduction. Input polarization of the signal at each amplifier was adjusted to be TE polarization by the manual polarization controller for maximum gain, and after each amplifier polarization-independent optical isolator (isolation>30dB) was inserted to prevent the unstable operation of the amplifier due to the undesired reflection. Input optical power to every amplifier was adjusted to be same using the optical attenuator after each isolator. Net gain of each amplifier is denoted in the figure.

We measured the BER characteristics of the system with different optical input power to the amplifier in case of 1, 5, and 10 stage amplifiers. Fig.13 shows the power penalty achieving the BER of 10^{-8}

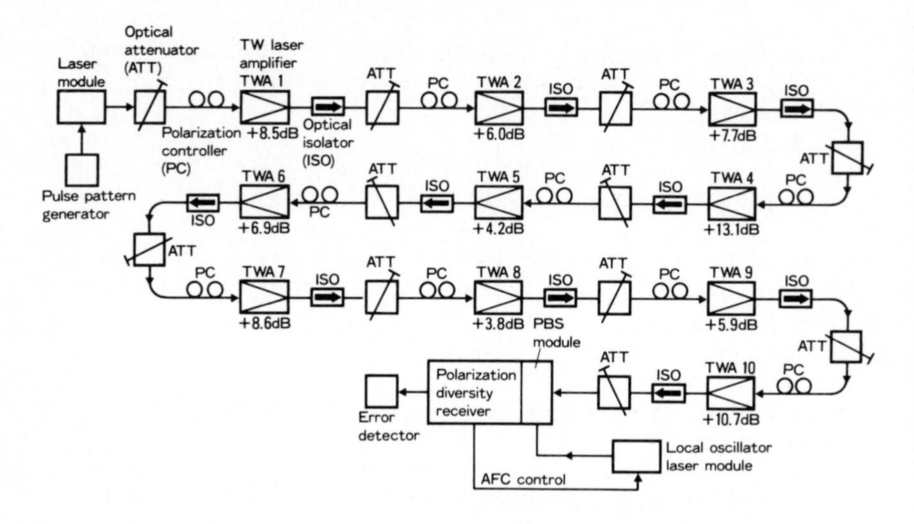

FIGURE 12
Experimental setup of the 10 stage TWA transmission.

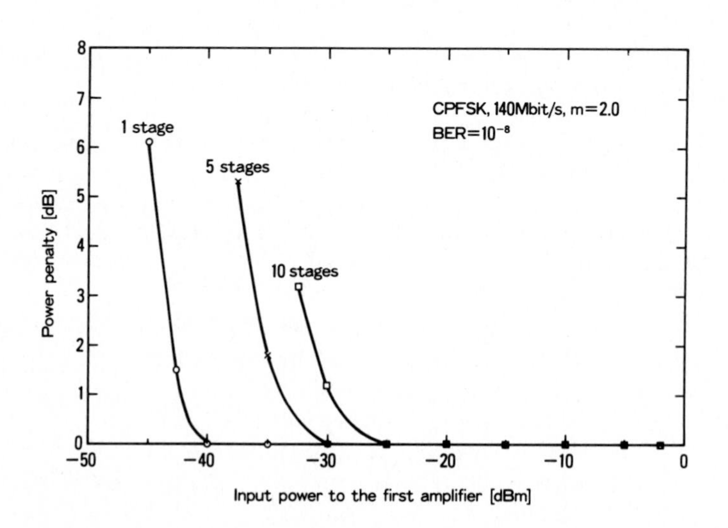

FIGURE 13
Power penalty of the 1, 5, and 10 stage amplifier system versus the input power to the first amplifier.

as the function of the input power to the first amplifier. From the figure, we can well understand the narrowing effect of the dynamic range in multistage amplifier systems because of the accumulation of the spontaneous emission noise.

Form these results, we can say that this important point should be taken into account in designing the all optical submarine cable network.

3.3.2. 546km Coherent FSK Transmission Experiment Using 10 Cascaded Semiconductor Laser Amplifiers

We carried out long-haul transmission experiments by using the single-mode optical fiber transmission lines instead of the optical attenuators in Fig.12.

Fig.14 shows the experimental setup. Net gain and the input power of each amplifier is shown in the figure. In this setup, total transmission length was 546km.

Fig.15 shows the BER performance of the system before and after 546km transmission. We observed no receiver sensitivity penalty by long-haul transmission because we adjusted the input power of each amplifier over around -20dBm for no transmission quality

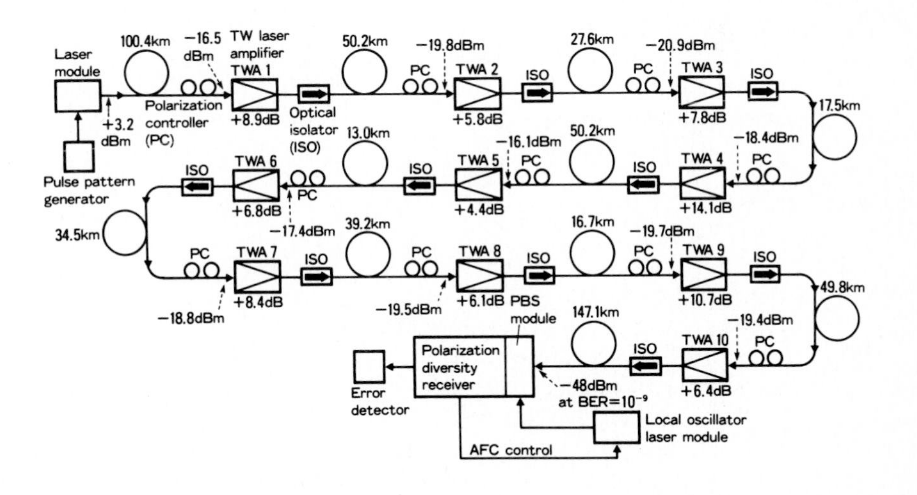

FIGURE 14
Experimental setup of the 546km FSK transmission.

degradation considering the results of Fig.13.

In the actual system design, we should pay attention to the optimum level diagram of the system for minimum transmission penalty.

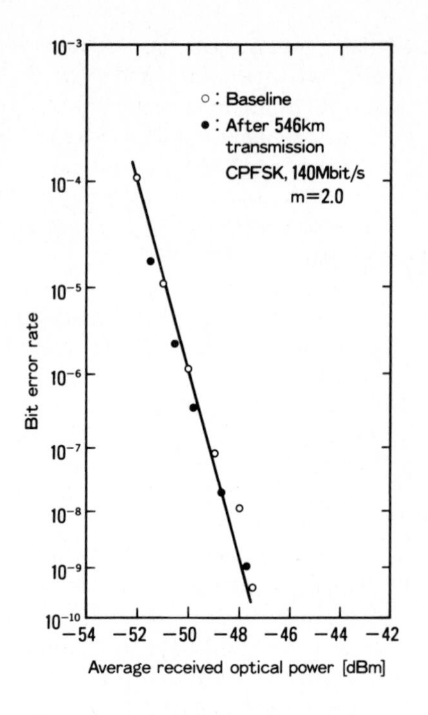

FIGURE 15
BER performance of the 546km system.

4. CONCLUSION

We investigated the several problems in realization of the coherent systems as well as the all optical network with FDM techniques and pointed out some important results.

ACKNOWLEDGEMENTS

The authors are grateful to Mr. K. Komuro, and Dr. T. Yamamoto of KDD Meguro R&D Laboratories for their encouragement and also express thanks to Dr. Y. Namihira, Dr. S. Yamamoto, and Mr. H. Taga for the fruitful discussion and cooperation in the experiments.

REFERENCES

[1] Ryu, S., Yamamoto, S., and Mochizuki, K., "Polarisation-Insensitive Operation of Coherent FSK Transmission System Using Polarization Diversity," Electron. Lett., Vol.23, No.25, pp.1382-1384, 1987.

[2] Okoshi, T., "Recent Advances in Coherent Optical Fiber Communication Systems," IEEE Jour. Lightwave Technol., Vol. LT-5, No.1, pp.44-52, 1987.

[3] Okoshi, T., Ryu, S., and Kikuchi, K., "Polarization-Diversity Receiver for Heterodyne/Coherent Optical Fiber Communications," Tech digest of IOOC'83, Tokyo, Japan, 1983, Paper 30C3-2.

[4] Glance, B., "Polarization Independent Coherent Optical Receiver," IEEE Jour. Lightwave Technol., Vol. LT-5, No.2, pp.274-276, 1987.

[5] Ryu, S., Yamamoto, S., Namihira, Y., Mochizuki, K., and Wakabayashi, H., "First Sea Trial of FSK Heterodyne Optical Transmission System Using Polarization Diversity," Electron. Lett., Vol.24, No.7, pp.399-400, 1988.

[6] Ryu, S., Mochizuki, K, and Wakabayashi, H., "Pattern Independent FSK Polarization Diversity Transmission Experiment Using a Laser Module with Built-in $LiNbO_3$ Waveguide Phase Modulator as External Cavity," Tech digest of IOOC'89, Kobe, Japan, 1989, Paper 18C2-2.

[7] Emura, K., Yamazaki, S., Fujita, S., Shikada, M., Mito, I., and Minemura, K., "Over 300km Transmission Experiment on an Optical FSK Heterodyne Dual Filter Detection System," Electron. Lett., Vol.22, No.21, pp.1096-1097, 1986.

[8] Agrawal, G. P., "Amplifier-Induced Crosstalk in Multichannel Coherent Lightwave Systems," Electron. Lett., Vol.23, No.22, pp.1175-1177, 1987.

[9] Darcie, T. E., Jopson, R. M., and Tkach, R. W., "Intermodulation Distortion in Optical Amplifiers from Carrier-Density Modulation," Electron. Lett., Vol.23, No.25, pp.1392-1394, 1987.

[10] Darcie, T. E. and Jopson, R. M., "Nonlinear Interactions in Optical Amplifiers for Multifrequency Lightwave Systems," Electron. Lett., Vol.24, No.10, pp.638-640, 1988.

[11] Ryu, S., Mochizuki, K., and Wakabayashi, H., "Influence of Nondegenerate Four-Wave Mixing on Coherent Transmission Systems Using In-line Semiconductor Laser Amplifiers," Technical digest of OFC'89, Paper THC4.

PART 2

COHERENT OPTICAL COMMUNICATION THEORY

INTRODUCTION TO THE EFFECT OF PHASE NOISE ON COHERENT OPTICAL SYSTEMS

Ian Garrett

British Telecom Research Laboratories,
Martlesham Heath, Ipswich IP5 7RE, U.K.

Laser phase noise, caused by spontaneous emission of photons at random instants, generally degrades the performance of coherent optical systems. However, by suitable design of the optical recevier, the degradation can be kept quite small, and coherent systems suitable for local loop applications can be designed around lasers with linewidths comparable with the data rate. This paper reviews the theoretical background for designing systems to cope with phase noise, which are now widely accepted as the future for optical communications. The last part of the paper takes a more distant look into the future and discusses the basis of non-classical states of light which may one day open the way to further increased performance in transmission or signal processing.

1. INTRODUCTION

Coherent optical communication systems have received considerable attentioin in recent years, both theoretically and experimentally. They offer two major advantages over direct detection optical systems: increased receiver sensitivity by 10 to 20 dB, and the possibility of adding very large numbers of further channels by frequency multiplexing on a single fibre. These properties make coherent optical systems attractive both for future long haul transmission and for broadband local networks. Much effort has gone into overcoming the technological problems associated with controlling the polarisation of the received signal, tracking its frequency or phase, and, most importantly, reducing the phase noise (linewidth) of semiconductor lasers so that they may serve as suitably coherent transmitters and local oscillators. These difficulties have all been overcome in the laboratory, but as yet there are no engineered solutions that may be used in operational coherent optical transmission systems.

Lasers for coherent transmission have linewidths covering the range from a few kHz to a few tens of MHz. Semiconductor lasers with external cavities have linewidths in the range from about 1 kHz to 1 MHz. DFB and DBR lasers with multiple regions have linewidths of typically a few MHz, while simple DFB lasers are usually in the regions of tens of MHz. With the development of high data rate systems and the understanding of how receivers may be designed to cope with phase noise, the linewidth is now regarded as perhaps less of a problem than tunability so far as single channel systems is concerned. However, the laser linewidth affects the channel spacing in multi-channel systems to an extent that is only beginning to be understood theoretically.

In recent years there have been several successful demonstrations of non-classical states of light, particularly so-called "squeezed states". These states of light have been predicted theoretically for about twenty years, but it is only now that laser technology and non-linear optics have developed to the point where such states can be generated reproducibly in the laboratory. The interest in these states of light stems from their unusual quantum noise properties. Classical states of light have equal noise contributions in both quadratures and these contributions are time-independent. Squeezed states and other non-classical states can be generated where the noise is predominantly in one quadrature, or where it cycles between the quadratures, so that an appropriate detection technique can allow measurements to be made below the semi-classical quantum noise limit. The quantum theory of light has several other predictions that differ from those of semi-classical theory and which may be important in such phenomena as detection of self-phase modulated signals and soliton propagation. This area of research looks to be the next horizon for optical communications and signal processing.

This paper is an introduction to these topics, intended to provide a background to the following five papers. The next two papers analyse the effect of phase noise in single channel systems, and the background to this problem is given in the next section. The two papers after that deal with phase noise effects in multi-channel systems, taking two rather different approaches. Section 3 gives some background to that problem. The fifth paper addresses generation and detection of some non-classical states of light, the basic ideas for which are discussed in section 4.

2. PHASE NOISE IN SINGLE CHANNEL SYSTEMS

2.1 The Origin of Phase Noise

The phase noise in lasers originates in the spontaneous emission of photons with phases that are random relative to the stimulated emission field in the laser. The spontaneous emission is, of course, essential to the laser action, which is an oscillation that build up from the spontaneous emission events in a way that is analogous to the beginning of any oscillation from noise. The spontaneous emission events are usually modelled as a Poisson point process, resulting in a phase for the laser field which is a Brownian motion Gaussian process. This random process is characterised by the variance $D\tau$ of the phase drift over a given measurement time interval τ. The quantity D is the phase diffusion coefficient. The Gaussian nature of the phase drift leads to a Lorentzian shape for the field power spectrum or lineshape, which is usually characterised by its full width at half maximum height (FWHM) $\Delta\nu$ (the linewidth), related to the phase diffusion coefficient by $2\pi\Delta\nu = D$. One can view the line broadening effect of the phase noise as the result of random phase modulation of the carrier.

2.2 Categories of Coherent System

It is useful to divide coherent systems into three categories, which correspond to the type of measurement that is made on the received field in order to detect the information.

The first category is truly coherent systems, where the phase of the received field is compared with that of a local oscillator laser which is phase-locked to the average signal phase. These systems are most demanding on laser linewidth because of the need for the optical phase locked loop to track the phase of the signal field. The requirement depends on the loop configuration. It is typically 0.01% for most types of phase-locked loop, but is an order of magnitude larger for a heterodyne Costas loop[1], implying the use of external-cavity controlled lasers. These systems have proved difficult to engineer, mainly because of problems with the optical phase-locked loop. They offer the best receiver sensitivity, however, with phase-shift keyed systems having a theoretical rceiver sensitivity of as little as 9 photons per bit.

The second category is differentially coherent systems, in which the receiver measures the phase change of the signal field between one symbol and the next. The local oscillator is not a phase reference, as in truly coherent systems, but serves as a means for translating the frequency range of the measurement from the optical to the electronic domain. Receivers for these systems use delay demodulation. The modulation format may be PSK or continuous-phase FSK. The linewidth requirements in these systems are less demanding than in truly coherent systems as the important quantity is the phase drift over the delay time of the demodulator. The receiver sensitivity is typically 3 to 5 dB worse than for homodyne PSK systems. Linewidths of less than 1% of the data rate are needed, and the receiver error rate exhibits a "floor", i.e. an error rate that cannot be improved by increasing the signal power. This floor represents the probabaility of the IF phase drifting by more than $\pi/2$ over the delay time of the demodulator. Its position depends only on the laser linewidths and on the delay time of the demodulator.

The third category is weakly coherent systems, in which the receiver measures the signal power within the bandwidth determined by the passband of the IF filter. Phase information is not used, and again the local oscillator serves as a frequency translation mechanism, not as a phase reference. Heterodyne detection of ASK and FSK signals with envelope or square-law demodulation are examples. Since the receivers for these systems do not measure phase, they are very tolerant to phase noise. However, phase noise does degrade the performance because it results in amplitude noise after bandpass filtering. One can see how this comes about by considering phase noise as random phase modulation. Before the IF filter, the random phase modulation results in an amplitude that is constant (if we may ignore quantum noise and excess laser noise). The IF filter passes a signal with a randomly varying spectrum that no longer results in a constant amplitude because some spectral components are filtered out. If the IF filter is wide, little of the randomly modulated signal is removed and so the amplitude noise is small. The amplitude noise is signal-dependent, and so results in an error rate floor if the decision threshold is fixed as some constant fraction of the mean signal.

2.3 The Basic Theoretical Problem

Truly coherent systems using optical phase-locked loops are comparatively well understood from the theory point of view, following the initial work of Okoshi, Kikuchi and co-workers[2] and the detailed and comprehensive analysis of Hodgkinson[1]. In the rest of this paper, we will concentrate on the other two categories of coherent systems where there are still theoretical problems to which there is as yet no really satisfactory and applicable solution.

Figures 1a and 1b are schematic diagrams of heterodyne receivers for PSK and for ASK modulated optical signals, using delay demodulation (differential detection) in the PSK receiver and square-law demodulation in the ASK receiver. These two receivers are chosen for discussion because they are of practical importance and because they illustrate the problems faced in analysing receiver performance. The object of the analysis is to calculate the error probability as a function of received signal power, taking into account the relevant receiver parameters such as the IF bandwidth and shape, the delay time in the DPSK demodulator, etc. This error probability can be calculated if one knows the probability density function for the output voltage $v(t)$ at the decision times. If one can ignore phase noise, then the problem is a well-known one in communication theory. Phase noise, however, introduces new difficulties.

In modelling the effect of phase noise in these systems, one meets a basic analytical problem in describing the effect of the IF filter on the signal corrupted by phase noise. If we use the equivalent base-band representation for simplicity, the output of the IF filter can be written as:

$$z(t) = e^{j\phi(t)} * h(t) \tag{1}$$

where $\phi(t)$ is the random phase of the signal from the photodiode, $h(t)$ is the impulse response of the IF filter, and "*" denotes convolution. Since the phase of the signal from the photodiode is the sum of the random phases of the transmitter and local oscillator lasers, it will also be a Brownian motion Gaussian process characterised by a linewidth $\Delta\nu$ equal to the sum of the linewidths of the transmitter and local oscillator lasers. The basic analytical problem is contained in equation (1): there are no well-developed general methods for dealing with this convolution with a random process in the exponent. All analyses published to date have resorted to some approximation or simplification in order to make progress.

As defined in equation (1), z is a complex quantity. In modelling DPSK receivers, it is primarily the phase of z which is of interest, while in modelling ASK receivers it is the amplitude or the squared amplitude.

2.4 A Survey of Published Work

For differentially coherent systems, Nicholson[3] decribed the effect of phase noise on a DPSK receiver and showed that there is an error rate floor. He attempted to take into account the effect of IF filtering on the statistics of the phase by computer simulation and analytical fitting. Salz[4] reviewed the performance of all three categories of system, laying a good basis for further analysis of differentially coherent systems. Patzak and Meissner[5] have studied the effect of IF filtering on the statistics of the signal phase by considering the most probable trajectory for the phase $\phi(t)$ which gives a zero output voltage from the receiver (i.e. the critical condition for errors). The then estimate the probability of trajectories in the vicinity of this critical trajectory, arguing that these are the cases which produce most errors.

In the case of weakly coherent systems, Salz recognised the problem contained in equation (1) but offered only an empirical rule for the IF bandwidth required to accomodate the phase noise, based on considerations of the signal spectrum. The first work to attempt to describe the effect of phase noise was that of Garrett and Jacobsen[6], [7], who gave a more explicit optimisation of the IF bandwidth and showed that there would be an error rate floor in a receiver with the decision threshold fixed as a constant fraction of the mean power. Their analysis made three approximations: (i) the phase noise results in a Gaussian random intermediate frequency, (ii) the signal amplitude is determined by the action of the IF filter on this random frequency, (iii) the decision gate samples this random process once every bit-time. The first approximation has been supported both theoretically and experimentally[8]. The second is intuitively appealing and leads to a predictive theory which has recently been shown to have a useful range of validity[9]. The third is a simplification which in fact helps to counteract the increasing inaccuracy of the second assumption when the linewidth is large.

Kazovsky and co-workers have taken a different approach, stemming from classical communication theory[10]. He calculates the second-order statistics of the output voltage of the receiver, including the effect of phase noise, and then assumes the output voltage to be a Gaussian process. This approach is flexible and has been applied to a variety of systems and receivers, and is reasonably accurate for describing effects which lead to reduction in receiver sensitivity of about 1 dB or less. At this point, the performance of the receiver is still determined largely by receiver noise such as local oscillator shot noise and thermal noise, which are Gaussian processes to a good approximation. Kazovsky's approach becomes less successful as the effects of phase noise become more important.

The most comprehensive and accurate analysis to date has been done by Foschini, Greenstein and Vannucci[11], [12]. They made the simplification of taking the IF filter as a finite-time integrator, so that equation (1) becomes:

$$x(\tau) = \int_0^{\tau} e^{j\phi(t)} d\tau$$

They derived an analytical approximation to the moment generating function of $z(t)$ which is good for small linewidth, and they also carried out a sophisticated numerical simulation of the process z to estimate its statistics, using a method based on the Radon-Nikodym Derivative

(RND). They calculated the receiver performance as a function of IF bandwidth and decision threshold setting for a heterodyne ASK receiver, and derived the optimum IF bandwidth and the resulting sensitivity for the receiver. They also derived the Forward (Fokker-Plank) equation for the probability density function of z.

Recently, Garrett and co-workers[13] have derived this Fokker-Plank equation by a somewhat simpler method and have solved it numerically both for the amplitude and for the phase of the filtered IF signal. Some of the results will be discussed in the third paper in this session (Jacobsen and Garrett).

2.5 Summary of results

For differentially coherent systems, the error rate floor is determined by the probability of a phase deviation of more than π during the demodulator delat time τ. Since the phase has a probability density function that is close to Gaussian, the error rate floor position is given by:

$$P_{floor} = 1/2\,\mathrm{erfc}\left(\frac{1}{4}\sqrt{\frac{\pi}{a\Delta\nu\tau}}\right) \tag{2}$$

where a is unity if the phase statistics are unaltered by the IF filter. Patzak and Meissner[5] have argued that a lies close to 0.68.

Figures 2a and 2b show the probability density function for the modulus of z defined by equation (2), obtained by three different methods. The solid lines were found by numerical solution of the Fokker-Plank equation[13], the broken lines by the numerical simulation using the RND theorem[11], and the dotted lines are the analytical approximation of Foschini and Vannucci[11]. Results are shown for values of Dt (normalised to the data rate) ranging from 0.125 to 8. It can be seen that the Fokker-Plank and RND results are in excellent agreement, and that the analytic approximation is good not only for small linewidth (Dt) but, perhaps more importantly in the present application, also for a range of probability densities covering about four orders of magnitude. This means that after multiple convolution to account for the action of the post-detection filter, the resulting probability density function will be accurate over many orders of magnitude and so provide a useful way of estimating receiver sensitivity.

2.5 Future Work

At present there are three reasonably accurate and convenient ways of obtaining numerical results for the probability density function of the amplitude of the filtered IF signal, as described above. To obtain the receiver sensitivity, one must take into account the effect of the post-detection filter. In differentially coherent systems, this filter usually serves only to remove the higher frequency components generated by the demodulation process, and has very little effect on the noise in the signal bandwidth. This is because the IF filter should be as narrow as is consistent with passing the signal without undue distortion in these systems. But in weakly coherent systems, the IF bandwidth may need to be several times the data rate if the ability of these systems to accomodate phase noise is to be exploited, and then the post detection filter has a significant part to play in noise filtering. The approach used by Greenstein and co-workers in modelling this filter is to regard it as summing a number of independent samples of the output of the square-law demodulator. They show that such samples are indeed independent if the interval between samples is at least as long as the integration time of the IF filter. Such an approach is good for the finite-time integrating IF filter. In the near future we shall see results for other filter shapes, from the simulation approach and from the Fokker-Plank equation, but unless the filter impulse responses are sensibly bounded in time, the summing model for the post-detection filter is not really applicable. A further problem is that the samples of the IF process are only independent if the receiver noise is white (strong local oscillator approximation). Other situations may be very hard to deal with theoretically, and it seems that the RDN numerical simulation approach may prove to be the most powerful.

3. MULTI-CHANNEL SYSTEMS

In addition to the phase noise effects described for single channel systems, the presence of a signal corrupted by phase noise in an adjacent channel can be expected to be noticeable to an extent that depends on the channel spacing as well as on the laser linewidths and other parameters. In the absence of phase noise, the channel spacing is determined by the acceptable amount of cross-talk arising from the overlap of the signal spectrum into the adjacent channel passband. Phase noise broadens the signal spectrum, of course, and the channel spacing must be increased to take that into account. However, simple considerations of the signal spectrum do not lead to correct calculations of phase noise effects in single channel systems, and one would expect the same to hold true in multi-channel systems. In a later paper in this session, Kazovsky[14] reports on more detailed calculations.

The more sophisticated numerical simulation and Fokker-Plank techniques have yet to be applied to ths problem. It is clear that they can be applied, with slight modification. For example, equation (1) becomes:

$$z(t) = e^{\Omega t + j\phi(t)} * h(t) \tag{3}$$

where Ω is the channel separation in frequency. An analogous Fokker-Plank equation can be derived, although its solution is somewhat more difficult than for the single channel case. Similarly, numerical simulation can be applied to this equation in much the same way as to equation (2).

One should point out that, in a multi-channel system, the IF filter is all-important in reducing cross-talk between the channels, and filters with a steep roll-off are likely to be used. Since the finite-time integrator is essentially a first-order filter, it is quite unsuited to the multi-channel case. It is important to apply methods that can deal with higher-order filter shapes.

There has been considerable work in recent years concerning optical amplifiers for use in coherent transmission systems. Optical amplifiers can now be made with very broad bandwidth, capable of amplifying signals in a multi-channel system. However, any non-linear process, either in the transmission path or in the amplifiers, will cause intermodulation distortion. Hodgkinson[15] will describe the intermodulation distorion produced in travelling wave amplifiers in multi-channel coherent systems.

4. NON-CLASSICAL STATES OF LIGHT

The semi-classical theory of light describes the emission of photons from a coherent source as a Poisson point process, in which the emission events are independent. Photons arrive at a detector as a Poisson process also. The quantum theory of light shows that other states of light are possible in principle, where the photons may be more bunched or they may be "anti-bunched".

Classically, the field in a one-dimensional cavity of "volume" V may be written in terms of in-phase and quadrature components as:

$$E(z,t) = 2\sqrt{2}E_o \sin kz\,(X\cos\omega t + Y\sin\omega t) \quad (4)$$

where $E_o = \sqrt{\hbar\omega/2\varepsilon_o V}$ is the "electric field per photon". The quantum mechnical representation of this field may be written in the same way, but now E, X and Y are quantum mechnical operators corresponding to the classical observable field and field components. The quadrature operators X and Y have variances which satisfy the uncertainity relation:

$$<(\Delta X)^2><(\Delta Y)^2> \geq 1/16$$

For a classical coherent field, the uncertainities in X and Y are equal, and each equal to 1/4. Figure 3 shows the quadrature components and their uncertainities for a coherent state with amplitude vector α. Note that the uncertainities in the quadrature operators X and Y are independent of the classical field amplitude, and are in fact the same for the vacuum state. Thus the quantum mechanical description of a coherent state may be regarded as the classical coherent state with added vacuum noise.

The uncertainities in photon number n and in phase ϕ may be derived from the geometry of Figure 3 when α is large in magnitude:

$$<(\Delta n)^2> = |\alpha|^2$$

$$\Delta\phi = \frac{1}{2|\alpha|}$$

The uncertainity principle implies that, for two canonical variables, increasing the measurement accuracy for one variable can only be done at the expense of accuracy in a simultaneous measurement of the other. This trade-off is not possible for classical coherent states of light. However, various non-classical states may be prepared with unequal variances so long as the product of the variances satisfies the uncertainity relation.

An important class of non-classical states of light is the class of "squeezed states", so called because the uncertainity circle of Figure 3 is squeezed into an elipse as shown in Figure 4. The major and minor axes of the elipse are of length $e^{s}/2$ and $e^{-s}/2$, where s is the squeezing parameter. Such states may be generated by a phase-sensitive non-linear optical process, and to date the greatest degree of squeezing has been achieved by below-threshold parametric amplification.

To see the importance of such states of light, consider homodyne detection of a beam of such light. Suppose that the local oscillator beam is strong, and that its phase χ relative to the signal may be varied. The measurement yields the projection of the signal vector on the horizontal axis of Figure 4, and varying χ corresponds to rotating the signal vector. The uncertainity in the detected amplitude is the projection of the uncertainity elipse on the horizontal axis, and is clearly dependent on the relative phase angle χ. If the degree of squeezing is large, and if the relative phase angle is near zero or π, the uncertainity in the measurement of amplitude may be smaller than the semi-classical quantum limit, and in principle may be made arbitrarily small.

The application of squeezed states in optical communication is not as clear-cut as the foregoing would suggest, however. The problem is that if the squeezed light is subjected to any random process, its degree of squeezing rapidly decreases. Such random processes include propagation in a lossy fibre, or normal detection with a photodiode of less than unity quantum efficiency. The reason is that any probabilistic process allows the mixing of the squeezed light with the normal vacuum state, which has equal uncertainities in the two quadratures. This is seen most clearly in an imperfect photo-detector, which can be modelled as a perfect photo-detector with a beam splitter in front of it to simulate the quantum efficiency of the detector. The beam splitter is a four-port device, and the fourth port allows in the vacuum state, which is partly reflected onto the photo-detector. However, the study of squeezed states and other non-classical states of light is at an early stage as yet. The theory has been applied almost exclusively to continuous-wave light, not to modulated or pulsed light. In the future, ways of retaining the squeezing by suitable interaction with the transmission medium may emerge, or amplifiers that increase rather than destroy the squeezing may be developed. The paper of Imoto[16] discusses the generation of these non-classical states, and a means of detecting them without destroying them.

REFERENCES

1. Hodgkinson, T.G., "Receiver analysis for synchronous coherent optical fibre transmission systems", IEEE J Lightwave Technol. vol LT-5 (1987) pp 573-586.

2. Kikuchi, K., Okoshi, T., Nagamatsu, M., and Henmi, N., "Degradation of bit-error rate in coherent optical communications due to Spectral spread of the transmitter and local oscillator", IEEE J Lightwave Technol. vol LT-2 (1984) pp 1024-1033.

3. Nicholson, G., "Optical source linewidth criteria for heterodyne communication systems with PSK modulation", Opt. Quantum Electron. vol 17 (1984) pp 399-410.

4. Salz, J., "Coherent lightwave communications", AT&T Tech. J., vol 64 (1985) pp 2153-2209.

5. Patzak, E., and Meissner, P., "Influence of IF filtering on bit error rate floor in coherent optical DPSK systems", IEE Proc J vol 35 (1988) pp 355-357.

6. Garrett, I., and Jacobsen, G., "Theoretical analysis of heterodyne optical receivers for transmission systems using (semiconductor) lasers with non-negligible linewidths", IEEE J Lightwave Technol. vol LT-4 (1986) pp 323-334.

7. Jacobsen, G. and Garrett, I., "Theory for heterodyne optical ASK receivers using square-law detection and post-detection filtering", IEE Proc J, vol 134 (1987) pp 303-312. (erratum: vol 135 (1988) p100).

8. Garrett, I., Jacobsen, G., and Pedersen, R. J. S., "Filtered laser beat-frequency fluctuations", IEE Proc. J. vol 135 (1988) pp 408-412.

9. Garrett, I. and Jacobsen, G., "Phase noise in weakly coherent systems", IEE Proc. J. vol 136 (1989) pp 159-165.

10. Kazovsky, L. G., Meissner, P., and Patzak, E., "ASK multi-port optical homodyne receivers", J Lightwave Technol. vol LT-5 (1987) pp 770-791.

11. Foschini, G. J. and Vannucci, G., "Characterising filtered light waves corrupted by phase noise", IEEE Trans. Inf. Theory vol IT- (1988).

12. Foschini, G. J., Greenstein, L. J., and Vannucci, G., "Noncoherent detection of coherent optical pulses corrupted by phase noise and additive Gaussian noise", IEEE Trans. Com. vol COM-36 (1988) pp 306-314.

13. Garrett, I., Bond, D. J., Waite, J. B., Lettis, D. S. L., and Jacobsen, G., "Impact of phase noise in weakly coherent systems - a new, accurate approach", to appear in IEEE J Lightwave Technol (1989).

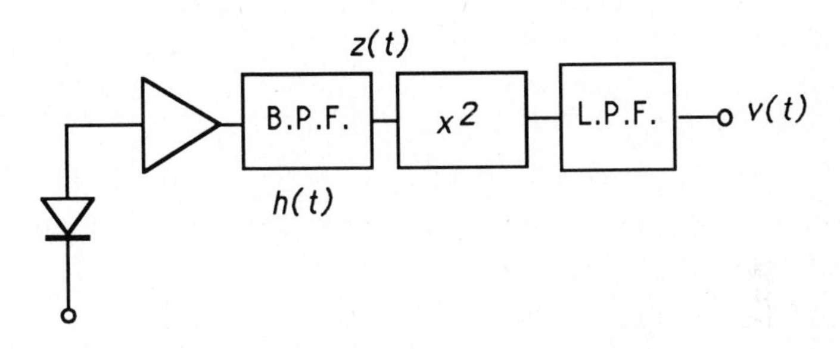

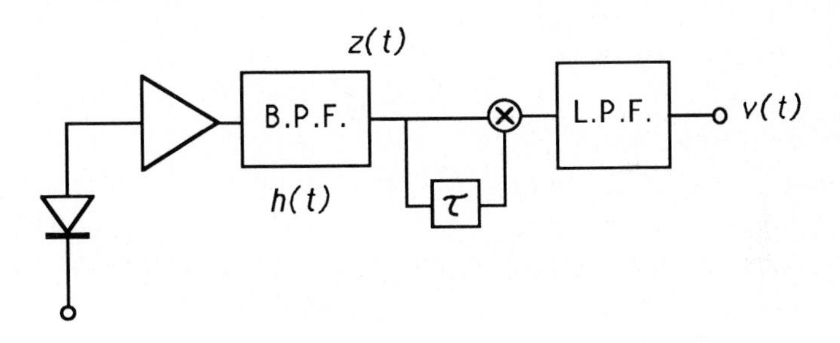

Fig 1: schematic diagrams of coherent optical receivers for (a) ASK signals, (b) DPSK detection.

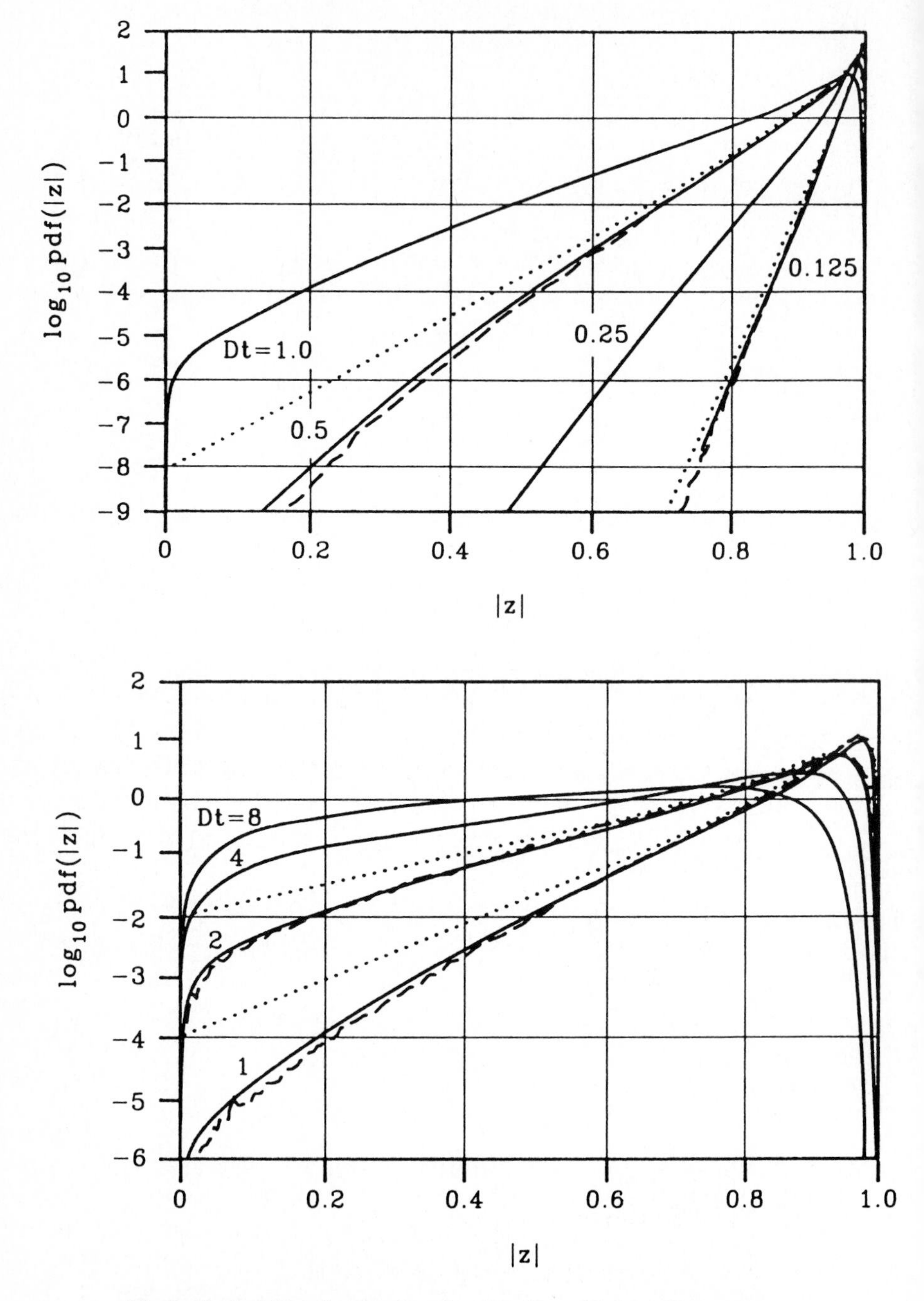

Fig 2. Probability density function for the modulus of $|z|$: Full lines: Fokker-Plank solution [13], broken lines: RND numerical simulation [11], dotted lines: analytic approximation [11].

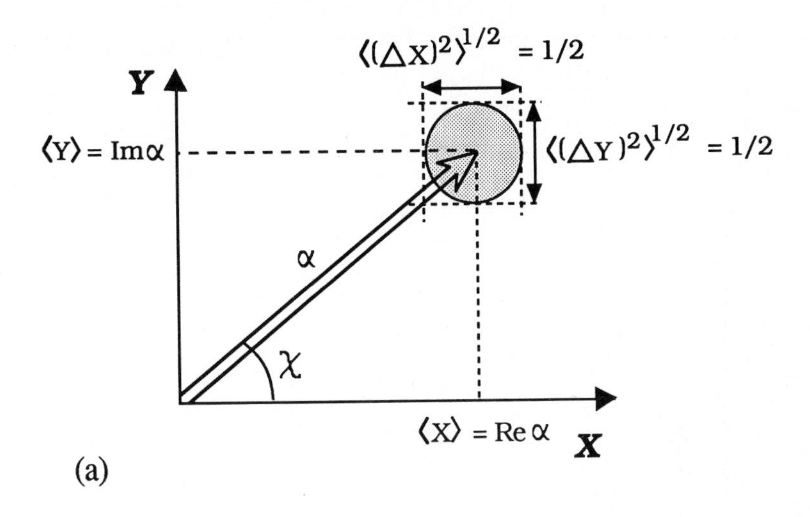

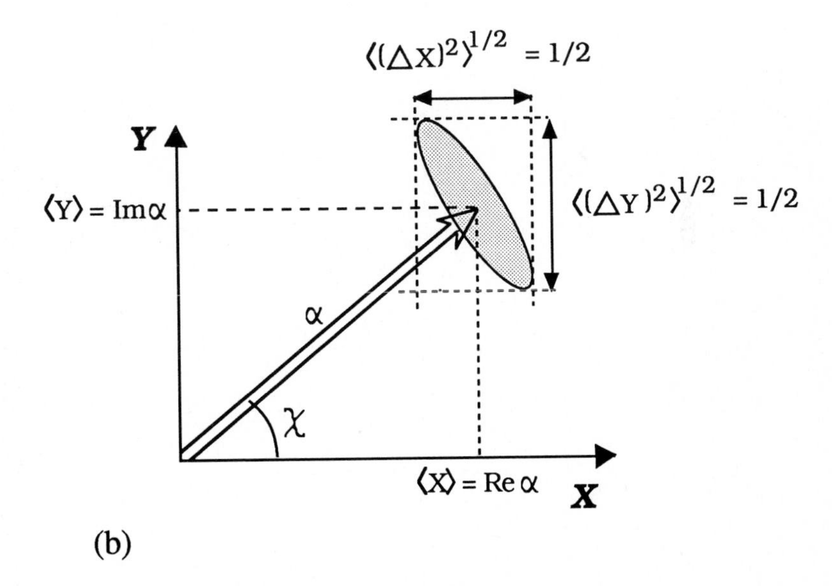

Fig 3. Phase-space description of the mean values and uncertainities of the quadrature operators X and Y. (a) for a coherent state; (b) for a squeezed state.

14. Kazovsky, L. G., "Channel spacing in coherent optical transmission systems", this volume.

15. Hodgkinson, T. G., "Analysis of intermodulation distortion in travelling wave laser amplifiers", this volume.

16. Imoto, N., "Quantum state control and non-demolition detection of photons", this volume.

The Detection and Analysis of Coherent Lightwave Signals Corrupted by Laser Phase Noise

G. Vannucci, L. J. Greenstein and G. J. Foschini

AT&T Bell Laboratories, Crawford Hill Laboratory, Holmdel, NJ 07733

ABSTRACT — **The detection of digital coherent lightwave signals impaired by phase noise is of great current interest. In this paper, we summarize our recent work on the subject.**

We examine the detection of waves modulated by *on-off-keying* modulation (OOK) that are further impaired by additive Gaussian noise. Our results show that, when the receiver parameters decision threshold and IF filter bandwidth are optimized, large amounts of phase noise can be accommodated with only minor increases in required signal-to-noise ratio.

We also show results for the theoretical minimum values of the required average optical power for achieving a bit error rate of 10^{-9}. We show that heterodyne detection for lightwave signals using OOK and FSK (frequency-shift keying) modulation can be made highly robust to phase noise, in contrast to the case of binary phase-shift keying.

Finally, we present an analytical method for estimating the *probability density function* (pdf) of the envelope of filtered optical signals with phase noise for different kinds of filter responses. We find that, for each of the several types of filters considered, the envelope pdf can be accurately fitted by an exponential function approximation, where the decay constant is related in a simple way to known system parameters.

We conclude with a brief description of work in progress.

I. INTRODUCTION

The detection performance of a coherent lightwave transmission link can be sharply degraded by laser phase noise. This problem has been the focus of many recent studies, a representative sampling of which is given by [1]-[10]. To see how phase noise can degrade receiver performance, consider the concrete example of an optical communication system employing *on-off keying* modulation (OOK). A binary on-off data stream with bit rate $1/T$ is used to externally modulate a single-mode laser tone containing phase noise. A local laser in the receiver, having its own phase noise, is mixed with the received signal in a photodetector, thereby transforming the optical pulse stream into an electrical one at some *intermediate frequency* (IF). This heterodyning operation not only increases the phase noise (because of the local laser) but also adds signal shot noise. In the absence of phase noise, the best way to detect the IF signal would be to extract a carrier reference from it and then perform coherent demodulation to baseband. The problem with this approach when phase noise is present is that the random phase can have large and rapid variations. Hence, a carrier recovery circuit might track it so imperfectly as to seriously degrade the demodulation process and the follow-on data detections.

An alternative approach is to use IF envelope detection. For a bit error rate of 10^{-9}, envelope detection is known to require only about 0.5-dB more signal power than does ideal coherent demodulation. This result, however, applies to signals with no phase noise, wherein the optimal IF filter is a *matched filter* with noise bandwidth $1/T$. In situations where the random phase is rapidly varying, significantly wider IF bandwidths are needed to spectrally capture the signal, and this has the effect of increasing the noise power at the output of the envelope detector. A useful tactic, then, is to offset this noise increase by following the envelope detector with a low-pass filter. In this paper we summarize our recent work [11]-[14] in which we analyze the preformance of this "double-filtering" scheme in a variety of circumstances.

We note that the double filtering strategy has been suggested or explored by other investigators, e.g., [5], [7], [9] and [10]. Indeed, some published results, such as those in [5] for FSK and in [10] for ASK, show numerical benefits for this detection approach that agree well with our results given in [12].

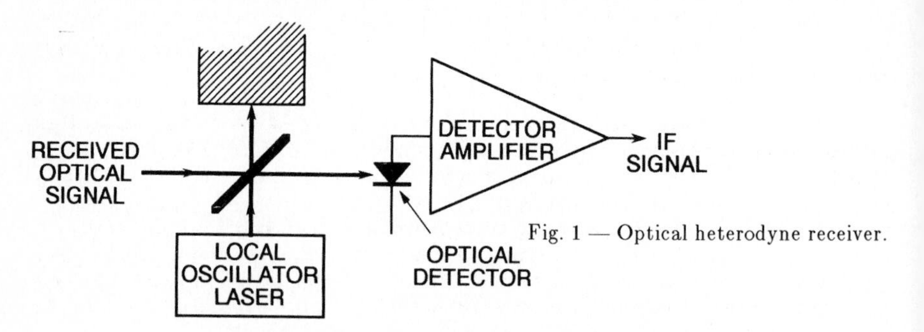

Fig. 1 — Optical heterodyne receiver.

II. SIGNAL, NOISE AND RECEIVER MODELS

A. The Optical Receiver

We assume an optical heterodyne receiver such as the one outlined in Figure 1. The light source used by the transmitter is assumed to be a single-mode laser and the receiver includes a similar laser used as a *local oscillator* (LO). The received optical signal is simply added to the LO signal as, for example, with a beam splitter, where the LO portion of the combined signal is much stronger than the received signal portion. The combined optical signal is then converted to an electrical signal by an optical detector, whose square-law amplitude response achieves the conversion to a microwave intermediate frequency. The value of the IF is assumed large enough to permit ideal envelope detection.

Gaussian noise is added during the conversion to an electrical current in two ways, namely, (1) shot noise produced in the process of photodetection; and (2) thermal noise introduced by the circuitry that follows the photodetector. We regard the additive Gaussian noise as white, which can be achieved by making the LO power high enough, for the chosen IF, to make shot noise dominant.

B. The IF Envelope

Let $s(t)\exp(j\theta(t))$ represent the complex envelope of the IF signal waveform containing phase noise, where $s(t)$ is the baseband pulse shape of amplitude S commencing at $t=0$; and $\theta(t)$ is the random function associated with the phase noise. Assume that this pulse is applied to a bandpass filter whose impulse response has a complex envelope $h(t)$. Both complex envelopes, $s(t)\exp(j\theta(t))$ and

$h(t)$, are referred to the same center frequency, f_c. The corresponding complex envelope of the filter output at a particular time t will be the convolution of the two, or

$$\mathsf{Z}(t) = \int_0^t s(\tau)h(t - \tau)e^{j\theta(\tau)}\,d\tau\,. \tag{1}$$

The *true* envelope of the filter output at time t, as obtained using an ideal envelope detector, is the modulus of this quantity, $M \triangleq |\mathsf{Z}(t)|$, hereafter called the output envelope. Note that $\theta(t)$ is the sum of the phase noises of the transmitting laser and the receiver's local oscillator laser; and $h(t)$ represents the impulse response of the IF filter, to be followed by an envelope detector (either linear or square-law). Again, the filter is made wide enough in bandwidth to pass the signal pulse corrupted by phase noise, but cannot be made arbitrarily wide; its purpose is to limit IF noise and, in some applications, also to select FDMA channels. The envelope-detected signal appearing at baseband is low-pass-filtered, to offset any increase in noise caused by widening the IF bandwidth.

C. The Noise Processes, n(t) and θ(t)

We model the additive noise, $n(t)$, as a complex Gaussian white noise process with two-sided spectral density N_0. A sufficient parameter for characterizing the additive noise is the signal strength, S^2, normalized by the noise power in a specified bandwidth. Following the usual convention, we specify that bandwidth to be equal to the bit rate. Thus, we define $\xi \triangleq S^2T/N_0$ and refer to it hereafter as the *IF signal-to-noise ratio* (or *IF SNR*). With OOK modulation, we should regard ξ as the *peak* IF SNR, the *average* IF SNR being $\xi/2$.

The phase variation, $\theta(t)$, is the difference between the independent phase variations, $\theta_1(t)$ and $\theta_2(t)$, of the transmitting and receiving lasers. An established model for these variations is that the associated radian frequencies (i.e., $\dot\theta_1(t)$ and $\dot\theta_2(t)$) are white and Gaussian,† [15]. Let the spectral densities of $\dot\theta_1(t)$ and $\dot\theta_2(t)$ be denoted by η_1 and η_2, respectively, in units of rps^2/Hz. It is straightforward to show [16] that the 3-dB linewidths of the laser sources are $\beta_1 = \eta_1/2\pi$ Hz and $\beta_2 = \eta_2/2\pi$ Hz, respectively, where the spectral shapes are Lorentzian (or first-order Butterworth) functions.

We thus see that the IF output of the photodetector will have a random frequency variation, $\dot\theta(t) = (\dot\theta_2(t) - \dot\theta_1(t))$, that is white and Gaussian with a spectral density of $(\eta_1+\eta_2)$ rps^2/Hz, resulting in a 3-dB linewidth $\beta = (\beta_1+\beta_2)$ Hz. The second important noise parameter of our study is this linewidth normalized by the bit rate, $1/T$. Thus, we define $\varsigma \triangleq \beta T$. All of our bit error rate results will be given as functions of ξ and ς.

D. Receiver Structure

Figure 2 shows a receiver block diagram for OOK detection. The key issue in specifying an IF filter is its bandwidth. There is a bandwidth, for any frequency response shape, for which the mean-square output noise is optimally

---

† Our model here differs from that of [7], which implicitly treats the instantaneous frequency as being random from bit to bit but constant over a bit period.

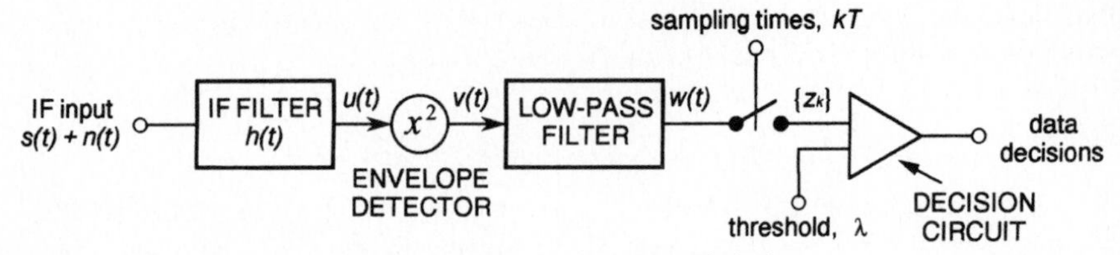

Fig. 2 — Double-filtering receiver structure for OOK modulation

balanced against the statistics of the output signal in the presence of phase noise.

In an early phase of our work [12] we considered a simplified situation where the IF filter is a finite-time bandpass integrator with integration time T'. Thus,

$$h(t) = \begin{cases} 1/T' & \text{for } 0 \leq t \leq T' \\ 0 & \text{elsewhere} \end{cases} \tag{2}$$

The associated frequency response is, except for a phase factor, $\sin \pi f T' / \pi f T'$, with f measured from f_c. The noise bandwidth of this filter is $1/T'$. In the absence of phase noise, the optimal design would be a *matched filter*, for which $T' = T$. Phase noise forces the choice of a shorter integration time (wider bandwidth) to secure decent statistics for the output signal amplitude.

In a later study [14] we examined the statistics of the envelope for different kinds of filter responses and for various combinations of phase noise severity and filter bandwidth. We found that, for each of the several types of filters considered, the shape of the filter response was a secondary factor in determining the envelope statistics: different filter shapes with individually optimized bandwidths will produce very similar detection results for the same signal and noise conditions. Accordingly, the results obtained in [12] and reproduced here for the simple bandpass integrator of (2) are indicative of the performance achievable in practice using practical filters.

The IF filter delivers its output to an envelope detector, which could be of either the *linear* or *square-law* type. For each kind of detector, the mathematical formulation leading to BER would be essentially the same. In [12] we noted that the details of detector design have little effect on receiver performance. Accordingly, we treated in detail only the ideal square-law detector whose baseband output is related to its (complex envelope) input by $v(t) = |u(t)|^2$.

We specified the post-detection low pass filter to be a finite-time integrator with integration time T''. Note that the IF input pulse is T seconds long, and that the composite impulse response of the IF and post-detection filtering has a time extent $T' + T''$. To avoid intersymbol interference, therefore, $T' + T''$ must be no greater than T; at the same time, for maximum noise averaging, T'' should be as large at this constraint allows. Thus, we chose $T'' = T - T'$. The results were expressed in terms of the ratio $m = T/T'$ which, in interesting practical cases, turns out to be much larger than unity. We took advantage of this by examining only integer values of m, which greatly simplified the analysis.

Under the above assumptions, the sampler output for the signal on $[(k-1)T, kT]$ is

$$z_k = w(kT) = \frac{1}{m} \sum_{i=1}^{m} v((k-1)T + iT'). \tag{3}$$

The key to our analysis lies in the fact that z_k is the average of m statistically independent, identically distributed random variables whose common *probability density function* (pdf) we can derive.

Finally, z_k is compared against a threshold λ to determine whether Data = 1 ($z_k \geq \lambda$) or Data = 0 ($z_k < \lambda$). The optimal λ is the one for which the sum of the error probabilities for the two data values is a minimum.

III. ANALYSIS

We denote the pdf of the generic term in the sum in (3) as $p(v\,|\,0)$ when Data = 0 and $p(v\,|\,1)$ when Data = 1. The BER can be obtained numerically in terms of these two conditional pdf's.

We recall the filter output signal given by (1), and the defining equation, (2), for the IF filter response. We can then describe the filter output at time $t = T'$ by

$$u(T') = \begin{cases} \mathsf{Z}(T') + n'; & Data = 1 \\ n'; & Data = 0 \end{cases} \tag{4}$$

where n' is a complex, zero-mean, Gaussian sample of variance $\sigma^2 = N_0/T' = mN_0/T$. For later convenience, we use X to denote the normalized magnitude of $\mathsf{Z}(T')$ i.e., $X \triangleq |\mathsf{Z}(T')|/S$, with $0 \leq X \leq 1$.

The pdf of $v = |u|^2$ when Data = 0 is well-known, i.e.,

$$p(v\,|\,0) = \frac{1}{\sigma^2} \exp(-v/\sigma^2),\ v \geq 0. \tag{5}$$

Note that an m-fold sum of i.i.d. exponential variates, such as v in (5), is a chi-square variate with $2m$ degrees of freedom. Using known relationships, we can obtain the probability of error given Data = 0. It is

$$P_0(Z) = \left[\sum_{i=0}^{m-1} \frac{1}{i!} \left(\frac{mZ}{\sigma^2}\right)^i\right] \exp\left(-mZ/\sigma^2\right) \tag{6}$$

When Data = 1, we can easily write the pdf for v conditioned on the value of X. Specifically, we invoke the results of Rice [17] for the envelope of a sinusoid plus narrowband Gaussian noise:

$$p(v\,|\,1,X) = \frac{1}{\sigma^2} I_0 \left(\frac{2SX\sqrt{v}}{\sigma^2}\right) \exp\left(-\frac{v + S^2X^2}{\sigma^2}\right),\quad v \geq 0. \tag{7}$$

To obtain $p(v\,|\,1)$, it is necessary to know the pdf for X, $p_x(X)$; then, $p(v\,|\,1)$ can be obtained through the integration

$$p(v\,|\,1) = \int_0^1 p(v\,|\,1,X) p_x(X) dX \tag{8}$$

We see, therefore, that the key to calculating the desired BER lies in the pdf for the quantity X, which is the envelope of a filtered sinusoid containing phase noise, the filtering being defined by (2). The difficulties associated with obtaining an expression for that pdf are discussed next.

IV. THE PDF OF X

The derivation of the pdf of X is treated in detail in [11]. Several different approaches are considered, each with its own advantages and disadvantages. The three most useful ones are now briefly described.

A. The Simulation Approach

The pdf of X depends strictly on $\beta T'$. A straightforward way to obtain this pdf is by using Monte Carlo simulation. Since $\theta(t)$ is a Wiener-Lévy process (also known as Brownian motion or continuous random walk), it is straightforward to simulate it through a computer program that generates sample paths for $\theta(t)$. (For this purpose the independent-increment feature of the Wiener-Lévy process is very useful.) It is also straightforward to numerically evaluate the integral $\int_0^{T'} \exp(j\theta(t))dt$ for a given sample path. Thus, we can obtain an estimate of $p_x(X)$ whose accuracy is limited only by the statistical fluctuations inherent in simulations. However, with standard simulation techniques, the accuracy (standard deviation) of the estimated probability of a quantile q, obtained with N independent simulation trials, is approximately $\sqrt{1/Nq}$. Therefore, if one wants a 10-percent accurate estimate of the 10^{-9} quantile probability, N must be of the order of 10^{11}, a prohibitively large number of simulation trials.

To deal with this limitation, a modified simulation program was devised that achieves a high degree of efficiency. The approach uses the Radon-Nikodym Derivative (RND) from probability theory [18] to transform the actual process into one in which the rare events occur more frequently. This procedure is akin to doing an integral using a change of variable, but the integral is over functions instead of numbers. It is related to the statistical technique known as "importance sampling". The result is that $p_x(X)$ can be accurately assessed down to quantile values as small as 10^{-9} with a total number of simulation trials on the order of 10^6. This is several orders of magnitude less than would be required using conventional simulation.

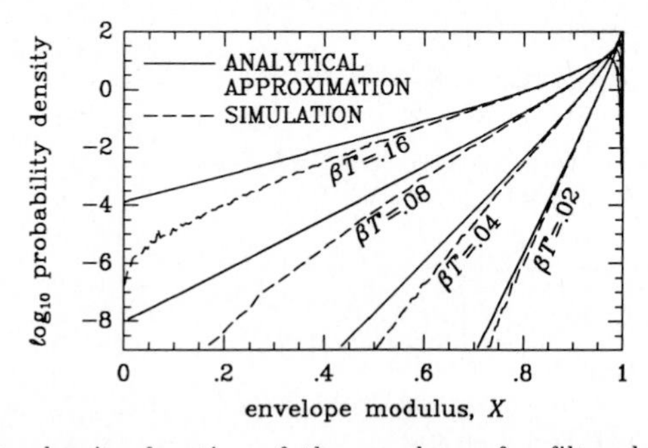

Fig. 3 — Probability density function of the envelope of a filtered sinusoid containing phase noise. The dashed curves represent results of computer simulations; the solid curves correspond to the analytical approximation of eq. (9). Results are shown for $\beta T' = .02, .04, .08$ and $.16$, where β is the linewidth due to phase noise and $1/T'$ is the IF filter bandwidth.

The effectiveness of this technique is evident in the dashed curves of Fig. 3. They are plots of $p_x(X)$ obtained through the modified simulation approach. The smoothness of the curves in the low-probability tails attests to the accuracy of the results.

B. The Analytical Approximation

The advantage of this approach is that it provides an explicit algebraic expression. Though only an approximation, it is asymptotically exact as $\beta T'$ tends towards zero. It uses perturbation theory to obtain the *moment generating function (MGF)* of the variate X. This MGF can then be inverted analytically to obtain the approximation $p_x(X) \cong -(d/dX)f((1-X)/(\pi\beta T'))$ where $f(\cdot)$ is given by eq. (A7e) of [11]:

$$f(y) = \frac{1}{\pi\sqrt{y}} \sum_{n=0}^{\infty} \Gamma(n+1/2)/[n!\Gamma(1/2)]$$
$$\times (4n+1)^{1/2} \exp[-(4n+1)^2/(16y)] K_{1/4}((4n+1)^2/(16y)) . \quad (9)$$

Here $K_{1/4}(\cdot)$ is the modified Bessel function. The function $f(\cdot)$ occurs in statistics in connection with the von Mises criterion for goodness-of-fit [20].

Despite its appearance, $f(\cdot)$ is very easy to calculate numerically. Some results of its use are shown as the solid curves in Fig. 3. Note that, where the analytical approximation departs significantly from the more exact simulation results, it is always in a pessimistic way, i.e., the BER obtained using the approximation will be an overestimate of the actual BER. Also, the difference becomes negligible as $\beta T'$ becomes very small. Therefore, one can use the approximate expression to upperbound BER for a given receiver, without the large amounts of computer execution time required by simulation.

Estimating the tightness of this bound is difficult. At BER levels around 10^{-9}, the analytical approximation may be in error by a large factor; however, because of the way $p_x(X)$ is used in (8), the final error in estimating receiver performance should turn out to be small. We verified this intuition by calculating receiver performance using both the analytical approximation and the more accurate simulation results for the *worst-case* data point (largest linewidth-to-bit-rate ratio and lowest BER). The difference between the two results was, indeed, negligible. We therefore confidently used the analytical approximation in obtaining our results for the double filtering receiver, which are discussed in the next Section.

C. The Fokker-Planck (FP) Method

This approach consists of modeling the processes of interest as a multidimensional Markov process. Specifically, if we denote the real and imaginary parts of $Z(t)$ as $r(t)$ and $q(t)$, respectively, the three-dimensional stochastic process $(r(t),\ q(t),\ \theta(t))$ can be easily shown to be a Markov process.

The well-developed theory of continuous-path Markov processes can be used to show that the pdf $p(r,q,\theta;t)$ satisfies the following partial differential equation (see Appendix B of [11]).

$$\pi\beta T' \frac{\partial^2}{\partial\theta^2} p - \cos\theta \frac{\partial}{\partial r} p - \sin\theta \frac{\partial}{\partial q} p = \frac{\partial}{\partial t} p . \quad (10)$$

Equation (10) is called the Fokker-Planck equation.

Since we are only interested in the joint statistics of $r(t)$ and $q(t)$, it would be helpful if we could write an equivalent equation without the θ variable. In [11] we presented elaborate mathematical manipulations to achieve the desired simplification. However, it has been recently pointed out to us [21] that the same result can be achieved in a much more straightforward way; namely, by observing that, with the alternative equivalent definition of $\mathsf{Z}(t)$ that we gave as eq. (B6) of [11],

$$\mathsf{Z}(t) = \int_0^t e^{j[\theta(\tau)-\theta(t)]}\, d\tau\,, \tag{11}$$

the two-dimensional stochastic process $(r(t),\ q(t))$ is also a Markov process. Using either approach, the resulting equation, in polar coordinates, is

$$\pi\beta T' \frac{\partial^2}{\partial\phi^2}p - \cos\phi\frac{\partial}{\partial\rho}p + \frac{\sin\phi}{\rho}\frac{\partial}{\partial\phi}p + \frac{\cos\phi}{\rho}p = \frac{\partial}{\partial t}p \tag{12}$$

with initial conditions

$$\lim_{t\to 0} p(\rho,\phi;t) = \delta(\rho,\phi)\,, \tag{13}$$

where ρ and ϕ are the modulus and argument of Z, respectively. A numerical solution of eq. (12) appears to be within the capabilities of present-day general-purpose PDE algorithms. In our work, we did not pursue such a numerical solution because the approximate expression given by (9), together with the simulation results, was quite adequate to achieve accurate results. However, in problems that require the full two-dimensional pdf of the complex quantity $\mathsf{Z}(t)$, a numerical solution of (12) appears to be the preferred approach to obtaining the necessary data.

V. *RESULTS*

A. *Results for BER*

A computer program was written that combines the analytical formulation of Section 3.1 with the results for $p_x(X)$ discussed in Section 4. In each program run, $\varsigma = \beta T$ is fixed, while the IF SNR, ξ, and the bandwidth ratio, m, are variables. For each combination of ξ and m, the program searches for the minimum of BER with respect to the threshold, λ, and outputs both the optimal λ and minimized BER.

Figure 4 shows curves of lowest BER vs. ξ for several values of ς. For each combination of ς and ξ, the BER result is for the discrete value of m that optimally balances mean square output noise against the output signal statistics. Thus, every point on each curve is minimized with respect to both threshold (λ) and bandwidth expansion factor (m).

Two curves in Fig. 4 bear discussion, starting with the dashed one. This is the result for BER that would arise if there were no phase noise *and* ideal coherent detection from IF were used. In this case, the optimal IF filter response would be (2) with $T' = T$; the optimal post-detection circuit would be just a sampler and a threshold at precisely half the signal amplitude, with no post-detection filtering; and the bit error rate would be $BER = 0.5\ \mathrm{erfc}(\sqrt{\xi/4})$. The second curve of interest is the one for $\varsigma = 0$. It represents the case of ideal noncoherent detection. In many texts, the formula presented for this case is equivalent to $BER = 0.5\ \exp(-\xi/4)$, which is slightly higher than the one shown here. This is because we have optimized λ instead of setting it at $S^2/2$, which is the usual approach. The optimal values we have obtained for λ/S^2 varied from about 0.14 to 0.40 over the interesting ranges of ς, m and ξ.

For each of the other four values of ξ, on each curve we have indicated the optimal discrete value of m. Note that the achievable detection degrades modestly as ς increases. For $BER = 10^{-9}$ and $\varsigma = 2.56$, for example, the required increase in ξ above the ideal coherent detection case is only 2.7 dB.

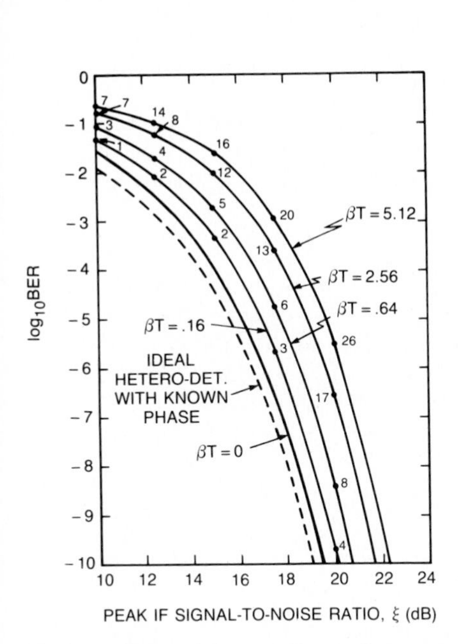

Fig. 4 — Optimized detection performance for OOK with double filtering receiver and square-law detection, with $\varsigma = \beta T$ (the linewidth-to-bit-rate ratio) as a parameter. The filled circles show the calculated points, with the corresponding optimum values of m indicated.

Fig. 5 — Example of bit error rate vs. IF signal-to-noise ratio, ξ, for OOK with double-filtering receiver. The curve shows, at each ξ, the minimum of BER with respect to integer values of IF bandwidth-to-bit-rate ratio, m. Other points show BER for suboptimal values of m.

B. Optimal and Suboptimal Bandwidths and Thresholds

The detection results are seen to be quite favorable when m is optimized. It is important to know, however, whether the bit error rate for a given SNR has a broad minimum with respect to m or is sharply dependent upon it. Consider, first, the curve for $\beta T = 0.27$, Fig. 5. As in Fig. 4, each point on the curve represents the minimum of BER with respect to integer m (and also with respect to the threshold, λ). More generally, at each of six SNR values spaced by 1 dB, BER points are shown for *several* values of m; the m-value giving the lowest point for a given SNR is optimal for that SNR, and the locus of the lowest points for all SNR defines the curve.

The important finding is that, for any SNR, the bit error rate increases only moderately as m ranges about its optimal value. For an SNR of 20 dB, for example, the optimal m is 5, but the degradation is slight as m decreases to 4 or increases to 8.

The sensitivity of OOK detection to the threshold setting, λ, is another matter. Figure 6 shows BER vs. λ for $\beta T = 0.27$ and an SNR of 20 dB. The optimal m for this case is 5, as noted above, and the optimal λ for that m is seen to be 0.320. As λ ranges 10 percent above and below this value, BER ranges over two orders of magnitude.

VI. EXTENSIONS

A. BER Results for FSK Modulation

The treatment outlined here for OOK modulation can also be applied with minor changes to *frequency-shift keying* (FSK) modulation in the limit of large separation between adjacent levels. Results were obtained in [12] for two-level FSK that are essentially equivalent to those of Figs. 4, 5, and 6, except for a 3-dB shift in the abscissa due to the definition of ξ as the *peak* SNR.

The results for two-level FSK can be easily extended to the case of an arbitrary number, M, of levels (M-FSK) as observed in [13]. Let the minimized bit error rate for 2-FSK be denoted by the two-variable function $B(\xi;\ \beta T)$. Then, for any M that is a power of 2, the bit error rate can be shown [22] to be closely approximated by $(M/2)B(\xi;\ \beta T)$, where ξ is the SNR as previously defined (peak and average SNR are the same with FSK) and T is now the *symbol* period.

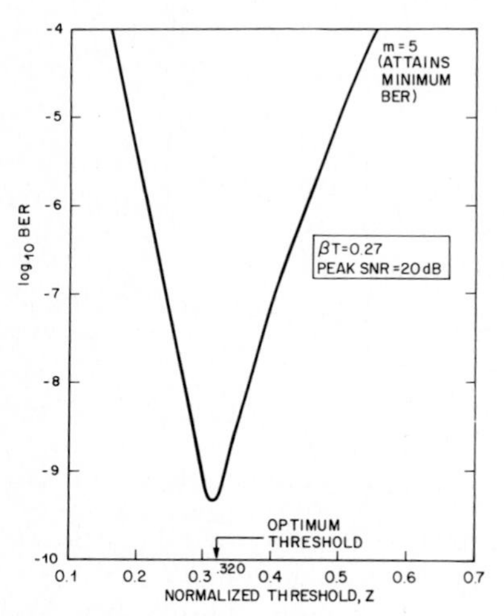

Fig. 6 — Example of bit error rate vs. threshold, λ, for OOK with double-filtering receiver. The curve is for optimal values of m.

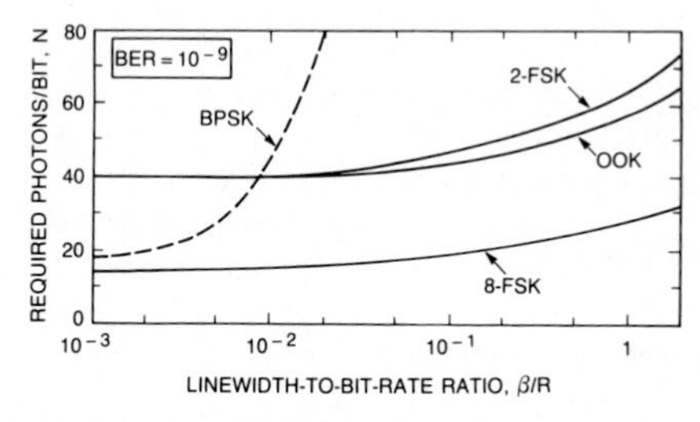

Fig. 7 — Minimum required average photons/bit vs. linewidth-to-bit rate ratio for several modulations. These results are for shot-noise-limited heterodyne detection and optimal design parameters (notably, the IF bandwidth). The dashed curve, for binary phase-shift keying, is taken from results in [5].

B. Optical Power Requirements

The results of Section V can be translated into a required optical level at the receiver. As discussed in [13], in the case where shot noise is dominant we can relate ξ quite simply to the average number of received photons per bit, N [23]:

$$N = \begin{cases} \xi/2; & \text{OOK Modulation} \\ \xi/(\log_2 M); & M\text{-FSK Modulation.} \end{cases} \tag{14}$$

Using curves like those in Fig. 4, it is easy to determine the values of ξ required to achieve a given bit error rate for given values of βT. By combining those curves with (14), we obtained the results shown in Fig. 7 for OOK and 2-FSK when $BER = 10^{-9}$. The abscissa represents the ratio of the (composite) laser linewidth, β, to the bit rate, R, which is defined as $R \triangleq \log_2 M/T$.

For *M*-FSK we can use (14) and the BER expression from the previous subsection, to obtain

$$BER(M\text{-FSK}) \cong (M/2)B(N \log_2 M;\ (\beta/R)\log_2 M) \tag{15}$$

Note that the first argument of $B(\cdot;\cdot)$ varies as $\log_2 M$, corresponding to the fact that the data period T (and, thus, the energy per data symbol) grows by this factor as M increases for the same bit rate. The required value of N for achieving a given BER is thus likely to be smaller by $\log_2 M$, at least for negligible amounts of phase noise. This is borne out by the curve for 8-FSK in Fig. 5.

As expected from classical communication theory, the results for OOK and 2-FSK are seen to be identical in the absence of phase noise. They diverge slightly, however, as β/R increases from zero (e.g., by about 0.5 dB at $\beta/R = 2$).

The dashed curve in Fig. 5, taken from results in [5], show the greater sensitivity to phase noise of heterodyne-detected binary phase-shift keying (BPSK). Thus, whereas this modulation is quite power-efficient when $\beta/R = 0$, both OOK and 2-FSK become significantly better for even modest values of β/R.

C. Other IF Filters

When one looks at Fig. 3, one immediately notices the peculiar fact that the pdf curves can be very closely approximated by straight lines. Since we used a log-linear scale, this means that a function that decays exponentially for decreasing X can be a useful empirical model for the pdf. This is not unique to the I&D filter defined in (2). In [14] we examined how the envelope pdf is affected by the shape of the response of the IF filter. For an arbitrary filter response, we developed a strict second-order upper bound to the probability distribution and a more accurate fourth-order approximation. In the case of the I&D filter, the approximate expression given by (9) is also available and its results are found to be essentially the same as the second-order upper bound. Fig. 8 shows a comparison of the results obtained by the various analytical techniques and by simulations for the I&D filter. The excellent accuracy of the fourth-order approximation was verified also for other filter responses by comparison to simulation results. The simulation algorithm was modified to allow the specification of an arbitrary filter response while retaining the advantages of the RND technique.

We considered three filter impulse responses beside the rectangular response of the integrate-and-dump filter. The first is that of a singly-resonant (or RC) filter; the second is that of a two-stage decoupled resonant (or 2-pole RC) filter; and the third is that of a second-order maximally flat (or Butterworth) filter. For all filters we found the same general behavior observed in Fig. 3: The curve for the pdf closely matches a straight line on a log-linear plot. Consequently, we were able to formulate the following empirical model for the pdf of X:

$$p(X) \approx \begin{cases} \alpha \exp(-\alpha(X_0 - X)); & 0 \leq X \leq X_0 \\ 0; & \text{elsewhere;} \end{cases} \tag{16}$$

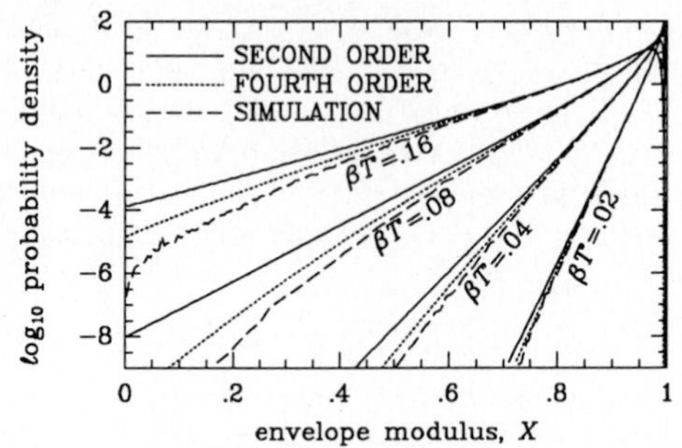

Fig. 8 — Probability density function of the envelope at the output of the I&D filter, as derived by both simulation and analysis. The analytical results are for both the second-order and fourth-order approximations. The second-order approximation is obtained from a strict upper bound to the probability *distribution* function; these results are essentially the same as the solid curves of Fig. 3.

where

$$X_0 = \int_0^{T'} h(t)dt \tag{17}$$

represents the maximum practical value of X in response to an optical pulse of duration T'. In most cases, T' is large enough to make X_0 approximately equal to the area under $h(t)$. In our calculations $h(t)$ was normalized to integrate to unity, so that we have $X_0 \approx 1$.

The slope of the exponential decay, α, is a function of the ratio of laser linewidth, β, to the *noise equivalent bandwidth* of the filter defined by

$$B \triangleq \int_{-\infty}^{\infty} |h(t)|^2 dt \Big/ \left| \int_{-\infty}^{\infty} h(t)dt \right|^2 ; \tag{18}$$

specifically,

$$\alpha = \frac{C}{\beta/B}[1+0.5\sqrt{\beta/B}] . \tag{19}$$

Note that for the I&D filter of (2), we have $\beta/B = \beta T'$, which is the same parameter that we have been using in the previous Sections.

The value of C depends on the details of the filter response. In our study we found that, while there were variations in the value of C from filter to filter, these were small; specifically, among the four types that we studied in [14], the Butterworth filter had the highest value, C=1.8, and the single-pole RC had the lowest value, C=1.3. Therefore, as observed in Section II.D, the optimized receiver performance given by Fig. 4 is, essentially, independent of the type of IF filter used.

Finally, one should note that $p(X)$ in (16) does not integrate precisely to unity because of the truncation at X=0. In numerical calculations where it is important that $p(X)$ integrate to unity, a scaling factor of $[1-\exp(-\alpha X_0)]^{-1}$ should be applied.

D. Direct-Detection Receivers with Optical Filtering

In the case of a direct-detection optical receiver preceded by an optical filter, the block diagram of Fig. 2 still applies. The IF filter now becomes the optical filter, which is typically used for channel selection in a multi-channel system. The envelope detector corresponds to the optical detector, which exhibits a square-law response as it delivers a current proportional to the instantaneous detected optical power. The rest of the receiver is unchanged.

The additive white Gaussian noise discussed in Section II.C is now replaced by optical background interference from unwanted channels. In situations where this is the dominant impairment and it can be modeled as white Gaussian noise, the results of Sections IV and V are directly applicable. This happens, for example, in systems employing optical Code-Division Multiple Access (CDMA) as discussed in [24] and [25].

VII. SUBJECTS FOR FURTHER STUDY

In this Section we briefly present two important extensions to this work that we are currently exploring in detail.

A. Direct-Detection with Optical Filtering - Full Noise Analysis

Consider a receiver like the one discussed in Section VI.D, but in a situation where it is not possible to neglect the noise introduced in the optical detection process (shot noise and receiver noise). Our treatment should be modified to include an additive Gaussian process introduced at the output of the square-law detector. Since this occurs *after* the non-linearity, it simply results in the addition of a Gaussian random variable to the sampled value denoted by z_k in Fig. 2. The variance of this random variable is easily computed. One can then modify the numerical calculations leading to Figs. 4, 5 and 6 to include this impairment.

B. Interference from Adjacent Channels

In a multi-channel system employing optical Frequency-Division Multiplexing (FDM), such as the one discussed in [26], the interference due to adjacent channels cannot, in general, be adequately modeled as white Gaussian noise. This is particularly true when the first filter is a Fabry-Perot interferometer (equivalent to the singly-resonant RC filter of Section VI.C). Indeed, assuming that the spectra of the interfering signals are broadened by phase noise to the same extent as the desired signal, it is easy to see how the resulting interference may have significantly non-Gaussian statistics, which must be accurately characterized to obtain valid BER performance results.

VIII. CONCLUSION

We have summarized our work to date on the subject of optical signal detection in the presence of laser phase noise. Our most important result is that, by using conventional designs with suitably chosen parameters, some coherent lightwave modulations can be efficiently detected despite considerable phase noise. Thus, the effective use of coherent lightwave transmission does not necessarily require bit rates far in excess of laser linewidths; indeed, good power efficiency is possible with laser linewidths in excess of the bit rate.

We are now working on refinements to our analysis that address the important practical problems of multi-channel systems based on optical frequency division.

REFERENCES

[1] T. Okoshi, "Heterodyne and Coherent Optical Fiber Communications: Recent Progress," IEEE Trans. on MTT, vol. *MTT-30*, no. 8, August 1982; pp. 1138-1149.

[2] M. Tamburrini, P. Spano and S. Piazzolla, "Influence of Semiconductor-Laser Phase Noise on Coherent Optical Communications Systems," Optics Letters, vol. *8*, no. 3, March 1983; pp. 174-176.

[3] T. Okoshi and K. Kikuchi, *Coherent Optical Fiber Communications*, Dordrecht/Boston/London/Tokyo: Kluwer Academic Publishers, 1988.

[4] J. Franz, "Evaluation of the Probability Density Function and Bit Error Rate in Coherent Optical Transmission Systems," Journal of Optical Communications, vol. *6*, no. 2, 1985; pp. 51-57.

[5] J. Salz, "Coherent Lightwave Communications," AT&T Technical Journal, vol. *64*, no. 10, December 1985; pp. 2153-2209.

[6] L. G. Kazovsky, "Balanced Phase-Locked Loops for Optical Homodyne Receivers: Performance Analysis, Design Considerations, and Laser Linewidth Requirements," Journal of Lightwave Technology, vol. *LT-4*, no. 2, February 1986; pp. 182-195.

[7] I. Garrett and G. Jacobsen, "Theoretical Analysis of Heterodyne Optical Receivers for Transmission Systems Using (Semiconductor) Lasers with Nonnegligible Linewidth," Journal of Lightwave Technology, vol. *LT-4*, no. 3, March 1986; pp. 323-334.

[8] J. Franz, C. Rapp and G. Söder, "Influence of Baseband Filtering on Laser Phase Noise in Coherent Optical Transmission Systems," Journal of Optical Communications, vol. *7*, no. 1, March 1986; pp. 15-20.

[9] L. G. Kazovsky, "Performance Analysis and Laser Linewidth Requirements for Optical PSK Heterodyne Communications Systems," Journal of Lightwave Technology, vol. *LT-4*, no. 4, April 1986; pp. 415-425.

[10] I. Garrett and G. Jacobsen, "The Effect of Laser Linewidth on Coherent Optical Receivers with Nonsynchronous Demodulation," Journal of Lightwave Technology, vol. *LT-5*, no. 4, April 1987; pp. 551-560.

[11] G. J. Foschini and G. Vannucci, "Characterizing Filtered Light Waves Corrupted by Phase Noise," IEEE Trans. on Inf. Theory, vol. *IT-34*, no. 6, Nov. 1988; pp. 1437-1448.

[12] G. J. Foschini, L. J. Greenstein and G. Vannucci, "Noncoherent Detection of Coherent Lightwave Signals Corrupted by Phase Noise," IEEE Trans. on Comm., vol. *COM-36*, no. 3, March 1988; pp. 306-314.

[13] L. J. Greenstein, G. Vannucci and G. J. Foschini, "Optical Power Requirements for Detecting OOK and FSK Signals Corrupted by Phase Noise," IEEE Trans. on Comm., vol. *COM-37*, no. 4, April 1989; pp. 405-407.

[14] G. J. Foschini, G. Vannucci and L. J. Greenstein, "Envelope Statistics for Filtered Optical Signals Corrupted by Phase Noise," IEEE Trans. on

Comm., to be published (Dec. 1989).

[15] F. G. Walther and J. E. Kaufmann, "Characterization of GaAlAs Laser Diode Frequency Noise," Sixth Topical Meeting on Optical Fiber Commun., New Orleans, February-March 1983; Paper TUJ5.

[16] W. A. Edson, "Noise in Oscillators," Proc. IRE, vol. *48*, Aug. 1960, pp. 1454-1466.

[17] S. O. Rice, "Statistical Properties of Sine-Wave Plus Random Noise," Bell System Technical Journal, vol. *27*, January 1948; pp. 109-157.

[18] L. A. Shepp, "Radon-Nikodym Derivatives of Gaussian Measures," Annals of Mathematical Statistics, vol. 37, 1966; pp. 321-354.

[19] D. Siegmund, "Importance Sampling in the Monte Carlo Study of Sequential Tests," Annals of Statistics, 4 (July 1976); pp. 673-684.

[20] T. W. Anderson and D. A. Darling, "Asymptotic Theory of Certain "Goodness of Fit" Criteria Based on Stochastic Processes," Annals of Mathematical Statistics, no. 23, 1952; pp. 191-192.

[21] J. Salz communicated to us this shorter derivation, which was identified by his student Amos Lapidot at the Technion, Haifa, Israel.

[22] S. Benedetto, E. Biglieri and V. Castellani, *Digital Transmission Theory*, Prentice-Hall, 1987; Section 5.5.1.

[23] W. K. Pratt, *Laser Communication Systems*, Wiley, 1969; Eq. (10-43).

[24] G. J. Foschini and G. Vannucci, "Using Spread-Spectrum in a High-Capacity Fiber-Optic Local Network," J. of Lightwave Tech., vol. *LT-6*, no. 3, March 1988; pp. 370-379.

[25] G. Vannucci, "Combining Frequency-Division and Code-Division Multiplexing in a High-Capacity Optical Network", IEEE Network magazine, vol. *3*, no. 2, March 1989; pp. 21-30.

[26] I. P. Kaminow, P. P. Iannone, J. Stone and L. W. Stulz, "FDMA-FSK Star Network with a Tunable Optical Filter Demultiplexer," J. Lightwave Technol., vol. *6*, no. 9, Sept. 1988; pp. 1406-1414.

Impact of phase noise on coherent systems - a Fokker-Plank approach

G. Jacobsen
TFL - Telecommunications Research Laboratory
Lyngsø Allé 2
DK-2970 Hørsholm, Denmark

I. Garrett
British Telecom Research Laboratories
Martlesham Heath
Ipswich IP5 7RE, England

Coherent optical systems for the future may use lasers with significant phase noise, manifest as broad linewidths. It is difficult to analyse the performance of a coherent optical receiver when the signals are corrupted by phase noise. The central theoretical problem arising from filtering a signal with phase noise is considered in a particular form which permits the derivation of the Forward or Fokker-Plank partial differential equation for the probability density of the output voltage of the I.F. filter of the receiver. We have developed a method for accurate numerical solution of this differential equation. Although the characteristic function of this probability density has been derived analytically in the limit of low phase noise, and the probability density has been simulated numerically, this is the first time that this problem has been solved accurately to low probability density. The results are used to discuss the I.F. bandwidth required for optical heterodyne receivers for ASK signals as well as to discuss the bit-error-rate floor arising due to the phase noise in DPSK and CP-FSK receivers. For DPSK or CP-FSK receivers with optimum I.F. filtering our analysis show that about 1.36 times as much phase noise can be tolerated than was predicted disregarding the effect of the filter.

1 Introduction

Coherent optical communication systems have received considerable attention in recent years, both theoretically and experimentally. They offer two major advantages over direct detection optical systems: increased receiver sensitivity by 10 to 20 dB, and the possibility of adding very large numbers of further channels by frequency multiplexing on a single fibre. These advantages make coherent optical systems attractive both for future long haul transmission and for broadband local networks. Much effort has gone into overcoming the technological problems associated with controlling the polarisation of the received signal, tracking its frequency or phase, and, most importantly, reducing the phase noise (linewidth) of semiconductor lasers so that they may serve as suitably coherent transmitters and local oscillators. These difficulties have all been overcome in the laboratory, but as yet there are no engineered solutions that may be used in operational coherent optical transmission systems.

Truly coherent systems make use of the phase information in the signal, such as in PSK modulation and in homodyne detection. Increasing the coherence of semiconductor lasers in order to be able to implement an optical phase locked loop is one of the most difficult of the engineering problems which has to be solved before truly coherent optical transmission can be used in the field.

Moderately coherent systems using delay demodulation, such as DPSK and CP-FSK systems, do not require an optical phase locked loop and are being considered seriously for operational use because the receiver sensitivity is only 3.5 to 5 dB worse than for truly coherent systems. These systems can also tolerate more phase noise on the lasers used for the source and for the local oscillator, and in the case of CP-FSK, can use direct modulation of the source laser. It is known [1-4] that laser phase noise results in an error-rate floor in these systems. The position of this floor depends on the phase drift over the demodulator delay time, τ_d, and the sum $\Delta\nu$ of the source and local oscillator laser linewidths. The measurement of the phase drift (i.e. the demodulation of the signal) takes place after bandpass filtering, however, and the effect of this on the phase noise is not well understood.

The alternative weakly coherent systems - heterodyne ASK and FSK systems with envelope or square-law demodulation of the I.F. signal - can be implemented using lasers with much broader linewidths with rather little reduction in receiver sensitivity. For example, in a weakly coherent system, using lasers with a total linewidth equal to half the data rate may, in theory, reduce the receiver sensitivity by only 2 or 3 dB relative to that obtainable for CP-FSK systems using lasers with no phase noise. Such a loss may be insignificant in many applications. Several workers [1-13] have investigated the theory of coherent receivers in which the phase noise is a significant factor. Their models have recently been compared and the similarities and differences of their predictions discussed [13]. The central theoretical problem is again that of analysing the result of filtering a signal which has significant phase noise. It is known that the filtering process results in amplitude noise, to an extent depending on the original phase noise (i.e. the linewidth) and on the filter bandwidth. The probability density function of this amplitude noise has been approximated to various degrees [11, 6] and simulated numerically [6], and the error rate performance of receivers for weakly coherent optical systems estimated on the basis of these results. However, this central theoretical problem has not yet been solved exactly.

In this paper, we present a partial differential equation (the Forward, or Fokker-Plank Equation) for the probability density of the amplitude noise, we discuss its numerical solution, and we give results for the performance of weakly and moderately coherent optical receivers based on the numerical solution. The partial differential equation is the same as that derived by Foschini and Vannucci [6].

The paper is organised as follows: section 2 defines the model of the receiver for a coherent heterodyne system, states the central theoretical problem in algebraic form, and describes the action of the I.F. demodulation on the signal. In section 3, we present the partial differential equation whose solution is the probability density of the amplitude noise, and we describe the method used to obtain a numerical solution. The results of this solution are related to receiver performance in section 4, and compared with the predictions of Foschini and co-workers [7]. Section 5 summarises our conclusions from this work.

2 The Central Theoretical Problem

2.1 The receiver model

A schematic diagram of the heterodyne receiver is shown in Fig 1. For an ASK receiver with square-law I.F. detection the delay τ_d is 0, whereas it is of the order of T, the bit-time, for a delay DPSK or CP-FSK receiver. The signal and local oscillator fields are incident on a photodiode. The resulting photocurrent is proportional to the squared modulus of the total incident optical field and is made up of three components, representing direct detection of the signal, direct detection of the local oscillator, and a beat component between the signal and local oscillator. This beat component is the difference frequency (I.F. carrier) between the signal and local oscillator optical frequencies, modulated by the signal information. The total photocurrent is amplified, resulting in a signal voltage which is then bandpass filtered to extract the modulated I.F. carrier $s(t)$ from the d.c. and baseband components, and from other channels if present.

For an ASK receiver the signal is demodulated from the I.F. carrier by an ideal square-law device followed by a low-pass filter. The function of this combination is often modelled as yielding the squared modulus of the modulated I.F. carrier as the baseband signal, an approximation which is good if the I.F. is high enough for the low-pass filter to separate baseband from double-frequency components. Alternatively the square-law device may be replaced by an envelope detector, which ideally yields the modulus of the modulated I.F. carrier as the baseband signal. We will concentrate on a discussion of the square-law device, since this is a good model of a simple, practical nonlinear demodulator. For a DPSK or a CP-FSK receiver the output of the delay-demodulator has a baseband component which is proportional to the cosine of the phase drift (due to phase noise and modulation) over the delay time.

2.2 Formulation of the central problem

We are concerned with the influence of phase noise on the signal as it is processed by the heterodyne receiver. If the transmitter laser is ideally modulated, there is no optical signal power during the ZERO symbols in an ASK receiver. For this reason, we confine our attention to the ONE symbols in what follows. For the DPSK or CP-FSK receiver our considerations apply for both symbols.

The output from the receiver front-end contains an I.F. signal component $s(t)$ at frequency ω_{IF}. This signal, which is corrupted by phase noise, may be represented as:

$$s(t) = e^{j(\omega_{IF}t+\phi(t))} \tag{1}$$

where $\phi(t)$ is the random I.F. phase, usually modelled as a Brownian motion Gaussian process. Here the signal has been normalised to unit amplitude. We will use the equivalent baseband representation of the I.F. signal, and write:

$$s(t) = e^{j\phi(t)} \tag{2}$$

If the I.F. filter has an equivalent baseband impulse response $h(t)$, the signal at the output of the I.F. filter is:

$$z_0(t) = s(t) * h(t) \tag{3}$$

where * denotes convolution. The ideal square-law detector yields:

$$Y(t) = |z_0(t)|^2 \tag{4}$$

This is filtered by the post-detection filter with impulse response $f(t)$ to yield the receiver output voltage $v(t)$:

$$v(t) = Y(t) * f(t) = \left| \int_{-\infty}^{\infty} h(\tau - t) e^{j\phi(\tau)} d\tau \right|^2 * f(t) \tag{5}$$

The central problem is the convolution integral in equation (5), which has the stochastic process $\phi(t)$ in the exponential. The subsequent convolution with $f(t)$ also presents an analytical problem when $T >> t$ as is the case for an ASK system. For a DPSK or CP-FSK system the phase noise has its effect at the I.F.-demodulation stage and we will not concern ourselves with details of the post-detection filter. General methods for handling integrals of random processes are poorly developed. For this reason, various researchers have sought to deal with the analytical difficulties by making approximate analyses of one kind or another.

2.3 A solvable general problem

We follow Foschini et al [7] in modelling the I.F. filter as a finite time integrator, and the low-pass filter as a discrete-time integrator, with integration times t and T' respectively, such that $t + T' = T$, the bit-time. Then the output voltage of the I.F. filter becomes:

$$z_0(t) = \int_0^t e^{j\phi(\tau)}\, d\tau \tag{6}$$

This is the problem considered by Foschini and Vannucci [6], and is the form of the problem which we address here. In section 3, we replace $\phi(t)$ by an equivalent phase process with the same statistics, which allows us to define a random variable $z(t)$ with the same probability density as $z_0(t)$ but which has the advantage of being Markov, which $z_0(t)$ is not. We can then use the powerful methods appropriate to Markov processes to derive the Forward (Fokker-Plank) Equation for the probability density of the output voltage of the I.F. filter. Our equation is the same as that derived by Foschini and Vannucci.

2.3.1 The effect of low-pass filtering for weakly coherent systems

The output voltage from the square-law detector, $Y(t)$, is equal to $|z(t)|^2$. This voltage is further filtered by the low-pass filter. Foschini et al [7] show that, if the I.F. filter is a finite-time integrator and the post-detection filter is a discrete-time integrator, then the resulting output voltage at the decision time, $v(T)$, is the sum of M independent, identically distributed samples of $Y(t)$, where M is the ratio T'/t of the filter integrating times. The probability density function (pdf) of $v(T)$ can therefore be found as the M-fold convolution of the pdf of $Y(t)$. In this way the influence of the I.F. bandwidth is represented. A broader bandwidth corresponds to a smaller integration time, t, hence to a smaller Dt, and also to a greater number M of convolutions.[1] If we normalise $Y(t)$ and $v(T)$ so that they each span a range from 0 to 1, increasing M has the effect of moving the mean nearer to 1, and of narrowing the pdf for $v(T)$ around the mean, thus decreasing the impact of the phase noise on the receiver output voltage. Increasing M implies increasing the I.F. bandwidth, allowing more of the receiver noise to pass, of course. This cannot be completely compensated for by the post-detection filter because of the non-linear demodulator following the I.F. filter. So there is an optimum value for the I.F. bandwidth which minimises the total output noise voltage.

2.4 The bit-error rate

2.4.1 The ASK receiver

The receiver output voltage $v(T)$ at the decision time is compared with a threshold voltage v_{th} and the symbol under decision is determined to be a ONE or a ZERO depending on whether $v(T)$ is above or below v_{th}. Since $v(T)$ is corrupted by noise, this decision is sometimes in error. The corrupting noise has two main origins: amplitude fluctuations caused by filtering the signal with phase noise, and receiver noise which includes the shot noise on the photocurrent due to the local oscillator field and various sources of thermal noise [10]. In a heterodyne ASK receiver, the receiver noise is the

[1] For moderately coherent systems we have $M \approx 1$ and the low-pass filter does not significantly influence the noise statistics due to phase noise.

same for both ONE and ZERO symbols. The amplitude noise resulting from the phase noise scales with the optical signal field, however, and is thus greater on ONE symbols. If the transmitter laser is ideally modulated there is no optical signal field during ZERO symbols and hence no amplitude noise resulting from phase noise. The decision threshold voltage v_{th} must be set so as to produce an optimum balance between errors on ONEs and errors on ZEROs, taking into account the different statistics of the output voltage for the different symbols. A second balance must be struck in determining the bandwidth of the I.F. filter. If it is too narrow, the effect of phase noise on the output voltage is greater, while if it is too large, the receiver noise contribution to the output voltage is increased. This balance must also take account of the spectral density of the receiver noise, which will only be white if the local oscillator shot noise is the overwhelming contribution. In this paper we will concentrate on the effects of phase noise on the received signal, and we will not complicate the picture by computing the optimum decision threshold setting. Instead we consider a range of fixed values between 0.25 and 0.4, which spans values of practical interest. We choose this simplified approach for several reasons: (i) the optimum threshold setting depends on the spectral shape of the receiver noise; (ii) the noise statistics on ZERO symbols depend on the received optical power level, which may not be zero in practice; (iii) the receiver model is somewhat idealised. We do not want to suggest that the results are exactly applicable to practical receivers. Rather it is the trend of the results which is of interest and importance. Choosing a fixed value for v_{th}, as might well be done in a practical coherent receiver, causes an error-rate floor [14, 11]; that is, an error-rate that cannot be lowered by increasing the signal power. The errors are caused predominantly by the amplitude noise resulting from filtering the signal with phase noise, and this noise scales with the signal field. The position of the floor depends on v_{th} and on the I.F. bandwidth (i.e. on M). By taking v_{th} as fixed, we are using our analysis to find how the error rate floor depends on M. In particular, we find the values of M (as a function of linewidth $\Delta\nu$) which place the floor at 10^{-12}, and use this as our basis for comparison with other models.

2.4.2 Moderately coherent systems

Previous analyses of DPSK and CP-FSK systems [1, 2, 4] have assumed that there is no significant change in the probability density function (pdf) of the phase at the output of the I.F. filter if the I.F. bandwidth is much greater than the total laser linewidth. In this case, the phase drift over the delay time of the demodulator is a Gaussian random process with zero mean and variance $2\pi\Delta\nu\tau_d$. The error-rate floor is then given by:

$$P_{floor} = \frac{1}{2}\operatorname{erfc}\left(\frac{1}{4}\sqrt{\frac{\pi}{a\Delta\nu\tau_d}}\right) \tag{7}$$

where $a = 1$ if the statistics are unaltered (we include a in equation (7) for later discussion). We have ignored the possibility of the phase drift θ over t being more than π. Filtering reduces the effect of the phase noise, so inasmuch as the complementary error function form is a good approximation, we expect $a \leq 1$. Patzak and Meissner [5] have argued that a lies between 0.5 and 0.75. Foschini and Vannucci [6] have derived an analytic approximation to the moment-generating function for the process at the output of the I.F. filter, which is valid in the limit of small $\Delta\nu\tau_d$. For the output phase, their equation (6) reduces to the moment-generating function for a Gaussian phase process with variance $2\pi\Delta\nu\tau_d/3$, one-third of the input phase variance, leading to a value of 2/3 of a in equation (7) in agreement with [5]. Our approach gives a similar result for a.

3 The Method of Solution

3.1 The Forward (Fokker-Plank) Equation

The method outlined here has been described in more detail by Bond [15] (see also [16]). The first step is to define a random variable $z(t)$ with the same probability density as $z_0(t)$ but which is Markov and can therefore be handled using the powerful techniques developed for such stochastic processes [17]. Then we obtain an expression for the time derivative of $z(t)$, which allows us to derive the Forward Equation for the density of z by well-established methods.

For fixed t, the phase process $\phi(t-\tau)-\phi(t)$ is normally distributed with mean zero and variance $D\tau = 2\pi\Delta\nu\tau$, where D is the phase diffusion coefficient and $\Delta\nu$ is the full width at half maximum of the I.F. signal spectrum. $\Delta\nu$ is the sum of the linewidths (full width at half maximum of the Lorentzian field power spectra) of the transmitter and local oscillator lasers. Since the increments (with respect to τ) of $\phi(t-\tau)-\phi(t)$ are independent, it is also a Brownian motion Gaussian process, with the same statistical properties as $\phi(t)$. The random variable

$$z(t) = \int_0^t e^{j(\phi(\tau)-\phi(t))}\, d\tau \tag{8}$$

has therefore the same probability density as $z_0(t)$ and differs from it only in the details of how it evolves in time. Clearly, the modulus of z is always less than or equal to t. We express $z(t)$ in cartesian and in polar form:

$$z(t) = x(t) + jy(t) = r(t)e^{j\theta(t)} \tag{9}$$

The probability density of $z(t)$ will be written as $p(x,y,t)$ or, with a slight abuse of notation, as $p(r,\theta,t)$. The Forward Equation for p can be derived as explained in [18, 15, 16]:

$$\frac{\partial p}{\partial t} = -\cos\theta\frac{\partial p}{\partial r} + \frac{\sin\theta}{r}\frac{\partial p}{\partial \theta} + \frac{D}{2}\frac{\partial^2 p}{\partial \theta^2} \tag{10}$$

with the initial condition that when $t=0$ the density p is concentrated as a delta function of unit weight at $r=0$.

It is possible to derive the moments of $z(t)$. Let

$$\mu_{m,n}(t) = E\{z(t)^m z^*(t)^n\} \tag{11}$$

where z^* is the complex conjugate of z. Then [15, 16]

$$\mu_{m,n}(t) = \int_0^t e^{-\frac{1}{2}D(m-n)^2(t-\tau)}(m\mu_{m-1,n}(\tau) + n\mu_{m,n-1}(\tau))d\tau \qquad (12)$$

This formula may be used to find all the $\mu_{m,n}$'s since $\mu_{0,0} = 1$ [15, 16]. We have used the moments up to fiftieth order to check the accuracy of the numerical solutions of the Forward Equation.

3.2 Numerical solution of the Forward Equation

Equation (10) represents advection with unit speed in the positive-x direction, and combined advection and diffusion in both positive and negative θ directions. The initial condition is a delta function in both r and θ. The solution of this equation possesses certain properties:

1. $|z(t)|$ cannot be greater than t, so that the solution is confined within a disc of radius t.
2. Since the phase process $\phi(t)$ has zero mean, the solution must be symmetric about the $\theta = 0$, $\theta = \pi$ axis.
3. The solution for $p(r, \theta, t)$ must be non-negative everywhere since p is a probability density function.
4. The probability density function $p(r, \theta, t)$ must integrate to unity over the $\{r, \theta\}$ domain.
5. Because of the convective behaviour of the Forward Equation and because the initial condition is a delta function, we expect the solution to have a very steep front near the circle $r = t$, at least for small time. For large time, diffusion will smooth out the influence of the initial condition.

We choose to develop the difference scheme from the conservative form of the Forward Equation (10), and we make a change in the dependent variable, defining $P = rp$. The Forward Equation (10) is then:

$$\frac{\partial P}{\partial t} = -\frac{\partial}{\partial r}(P\cos\theta) + \frac{\partial}{\partial\theta}\left(\frac{P}{r}\sin\theta + \frac{D}{2}\frac{\partial P}{\partial\theta}\right) \qquad (13)$$

We can write this as:

$$\frac{\partial P}{\partial t} = L_r(P) + L_\theta(P) \qquad (14)$$

where L_r and L_θ are radial and angular operators. We can then use an operator splitting method for the difference scheme. The scheme that we used has been reported in [19]. For the radial step, we used a flux-limiting (total variation diminishing) scheme due to Roe [20] to preserve the steep front of the solution near $r = t$. A typical contour plot of the numerical solution of the Forward Equation, $p(r, \theta, t)$ is shown in Figure 2 for Dt

having the value 8. It can be seen that this plot possesses properties 1, 2, 3 and 5 listed above.

The solutions were checked against the moments derived from equation (12) which were computed up to fiftieth order for x,y and $r^2 = |z(t)|^2$ as well as for for the central moments. The results are summarized in Tables 1-4 for $Dt = 0.125$, 0.25 and 2. Table 1 shows that zero order order moments check to better than 6 significant digits and that the means of x and r^2 are given with a precision better than 5.2×10^{-2} %. Tables 2-4 show central moments up to fiftieth order. The tables are given up to a moment order n, where results derived from equation (12) are no longer accurate with more than 1 % relative precision. Accordingly, the deviation listed specifies the accuracy of the Fokker-Plank numerical solution. The accuracy gets poorer for higher order moments as expected. It is, however, remarkable that even for small absolute values of the central moments we have in general agreement within a few percent. This indicates that the Fokker-Plank results can be trusted even in the lower parts of the pdf's.

The pdf of $r = |z(t)|$ is obtained from $p(r, \theta, t)$ for given t by integrating with respect to θ over $\{0, 2\pi\}$. The pdf of the output phase, θ, is obtained by integrating over r.

4 Results and Discussion

4.1 The ASK receiver

Representative examples of the pdf of $r = |z(t)|$, obtained by numerical solution of the Forward Equation (10) and subsequent integration over θ, are shown in Figure 3 for different values of the variance Dt of the phase noise. The general form of these curves shows that the envelope of the output voltage from the I.F. filter, $|z|$, is most probably near unity on the normalised voltage scale, but has non-zero probability for all values between 0 and 1. The peak of the distribution is below unity, as expected, since any phase fluctuation will reduce $|z|$. Increasing the phase noise parameter Dt reduces the mean and broadens the distribution. Shown for comparison are the analytic approximation (dotted line) and numerical simulation (broken line) from Foschini and Vannucci [6] for Dt of 1.0 and 2.0. We see that our results are in excellent agreement with the numerical simulations.

The cumulative distribution function (cdf) of the receiver output voltage $v(T)$ at the decision time is calculated as follows. First the pdf of $r^2 = |z|^2$ is found from the pdf of $r = |z|$. Then this new pdf is convolved with itself $M = T'/t$ times, using a numerical Discrete Fourier Transform routine, to yield the pdf of the output voltage at the decision time, $v(T)$, as discussed in section 2.3.1. Integrating this pdf gives cdf curves [16].

Figure 4 shows the estimated IF bandwidth required for error rate floors of 10^{-12}, as a function of linewidth, for threshold values of 0.25, and 0.40 [16]. We show curves from this work (full lines) and from the analytic approximation of Foschini and Vannucci [6] (broken lines). These curves show that any linewidth can be accomodated by a sufficiently broad IF filter, while still achieving an acceptable error rate floor performance. It can be seen that the results agree rather well for all linewidths. Our numerical solution of the Forward Equation always predicts the lower IF bandwidth requirement. The optimum IF bandwidth is clearly very dependent on the threshold setting. The threshold setting may be optimised (to minimise the BER), taking into account the combined effect of receiver noise and phase noise, and this will determine the required IF bandwidth. This optimisation has been done for additive white Gaussian receiver noise by Foschini et al [7]. It is hard to make a detailed comparison between their results and the results shown in Fig 4 because of the effect of receiver noise, but it appears that where comparison can be made, the criterion of an error rate floor at 10^{-12} slightly overestimates the linewidth that can be accomodated by a receiver optimized for a BER of 10^{-9}.

Foschini et al's analytical approximation [7] is very suitable for practical receiver design, since it is easily implemented on a computer, and predicts results which are reasonably close to the numerical solution of the Forward Equation.

4.2 The moderately coherent systems

Typical pdf's for θ for different linewidths can be seen to be very close to Gaussian with a variance of $0.363 \times Dt$ [21]. Fig 5 shows the error-rate floor, for the case when the demodulator delay time τ_d is the same as the filter integration time t. In this case the output of the low-pass filter of the receiver is simply cosine of the difference between two identically distributed independent θ-samples. The resulting phase process is therefore very closely Gaussian with variance $0.726 \times Dt$, corresponding to an error-rate-floor given by equation (7) with $a = 0.726$. The curves in the figure correspond to equation (7) with $a = 1$ (no I.F. filter), 0.726 (our result), 0.680 ([5]) and 2/3 (implied from [6]) for comparison. Our result implies that if τ_d is the same as t, about 1.36 times the laser linewidth may be tolerated compared with a prediction that ignores the effect of I.F. filtering on the phase noise.

In practical moderately coherent receivers, the equality of time-constants is only approximately satisfied. The situation $t > \tau_d$ is avoided because of the resulting intersymbol interference. The opposite situation, with $t < \tau_d$, is of practical interest. From the form of the Fokker-Plank equation, we can consider the delay time τ_d to cover two periods: an integration time of duration t, and a phase diffusion period of duration $\tau_d - t$. Since these periods are disjoint, phase fluctuations during these periods are independent, so that the variance of the overall phase drift is the sum of the variances of the drifts in the two periods. Since the phase diffusion is a Gaussian process with variance $D(\tau_d - t)$, and the phase drift during the integration period is nearly Gaussian with variance $0.363 \times Dt$, the resulting process is also nearly Gaussian, with variance $D\tau_d(1 - 0.637t/\tau_d)$. The output phase of the delay demodulator is the difference between the phase sample from the delayed arm and an independent θ-sample from the direct arm. The resulting phase process is thus nearly Gaussian with variance:

$$\sigma^2 = 2\pi\Delta\nu_{eff}\tau_d = 2\pi\Delta\nu\tau_d(1 - 0.274t/\tau_d) \qquad (15)$$

The effective linewidth tends to the actual linewidth as t tends to zero (I.F. bandwidth tends to infinity), as it must. Equation (15) implies that the I.F. bandwidth should be reduced as far as possible so that t approaches τ_d. This strategy reduces the effects of both receiver noise and phase noise, in contrast with weakly coherent receivers where a wide I.F. bandpass is beneficial in reducing the amplitude noise arising from filtering the phase noise.

5 Conclusions

We have derived the Forward (Fokker-Plank) Equation for the probability density function of the modulus of the output voltage of a passband filter fed by a signal with phase noise. We have carried out the first accurate numerical solution of this equation and have checked the accuracy using iteratively derived moments up to fiftieth order. We have used the results to predict, more accurately than has been possible before, the I.F. bandwidth required by a heterodyne optical receiver to accomodate laser phase noise. For delay demodulation receivers our analysis shows that the pdf of the output phase of the I.F. filter is very nearly Gaussian for linewidths of up to 5% of the filter bandwidth. The variance of the phase fluctuations over the filter integration time is reduced by a factor of 0.363, close to Foschini and Vannucci's approximate result of one-third. After a delay time τ_d greater than the integration time t the pdf is close to Gaussian, with variance $2\pi\Delta\nu\tau_d(1 - 0.274t/\tau_d)$. This result is close to Patzak and Meissner's result for $t = \tau_d$ of $0.68 \times 2\pi\Delta\nu\tau_d$.

Acknowledgements

We thank the Director of Research of TFL - Telecommunications Research Laboratory and the Director of Research and Technology of British Telecommunications plc for permission to publish this work.

References

[1] Jacobsen, G., and Garrett, I., *J Lightwave Technol.*, 1987, **LT-5,** pp. 478-484
[2] Garrett, I., and Jacobsen, G., *J Lightwave Technol.*, 1988, **LT-6,** pp. 1415-1423

[3] Nicholson, G., *Opt. Quantum Electron.*, 1985, **17**, pp. 399-410
[4] Kazovsky, L. G., *J. Opt. Commun.*, 1986, **7**, pp. 66-78
[5] Patzak, E., and Meissner, P., *IEE Proc. J*, 1988, **135**, pp. 355-357
[6] Foschini, G. J., and Vannucci.*IEEE Trans. Inf. Theory*, 1988, **IT-34**,, pp. 1437-1448
[7] Foschini, G. J., Greenstein, L. J., and Vannucci. G., *IEEE Trans. Commun.*, 1988, **COM-36**, pp. 306-314
[8] Kazovsky, L. G., Meissner, P., and Patzak, E., *J. Lightwave Technol.*, 1987, **LT-5**, pp. 770-791
[9] Salz, J., *AT&T Tech. J.*, 1985, **64**, pp. 2153-2209
[10] Garrett, I., and Jacobsen, G.: , *J. Lightwave Technol.*, 1986, **LT-4**, pp. 323-334
[11] Jacobsen, G., and Garrett, I., *IEE Proc. J.*, **134**, pp. 303-312 (*Erratum* 1988, **135**, p. 100)
[12] Franz, J., *J. Opt. Commun.*, 1987, **8**, pp. 57-66
[13] Garrett, I., and Jacobsen, G., *IEE Proc. J*, 1989, **136**, pp. 159-165
[14] Jacobsen, G., and Garrett, I., *Electron. Lett.*, 1985, **21**, pp. 268-270
[15] Bond, D. J., to appear in *British Telecom Technol. J.*, 1989
[16] Garrett, I., Bond, D. J., Waite, J. B., Lettis, D. S. L., and Jacobsen G., to appear in *J. Lightwave Technology*, 1990
[17] Papoulis, A.: 'Probability, random variables, and stochastic processes' (McGraw-Hill 1984)
[18] Arnold, L.: 'Stochastic Differential Equations' (Wiley-Interscience, New York, 1974)
[19] Waite, J. B., and Lettis, D. S. L., to appear in *British Telecom Technol. J.*, 1989
[20] Roe, P. L.: 'Some contributions to the modelling of discontinuous flows', Proc. AMS/SIAM Summer Seminar, La Jolla, 1983; 'Lectures in Applied Mathematics', 1985, **22**
[21] Jacobsen, G., Jensen, B., Garrett, I., and Waite, J. B., submitted for publication

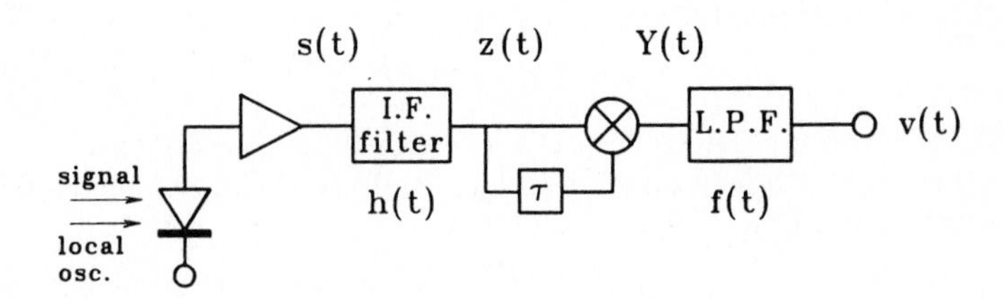

Fig 1 Schematic diagram of a heterodyne optical receiver. LPF: post-detection filter. Time responses $s(t)$, $h(t)$, $z(t)$, $Y(t)$, $f(t)$ and $v(t)$ are explained in section 2.2.

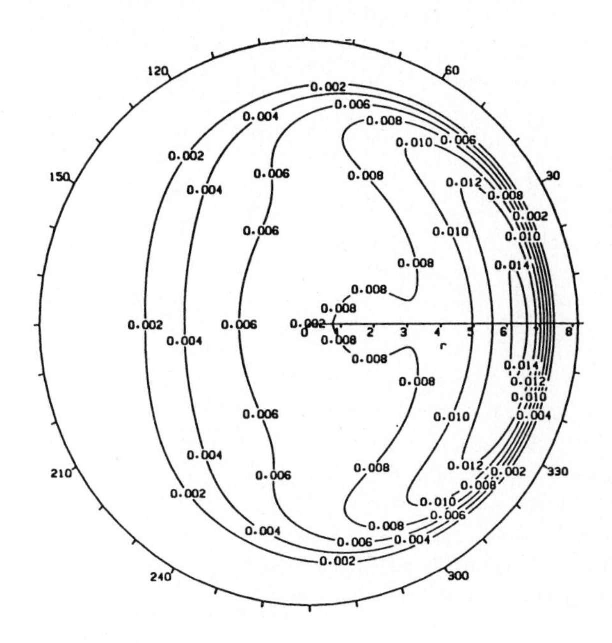

Fig 2 Contour plot of the numerical solution of the Forward Equation, $p(r, \theta, t)$, for $Dt = 8$.

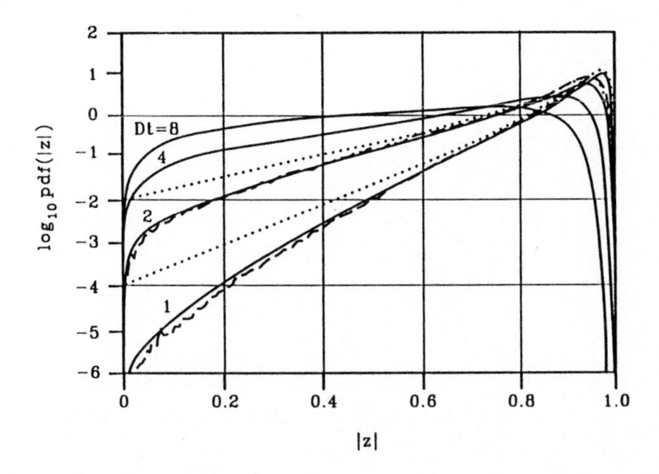

Fig 3 The probability density function of $|z|$, for different values of the variance $Dt = 2\pi\Delta\nu t$ of the phase noise. Full curves : this work. Broken curves and dotted curves (for Dt of 1 and 2): numerical simulation and analytic approximation [6].

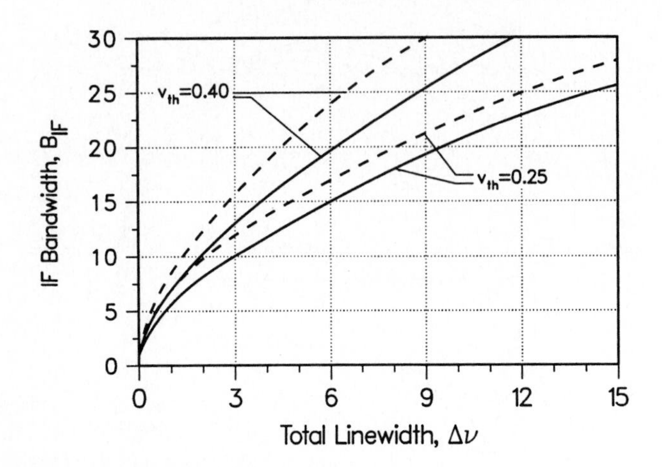

Fig 4 The I.F. bandwidth required to achieve an error rate floor of 10^{-12}, as a function of linewidth and for thresholds v_{th} of 0.25 and 0.40. Full curves: this work. Broken curves: using the analytic approximation [6].

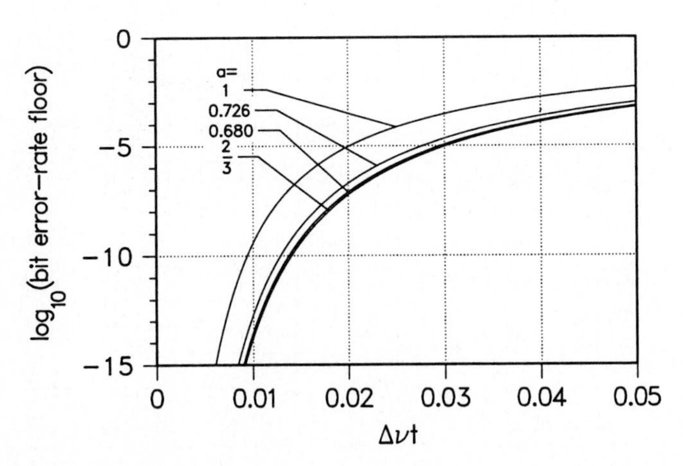

Fig. 5 The error-rate floor for $t = \tau_d$. Lines: equation (7) with values of a as marked.

moments		$Dt = 0.125$	$Dt = 0.25$	$Dt = 2.0$
Zero order	value	1.0000	1.0000	1.0000
	accuracy (%)	9×10^{-7}	1×10^{-9}	7×10^{-5}
r^2	value	0.97949	0.95960	0.73576
	accuracy (%)	2.3×10^{-2}	1.6×10^{-2}	6×10^{-3}
x	value	0.96939	0.94002	0.63212
	accuracy (%)	3×10^{-3}	1.3×10^{-2}	5.2×10^{-2}

Table 1

n		Central Moments, $Dt = 0.125$		
		x_C^n	y_C^n	r_C^{2n}
2	value	1.1939×10^{-3}	3.8575×10^{-2}	3.2343×10^{-4}
	accuracy (%)	2.6	0.7	3.0
6	value	9.014×10^{-7}	6.8063×10^{-4}	1.6182×10^{-8}
	accuracy (%)	11	3.2	14
8	value		1.4660×10^{-4}	3.69×10^{-10}
	accuracy (%)		5	25
10	value		3.7968×10^{-5}	
	accuracy (%)		6	
20	value		2.131×10^{-7}	
	accuracy (%)		15	

Table 2

n		Central Moments, $Dt = 0.25$		
		x_C^n	y_C^n	r_C^{2n}
2	value	4.3848×10^{-3}	7.1572×10^{-2}	1.2061×10^{-3}
	accuracy (%)	1.7	0.4	2.1
4	value	2.2369×10^{-4}	1.3236×10^{-2}	1.6125×10^{-5}
	accuracy (%)	4.5	1.0	5.7
8	value	7.9814×10^{-6}	1.1983×10^{-3}	4.6842×10^{-8}
	accuracy (%)	11	2.6	14
10	value	2.6138×10^{-6}	4.6411×10^{-4}	4.8626×10^{-9}
	accuracy (%)	21	3.4	19
20	value		1.2696×10^{-5}	
	accuracy (%)		7.0	

Table 3

n		Central Moments, $Dt = 2.0$		
		x_C^n	y_C^n	r_C^{2n}
2	value	9.7203×10^{-2}	2.3898×10^{-1}	3.1424×10^{-2}
	accuracy (%)	0.3	0.02	0.5
10	value	2.9650×10^{-2}	1.7694×10^{-2}	1.5240×10^{-4}
	accuracy (%)	1.8	0.4	1.7
20	value	0.11081	2.4153×10^{-3}	2.6049×10^{-6}
	accuracy (%)	11	1.1	2.2
30	value	0.9404	5.2056×10^{-4}	7.1069×10^{-8}
	accuracy (%)	107	2.0	2.4
50	value		4.5424×10^{-5}	
	accuracy (%)		4.1	

Table 4

A Coherent Phase-Diversity Optical Receiver with Double Frequency Conversion

G. Vannucci

AT&T Bell Laboratories, Crawford Hill Laboratory
Holmdel, NJ 07733 U.S.A.

ABSTRACT — **A conventional "phase-diversity" optical receiver uses homodyne detection on two separate branches to provide two orthogonal baseband outputs. Together, these outputs provide a complete characterization of the received optical signal. We present a novel design for an equivalent receiver that achieves some of the advantages of both heterodyne and homodyne detection while avoiding the need for the optical quadrature hybrid (a difficult-to-realize component) required by the conventional design. The new design has disadvantages, in that receiver sensitivity is reduced by approximately 6 dB, and in that it requires the use of electro-optic modulators. However, in applications where ultimate sensitivity is not sought, the advantages of this design make it a very interesting alternative.**

I. INTRODUCTION

Coherent detection of optical signals has become a very important area of investigation for future fiber-optic communication systems. Heterodyne detection schemes have been demonstrated in the laboratory by several groups [1-4]. Homodyne detection, however, has turned out to be more difficult to achieve, primarily because of the more stringent constraints on laser stability and control; reference [5] reports the best results to date.

Homodyne detection offers several advantages over heterodyne detection, including a) greater receiver sensitivity (by 3 dB); b) closer channel packing in multi-channel systems; and c) reduced receiver noise, due to the quadratic noise spectrum introduced by the detector amplifier [6]. Despite these advantages, the heterodyne technique is usually favored because of its reduced requirements on the frequency stability of the local-oscillator laser.

The so-called "phase-diversity" receiver attempts to circumvent some of the difficulties of homodyne detection by providing two (or more) baseband outputs that, together, give a complete representation of the received signal [7,8]. However, it requires a component called an "optical quadrature hybrid," which has proven difficult to realize in practice [9,10]. In this paper we present a design for a phase-diversity receiver that achieves some of the advantages of both heterodyne and homodyne detection while avoiding the need for an optical quadrature hybrid. Its disadvantages include a reduction in receiver sensitivity of a few dB and the use of electro-optic modulators. It is similar to the heterodyne scheme in that it includes two stages of frequency conversion (from optical carrier to IF and from IF to baseband); however, both occur in the optical domain, *prior* to optical detection.

II. "PHASE-DIVERSITY" OPTICAL RECEIVERS

It is well known that, through the use of a "quadrature hybrid" (also known as a 90° hybrid), it is possible to implement a homodyne receiver that provides both the in-phase and quadrature components of the input signal [7,10]. These components provide a complete characterization of the received signal, which can be represented as

$$r(t) = I(t)\cos\omega_0 t + Q(t)\sin\omega_0 t . \tag{1}$$

Here ω_0 is the carrier angular frequency, $I(t)$ is the in-phase component and $Q(t)$ is the quadrature component. If $r(t)$ is a passband signal with bandwidth W, the $I(t)$ and $Q(t)$ are low-pass signals with bandwidth $W/2$.

Such a receiver is known as a "phase-diversity" receiver. Figure 1 shows a possible implementation using an optical quadrature hybrid which takes advantage of the "balanced optical receiver" structure [6] for the optical detectors. Since the two output components (or rails) provide a complete representation of the received signal, any modulation scheme can be decoded by processing the two rail outputs with the appropriate electronic hardware. For example, Figure 2 shows the block diagram for an incoherent *on-off keying* (OOK) detector.

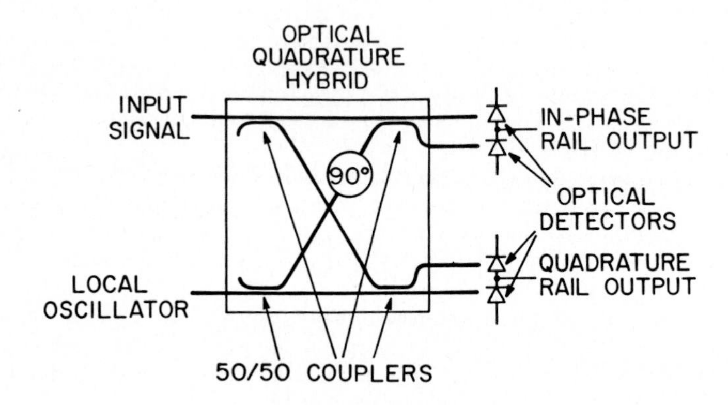

Fig. 1 — Phase-diversity (homodyne) optical receiver (Receiver A). The received optical signal is combined with the output of the local-oscillator laser into two branches, with a relative phase shift of 90°. The detected outputs represent the two orthogonal components of the received optical field.

There are two advantages in using a phase-diversity receiver instead of an optical heterodyne receiver. First, the speed required of the receiver electronics is considerably reduced, as the top frequency component to be handled is $W/2$, in contrast to a heterodyne receiver where the top IF frequency component cannot easily be made less than about $(3/2)W$. Second, because the noise introduced by the optical detector-amplifier has a quadratic spectrum [6], by operating near baseband instead of about the IF frequency the resulting total noise power is greatly reduced.

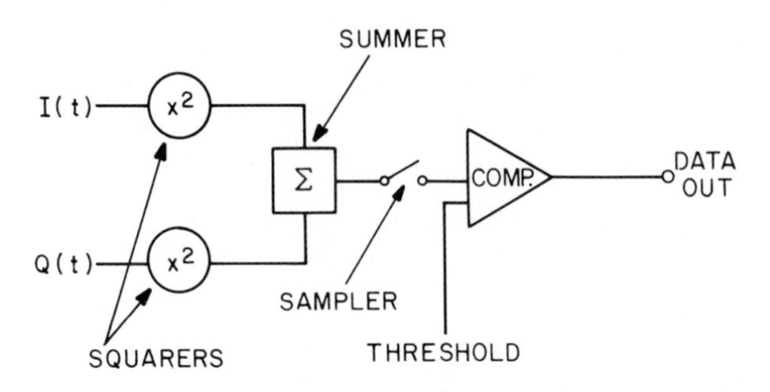

Fig. 2 — Incoherent OOK detector. The sum of the squares of the two components is proportional to the received optical power.

On the negative side, the phase-diversity receiver outlined in Fig. 1 is difficult to realize in practice. The optical quadrature hybrid, although demonstrated experimentally in the laboratory, is a delicate device that requires great accuracy and stability [9,10]. Path-length differences among the various branches must be maintained to an accuracy of a fraction of a wavelength. In the next section, we describe a structure for a phase-diversity receiver that avoids the need for an optical quadrature hybrid by implementing heterodyne frequency conversion *prior* to optical detection. Many of the advantages of the more conventional phase-diversity receiver are retained; however, there is a price to be paid in terms of reduced receiver sensitivity, by a factor of about 6 dB. In applications where great receiver sensitivity is not of paramount importance, this receiver structure should be seriously considered. In our analysis we shall refer to the receiver structure of Fig. 1 as "Receiver A"; the new structure will be "Receiver B".

III. NEW RECEIVER STRUCTURE

Frequency conversion of a signal requires the use of a nonlinear component. In conventional optical heterodyne or homodyne detection, the nonlinearity is provided by the quadratic relationship between the amplitude of the incident electromagnetic field and the output current of the detector diode. In our scheme, the required nonlinear component consists of the three-port block outlined in Fig. 3. The optical signal is first passed through an electro-optic switch, i.e., a device that routes the optical power to one of two output ports in response to a control voltage. It can be realized in Lithium Niobate as a directional coupler whose coupling coefficient varies smoothly (although not in a linear fashion) as a function of applied voltage [11]. What follows the switch is the familiar balanced receiver [6] design: The two detector diodes are connected with opposite polarities so that the output voltage will be positive or negative, depending on which of the two branches receives the optical signal.

This block can be modeled as a multiplier of sorts, in that the output voltage from the balanced optical receiver can be written as the product of the

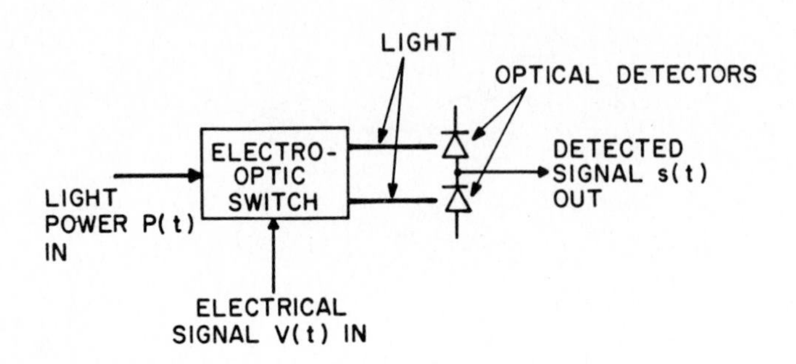

Fig. 3 — Electro-optic "multiplier". The output voltage is proportional to the input power level multiplied by either $+1$ or -1, depending on the state of the switch.

input *optical power*, $P(t)$, and a function of the applied voltage, $V(T)$:

$$s(t) = \alpha P(t) f(V(t)) . \tag{2}$$

Here $f(\cdot)$ accounts for the instantaneous nonlinearity in the response of the electro-optic switch to the applied voltage $V(t)$, and α is a constant of proportionality.

The careful reader will notice that in the previous section we referred to the optical *field* itself as being our "signal", while here we regard the optical *power* as the "signal". The latter is the usual convention when dealing with *incoherent* optical systems while the former is more common with *coherent* optical detection. The shift in nomenclature is equivalent to applying a "virtual" square-law detector to the signal while regarding the optical detector as a linear device. Thus, the coherent-communication expert will say that the 50/50 couplers in Fig. 1 simply add the received-signal and local-oscillator fields linearly, with the square-law response of the optical detectors being responsible for the frequency conversion to IF. By contrast, someone more familiar with *incoherent* communications will regard the output of the couplers as already containing the IF signal in the form of power-level variations, with the optical detector simply acting as a linear converter of optical power into electric current. Of course, the two points of view are perfectly equivalent and lead to the same final result. In our treatment we shall adopt the second point of view, which makes the operation of the receiver easier to understand.

The complete block diagram of Receiver B is shown in Figure 4. The received optical signal is superposed with the local oscillator signal through a 50/50 optical coupler. As in a conventional heterodyne receiver, the instantaneous optical *power* after the coupler is the desired IF signal, downconverted to an IF center frequency in the microwave range. The two coupler outputs provide two identical replicas (except for a 180° phase difference) of the IF signal. Unlike a conventional heterodyne receiver, the IF signal (in each branch) is downconverted to baseband *prior* to optical detection. To achieve the downconversion the signal is mixed with the waveform from an electronic oscillator through the multiplier block of Fig. 3. The frequency stability and control requirements for the electronic oscillator are not different from what they would be if the conversion were done on an *electrical* IF signal, as in a conventional heterodyne receiver. The advantage of this scheme lies in the fact that the optical detectors only "see" the baseband signal, so that their bandwidth requirements are substantially reduced.

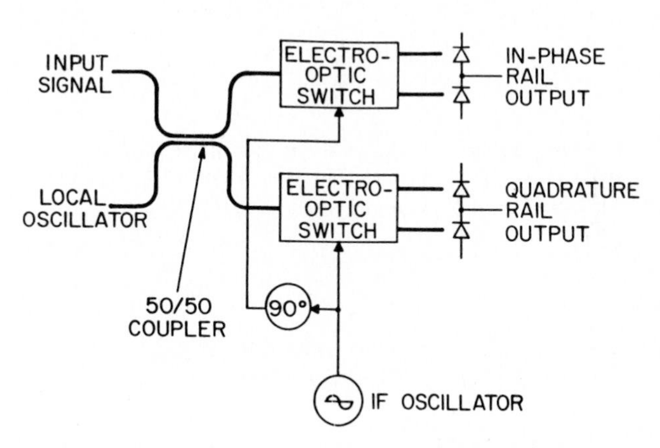

Fig. 4 — Novel phase-diversity heterodyne optical receiver (Receiver B).

The IF optical signals at the two outputs of the 50/50 coupler have an IF carrier frequency that is many orders of magnitude less than the optical carrier frequency. As a result, the required path-length accuracy for both the optical and electronic paths is correspondingly less stringent (it is of the order of a fraction of the wavelength of the IF carrier). This removes the major difficulty associated with the design of an optical quadrature hybrid.

IV. RECEIVER PERFORMANCE

In this section we analyze the performance of Receiver B. In the previous section we have discussed how this receiver performs the same function as the more conventional Receiver A; therefore, we shall characterize the performance of the former by comparison with the latter.

The relevant performance parameter is, of course, the Signal-to-Noise Ratio (SNR) available at the In-phase and Quadrature outputs. The two main sources of noise are 1) the detector/amplifier electronics and 2) the detection process itself, which produces shot noise of a level proportional to the strength of the local oscillator [12]. When the local oscillator is strong enough that shot noise becomes the dominant impairment, the receiver is said to operate in the "shot-noise limit".

Note that the amount of shot noise is the same for both receivers. This is because (if we neglect the insertion loss of the electro-optic switches) the total optical power incident on each of the two diode pairs is approximately $P_\ell/2$ in both cases, with P_ℓ being the local-oscillator optical power. More specifically, in Receiver A each diode of each pair sees, essentially, a constant power level of

$P_\ell/4$ while in Receiver B the power level seen by each diode varies depending on the voltage applied to the electro-optic switch, but the total for each diode pair stays constant at $P_\ell/2$. The amount of shot noise appearing at the output of the balanced receiver is not affected by how the optical power is subdivided between the two diodes; hence, shot noise level is the same for both receivers. With this in mind, we proceed to calculate the sensitivity of Receiver B and compare it to the sensitivity of Receiver A, which will give us a direct comparison of shot-noise-limited performance.

The received optical *field* can be represented as

$$\sqrt{2P_s}\,A(t)\cos(\omega_c t+\phi(t)) \tag{3}$$

where P_s is the average received power, ω_c is the optical carrier angular frequency, $\phi(t)$ represents possible phase modulation and $A(t)$ represents possible amplitude modulation. The local oscillator signal can be represented as†

$$\sqrt{2P_\ell}\,\cos(\omega_c+\omega_i)t \tag{4}$$

where ω_i is the IF angular frequency. The optical *field* after the 50/50 optical coupler can be represented as

$$\sqrt{P_s}\,A(t)\cos(\omega_c t+\phi(t)) \pm \sqrt{P_\ell}\,\cos(\omega_c+\omega_i)t\,, \tag{5}$$

with the '+' sign for the signal on one branch and the '−' sign for the other branch. The instantaneous power in (5) can be closely approximated as

$$P_\ell/2 \pm \sqrt{P_s P_\ell}\,A(t)\cos(\omega_i t-\phi(t))\,. \tag{6}$$

Next, this signal is multiplied by the IF oscillator waveform through the multiplier block of Fig. 3, according to eq. (2). The resulting detected signal depends on the functional form of $f(\cdot)$ as well as on the waveform generated by the IF oscillator; however, maximum receiver sensitivity is achieved when the electro-optic switch is operated as a true switch, i.e., the multiplier $f(V(t))$ in eq. (2) takes only the values ± 1. For this to occur, the waveform produced by the IF oscillator must be a perfect square wave. Under these conditions, the multiplier in eq. (2) can be expressed as

$$f(V(t)) = \mathrm{sgn}(\cos\omega_i t) \tag{7}$$

and the detected signal on, for example, the In-phase rail can be easily found to

† Here we neglect oscillator phase noise, whose possible presence does not alter our final results.

be

$$I(t) = \eta[\frac{1}{2\pi}\int_{2\pi}^{0}\cos\phi\ \mathrm{sgn}(\cos\phi)d\phi]\ \sqrt{P_s P_\ell}\ A(t)\cos\phi(t) + [\text{out-of-band terms}]$$

$$= \eta\frac{2}{\pi}\sqrt{P_s P_\ell}\ A(t)\cos\phi(t) + [\text{out-of-band terms}]\ , \tag{8}$$

where η is the responsivity of the optical detector.

In actuality, the waveform produced by the IF oscillator is most likely to be a sinusoid. If we make the simplifying assumption that $f(\cdot)$ is linear, we can write

$$f(V(t)) = \cos\omega_i t\ , \tag{9}$$

and the detected signal becomes

$$I(t) = \eta[\frac{1}{2\pi}\int_{2\pi}^{0}\cos^2\phi d\phi]\ \sqrt{P_s P_\ell}\ A(t)\cos\phi(t) + [\text{out-of-band terms}]$$

$$= \eta\frac{1}{2}\sqrt{P_s P_\ell}\ A(t)\cos\phi(t) + [\text{out-of-band terms}]\ . \tag{10}$$

We see that receiver sensitivity is degraded with respect to eq. (8) by the factor $\pi/4$, corresponding to 2.1 dB.

The actual form of $f(\cdot)$ is given, for the most basic switch structure, by eq. (3) of [11]:

$$f(V) = \sin^2[\phi_0\sqrt{1+(V/V_0)^2}\,]/[1+(V/V_0)^2] \tag{11}$$

where ϕ_0 is a design parameter that can take anyone of the values, $\pi/2$, $3\pi/2$, $5\pi/2$, ... and V_0 depends on the physics of the material as well as on device design. Fig. 5 shows what $f(V)$ looks like for $\phi_0 = \pi/2$ and $3\pi/2$. The general character of $f(V)$ (zero-derivative at the end points as it changes from one to zero or vice versa) will be the same for any practical switch design. The waveform that results, when a sinusoidal oscillator is used to drive the device, is intermediate between a sinusoid and a square wave; correspondingly, the actual receiver sensitivity will be intermediate between eq. (8) and eq. (10).

For comparison, the detected signal on the in-phase rail of Receiver B can be easily found to be

$$\eta\sqrt{P_s P_\ell}\ A(t)\cos\phi(t)\ . \tag{12}$$

The corresponding sensitivity is better than eq. (10) by a factor 2 (or about 6 dB) and better than eq. (8) by a factor $\pi/2$ (or about 4 dB). Therefore, the actual reduction in sensitivity lies somewhere between 4 dB and 6 dB, assuming no

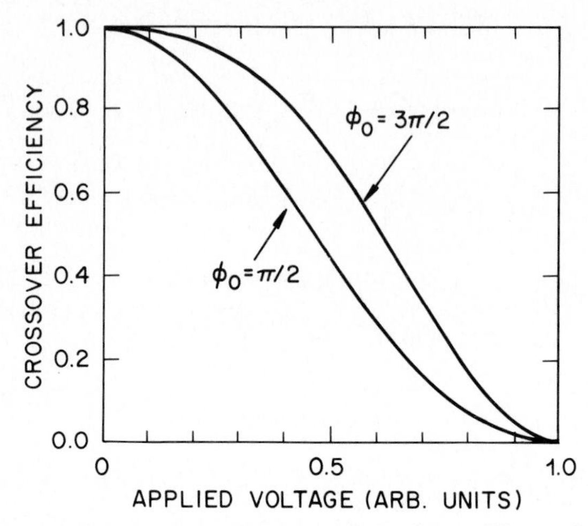

Fig. 5 — Response of electro-optic switch to applied voltage. The vertical axis shows the proportion of optical power going to the "crossover" output port. The rest of the optical power goes to the other (direct) output port.

insertion loss in the electro-optic switch.

V. PRACTICAL CONSIDERATIONS

In the previous Section we found that Receiver B requires a signal level 4-6 dB larger than Receiver A in the idealized situation where electronics noise (thermal or otherwise) is negligible and the electro-optic switch has zero insertion loss. In practice, the non-negligible insertion loss of the switch will attenuate both the received signal and the local oscillator by a certain factor. Correspondingly, the sensitivity of Receiver B will be further reduced by that factor (assuming operation in the shot-noise limit).

VI. CONCLUSION

We have presented an alternative design for a phase-diversity optical receiver based on double frequency conversion. It can be regarded as intermediate between a heterodyne receiver (where the local-oscillator laser is at a frequency different from that of the signal) and a homodyne receiver (where the optical detector produces the baseband signal directly). As such, it combines some of the advantages (and disadvantages) of both techniques. Its main advantage is that it does not require an optical quadrature hybrid, a component that has proven difficult to realize. In applications where somewhat-reduced receiver sensitivity is not a problem, the new design provides an interesting receiver alternative.

Several variations on the structure proposed here are also possible. One particularly interesting variation consists of realizing the three-port nonlinear block of Fig. 3 with the technique described by T. E. Darcie et. al. in [13], using a modulated optical amplifier. The calculations presented in the previous section can be easily reproduced for this modified design. Potential advantages of this variation include somewhat improved sensitivity, through the replacement of the switches' insertion loss with amplifier gain, and a reduced role for electronics noise.

The fact that electro-optic devices are needed for this scheme is, admittedly, a weakness; however, the devices are driven by a CW waveform, as opposed to a broadband signal (as is the case when the switch is used as a digital modulator) and this makes the drive circuitry somewhat simpler, especially considering that a resonant design for the electro-optic switch [14,15] may substantially reduce drive-power requirements, as compared to broadband applications.

Finally, this receiver design can also be used to realize a single-rail homodyne receiver, as discussed in ref. [16]; but in this case the primary advantage of the scheme — the elimination of the quadrature hybrid — is no longer there.

VII. ACKNOWLEDGMENT

The author would like to thank J. M. Kahn for several helpful suggestions.

REFERENCES

[1] T. Okoshi, "Heterodyne and Coherent Optical Fiber Communications: Recent Progress," IEEE Trans. on MTT, vol. *MTT-30*, no. 8 (Aug. 1982) pp. 1138-1149.

[2] A. H. Gnauck, R. A. Linke, B. L. Kasper, K. J. Pollock, K. C. Reichmann, R. Valenzuela, R. C. Alferness, "Coherent Lightwave Transmission at 2 Gbit/s over 170 km of Optical Fibre using Phase Modulation," Electron. Lett., vol. *23*, no. 6 (12 March 1987) pp. 286-287.

[3] P. S. Henry, "Lightwave Primer," IEEE J. Quantum Electron., vol. *QE-21* (1985) pp. 1862-1879.

[4] R. A. Linke and A. H. Gnauck "High-Capacity Coherent Lightwave Systems," J. of Lightwave Technology, vol. *6*, no. 11 (Nov. 1988) pp. 1750-1769.

[5] J. M. Kahn, "1 Gbit/s PSK Homodyne Transmission System Using Phase-locked Semiconductor Lasers," IEEE Photonics Technol. Lett., October 1989 (to be published).

[6] S. E. Miller and I. P. Kaminow, eds. Optical Telecommunications II, Chapter 18 by B. L. Kasper: "Receiver Design", Academic Press, New York (1988).

[7] A. W. Davis, M. J. Pettitt, J. P. King and S. Wright "Phase Diversity Techniques for Coherent Optical Receivers," J. of Lightwave Technology, vol. *LT-5*, no. 4 (April 1987) pp. 561-572.

[8] L. G. Kazovsky, "Phase- and Polarization-Diversity Coherent Optical Techniques," J. of Lightwave Technology, vol. 7, no. 2 (Feb. 1989) pp. 279-292.

[9] L. G. Kazovsky, L. Curtis, W. C. Young, N. K. Cheung, "All-Fiber 90° Optical Hybrid for Coherent Communications," OFC/IOOC '87, Reno, NV (Jan. 1987).

[10] T. Hodgkinson, R. A. Harmon, D. W. Smith "Demodulation of Optical DPSK using In-Phase and Quadrature Detection," Electron Lett., vol. *21*, (1985) p. 867.

[11] R. C. Alferness, "Guided-Wave Devices for Optical Communication," IEEE J. Quantum Electron., vol. *QE-17* (June 1981) pp. 946-959.

[12] W. K. Pratt, *Laser Communication Systems*, Wiley, 1969; Eq. (10-43).

[13] T. E. Darcie, S. O'Brien, G. Raybon, C. A. Burrus, "An Optical Mixer-Preamplifier for Lightwave Subcarrier Systems," Electron. Lett., vol. *24*, no. 3 (4 Feb. 1988) pp. 179-180.

[14] M. Izutsu, H. Murakami, T. Sueta, "Standing-Wave Structure Optical Waveguide Modulator Using a Resonant Electrode at 10 GHz," OFC/IOOC '87, Reno, NV, Poster Paper TUQ32 (Jan. 1987).

[15] L. A. Molter-Orr, H. A. Haus, F. J. Leonberger, "20 GHz Optical Waveguide Sampler," IEEE J. Quantum Electron., vol. *QE-19*, No. 12 (Dec. 1983).

[16] W. J. Stewart, U.S. Patent No. 4,648,134, March 3, 1987.

CHANNEL SPACING IN COHERENT OPTICAL TRANSMISSION SYSTEMS

Leonid Kazovsky

Bellcore
331 Newman Springs Road
Red Bank, NJ 07701–7040
USA

The following effects can deteriorate the performance of multichannel coherent optical transmission systems: (a) direct–detection and signal–cross–signal interference; (b) image–channel interference; (c) adjacent channel crosstalk; and (d) excess shot noise. All these phenomena degrade the performance of single–detector receivers; balanced receivers eliminate interference (a) while image–rejection receivers eliminate interference (b). The additional complexity and the need for single–polarization fibers make practical applications of image–rejection receivers unlikely. Balanced receivers offer a reasonable compromise between performance and complexity. In high–speed systems, the intermediate frequency IF is smaller than the optical–domain channel spacing D_{op}. As a result of spectrum folding, the electrical–domain channel spacing D_{el} is smaller than D_{op} by $2 \cdot IF$. The minimum channel spacing is achieved in homodyne systems (IF = 0) where $D_{op} = D_{el}$ can be theoretically as small as the bit rate R_b for all binary modulation formats (ASK, FSK and PSK) under ideal conditions: perfect *sinc* pulses and brick–wall filters. With more practical filters and pulse shapes, D_{el} increases to[1] about $3.6R_b$ for ASK, $2.7R_b$ for PSK, and $2.0R_b$ for FSK. Finite laser linewidth $\Delta\nu$ further increases D_{el}. For example, in FSK systems with $\Delta\nu/R_b = 0.25\%$, D_{el} increases to $2.1R_b$; in ASK systems with $\Delta\nu/R_b = 0.25$, D_{el} increases to $8R_b$. Even larger channel spacing is required if line coding is used and/or if the signal power is different in different channels. As a result, D_{op} as large as $58R_b$ had to be used in some recent multichannel experiments. Large channel spacing, in conjunction with a limited tuning range of local oscillator lasers, limits the maximum number of channels in coherent transmission systems.

1. INTRODUCTION

A major advantage of coherent optical fiber communications systems is their ability

[1]With 1 dB sensitivity penalty.

to transmit many closely spaced channels simultaneously through a single fiber. Thus, serious efforts have been devoted recently to multichannel coherent systems research. One important problem in this area is to establish the minimum permissible channel spacing and to understand its dependence on main system parameters: bit rate; intermediate frequency; laser linewidth; modulation format; line coding; receiver structure; and (for FSK systems) frequency deviation. In turn, channel spacing has a direct impact on such important device and subsystem parameters as the tunability range of local oscillator lasers and the accuracy of transmitter frequency stabilization. This paper is devoted to the channel spacing for binary coherent transmission systems; we consider several important cases ranging from idealized situations (leading to fundamental limits) to experimentally demonstrated systems. The paper is organized as follows. Mechanisms of performance degradations in multichannel coherent systems are discussed in Section 2. In Section 3, we discuss electrical–domain interference for three important types of coherent receivers: single–detector; balanced dual–detector; and image–rejection. Section 4 deals with idealized coherent systems, and presents sensitivity penalty for ASK, FSK and PSK modulation formats. The impact of phase noise on channel spacing is discussed in Section 5 while Section 6 deals with the impact of line coding. Section 7 summarizes the results of recent experimental measurements of channel spacing while Section 8 contains conclusions of this paper.

2. MECHANISMS OF PERFORMANCE DEGRADATION IN MULTICHANNEL COHERENT SYSTEMS

The following physical mechanisms deteriorate the performance of multichannel coherent systems: excess shot noise; direct–detection and channel–cross–channel interference; adjacent channel crosstalk; and image–channel interference. In this section, we define the foregoing interferences and discuss their relative importance.
The complex amplitude of a multichannel optical signal is

$$E_s = \sum_k E_k \quad , \quad 1 \leq k \leq K \tag{1}$$

where E_k is the complex amplitude of channel k, and K is the total number of optical channels[2]. Let $E_{lo} \equiv \sqrt{P_{lo}}$ be the complex amplitude of the laser local oscillator (LO), where P_{lo} is the LO power. Further, let N be the subscript of the desired channel. Then [1]

$$E_k = m_k \sqrt{P_s} \exp\{j \cdot [2\pi \cdot f_{if} + 2\pi \cdot (k-N) \cdot D_{op} \cdot t + \varphi_k]\} \tag{2}$$

where m_k and φ_k reflect the possible amplitude and angle modulation of the k–th channel, P_s is the peak signal power (assumed to be the same in all channels), f_{if} is the intermediate frequency, and D_{op} is the optical domain channel spacing.
The optical hybrid of a coherent receiver generates one or more linear combinations of E_s with E_{lo}. For simplicity, let us assume in this section that the hybrid has only

[2]If an optical filter is used, then K is the number of optical channels in its passband.

one output:

$$E_{out} = E_s + E_{lo} \qquad (3)$$

In Section 3, we will discuss practical hybrids, and see how they are different from our simple model (3); but first, we discuss the interferences induced by the presence of many channels using a simplified model (3).

EXCESS SHOT NOISE

The power spectral density (PSD) of the shot noise in a single–channel receiver is

$$\eta = 2\, q\, R\, P_{lo} \qquad (4)$$

where q is the electron charge, and R is the detector responsivity. In a multichannel receiver, the PSD of the shot noise is

$$\eta = 2\, q\, R\, (P_{lo} + K\, P_s) \qquad (5)$$

Comparison of expr. (5) with expr. (4) reveals that the multichannel receiver suffers from an excess shot noise with PSD of $2qRKP_s$. To estimate the resulting sensitivity penalty, assume that a K–by–K star coupler is used for channel combining; also assume that losses in the system components are negligible. In this case, if local oscillators and transmitter lasers have the same power, then $KP_s \approx P_{lo}$, leading to a 3 dB sensitivity penalty. In practice, optical signals are substantially attenuated in connectors, fibers and other components, so that $KP_s \ll P_{lo}$. As a result, the penalty is less than 1 dB when as many as 2000 channels are transmitted [1]. Since this number is independent from the channel spacing, the excess shot noise is not discussed further in this paper.

DIRECT–DETECTION AND CHANNEL–CROSS–CHANNEL INTERFERENCE

When the optical signal (1) is detected, the resulting current is equal to

$$i = R \sum_k |E_k|^2 + R \sum_k \sum_l E_k E_l^* \ , \quad k \neq l \qquad (6)$$

where the first term represents the direct–detection interference, and the second term represents the channel–cross–channel interference.
Direct–detection currents have low–pass spectra; in fact, for ideal angle–modulated signals the PSD of $|E_k|^2$ is just a delta–function, and can be easily filtered out. For ASK signals, the IF has to be sufficiently large to keep the desired signal spectrum away from the direct–detection spectrum. Alternatively, a balanced receiver can be used to suppress the direct–detection and channel–cross–channel interference [1].
Now consider the spectrum of the channel–cross–channel interference. The spectrum of the beat between the k–th and the l–th channel is centered around $|k-l| \cdot D_{op}$. Therefore, the total spectrum of the channel–cross–channel interference occupies a

broad frequency range from zero to $K \cdot D_{op}$ [Hz], and consists of many components spaced D_{op} [Hz] apart. Placing the IF beyond $K \cdot D_{op}$ to avoid the channel–cross–channel interference is impractical because of receiver bandwidth limitations. Fortunately, the PSD of the channel–cross–channel interference is fairly small comparatively to the signal PSD; it can be further attenuated using balanced receivers.

ADJACENT–CHANNEL CROSSTALK AND IMAGE–CHANNEL INTERFERENCE

The main components of the photodetector current are due to the beat between the received signal and LO fields. The desired signal component is equal to $2 \cdot R \cdot \sqrt{P_{lo} P_s} \cdot m_N \cdot \cos\,(2\pi \cdot f_{if} \cdot t\ + \varphi_N)$. Fig. 1 illustrates the adjacent channel crosstalk and the image–channel interference for IF > D_{op}; this condition can be

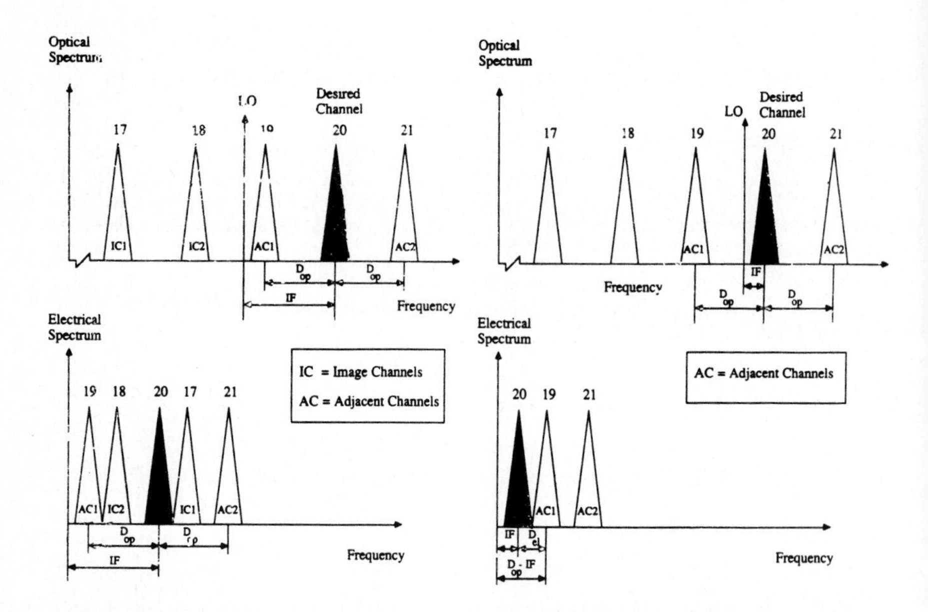

Fig. 1. Adjacent–channel crosstalk and image–channel interference for IF > D_{op}; numerical data: N=20; IF=1.3D_{op}.

Fig. 2. Adjacent–channel crosstalk and image–channel interference for IF < D_{op}; numerical data: N=20; IF=0.3D_{op}.

satisfied in low–speed narrow–linewidth systems. In this case, the adjacent channel crosstalk is generated by the channels number N–1 and N + 1. The image–channel interference is generated by the channels number N–2 and N–3 in the example shown, since the LO frequency is between channels N–1 and N–2. Inspection of Fig. 1 reveals that the strongest interference in this case is the image channel interference due to channel N–3; the impact of adjacent channel crosstalk is negligible.

Fig. 2 illustrates the case IF < D_{op}; this condition is likely to be satisfied in high–speed systems and/or in systems employing wide–linewidth lasers. In this case, the IF should be kept minimum since the equalized receiver noise is proportional to f^2, and becomes an increasingly difficult problem at high frequencies. Referring to Fig. 2, we notice that the strongest interference is generated by the channel number N–1 if $f_{lo} < f_N$; the term *adjacent–channel crosstalk* is normally used to describe this interference. Note that the electrical–domain channel spacing D_{el} is smaller than D_{op}:

$$D_{el} = D_{op} - 2 \cdot f_{if} \tag{7}$$

Since the sensitivity penalty depends on the electrical–domain spacing D_{el}, expr. (7) indicates that the IF should be kept minimum in multichannel systems if small channel spacing is desired. The physical reason for deterioration of channel spacing described by expr. (7) is the folding of the optical spectrum around the LO frequency during heterodyning.

3. ELECTRICAL–DOMAIN INTERFERENCE IN COHERENT RECEIVERS

As discussed in Section 2, a multichannel receiver suffers from several interferences. Some receivers can cancel (or at least attenuate) certain interferences. In the remainder of this section, we discuss three important receiver types and their interference–rejection capabilities assuming IF < D_{op}; this is the case in high–speed and/or wide–linewidth coherent systems.

3.1. SINGLE–DETECTOR RECEIVER

A single–detector receiver is shown in Fig. 3; it is described by eq. (3) to within a

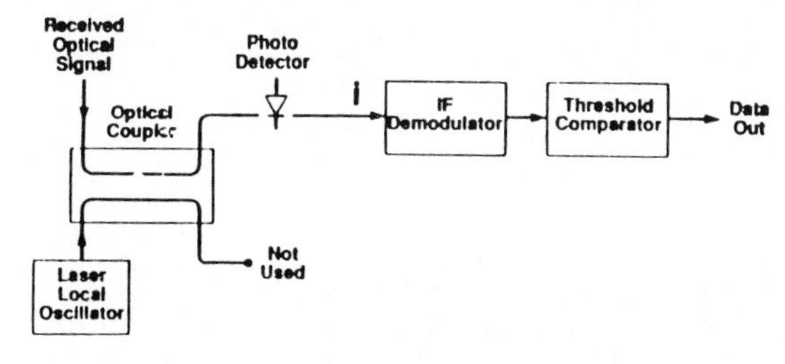

Fig. 3. A single–detector coherent receiver.

constant. Thus, this receiver suffers from all the interferences described in Section 2. ASK homodyne detection should be avoided in this case because of the strong impact of direct–detection interference. Heterodyne detection is a better choice provided the IF is sufficiently large to prevent an overlap with direct–detection interference. This receiver is wasting the signal and the LO power at the unused output port of the coupler; it is also susceptible to the LO intensity noise and to direct–detection and channel–cross–channel interferences. Hence, its use is not recommended for multichannel applications.

3.2. DUAL–DETECTOR BALANCED RECEIVER

Fig. 4 shows a block–diagram of a dual–detector balanced receiver. This receiver has

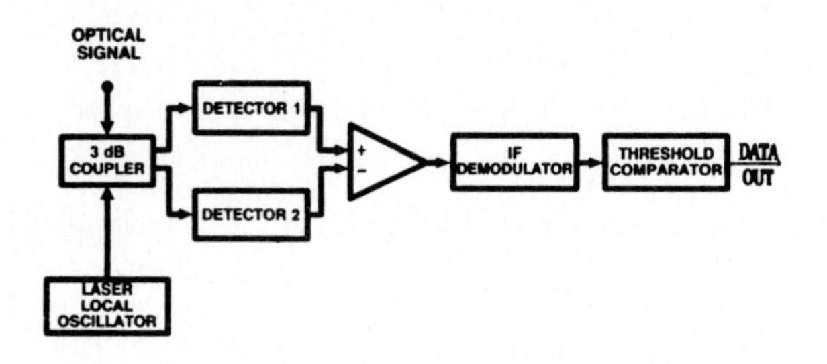

Fig. 4. A dual–detector balanced coherent receiver [1].

several desirable features: it uses efficiently the signal and the LO powers, and cancels the LO intensity noise, direct–detection interference, and channel–cross–channel interference [1]. When combined with a polarization–diversity approach, this receiver becomes an excellent choice for multichannel systems.

3.3. IMAGE–REJECTION RECEIVER

An image–rejection receiver [2] is shown in Fig. 5. This receiver rejects the image–channel interference and, therefore, leads to a closer channel spacing than balanced receivers. However, image–rejection receivers are fairly complex and require the state of polarizations of *all* channels to be aligned with that of the LO. Hence, this receiver is unlikely to be used in practice.

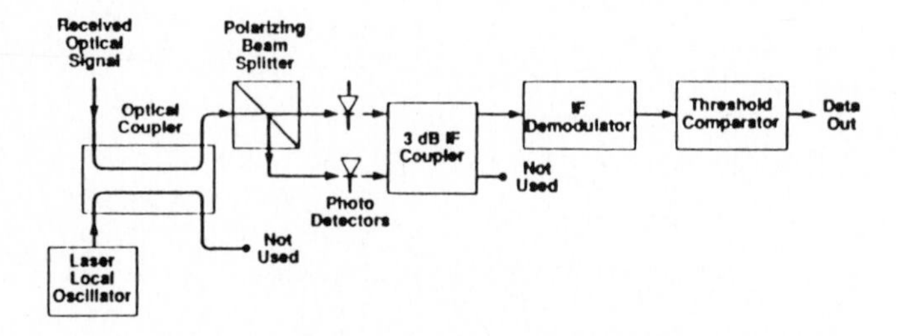

Fig. 5. An image–rejection coherent receiver [2].

To summarize, balanced receivers (particularly balanced polarization–diversity receivers) are the most reasonable choice for multichannel coherent systems. They suppress the direct–detection and the channel–cross–channel interferences, and, if $D_{op} > IF$, leave the adjacent channel crosstalk as the only important interference. Hence, only balanced receivers are considered in the remainder of this paper.

4. SENSITIVITY PENALTY IN IDEALIZED MULTICHANNEL SYSTEMS

In this section, we discuss the sensitivity penalty arising from the presence of additional channels in idealized multichannel coherent systems; we neglect temporarily several factors affecting the channel spacing in practical systems, such as the phase noise and line coding. These factors are addressed in Sections 5 and 6, respectively.
As discussed in Sections 2 and 3, the main source of performance degradation in a multichannel environment is the adjacent channel crosstalk. The crosstalk sensitivity penalty depends on the electrical domain channel spacing D_{el}. Theoretically, D_{el} can be as small as the bit rate R_b with no sensitivity penalty [3] for all three binary

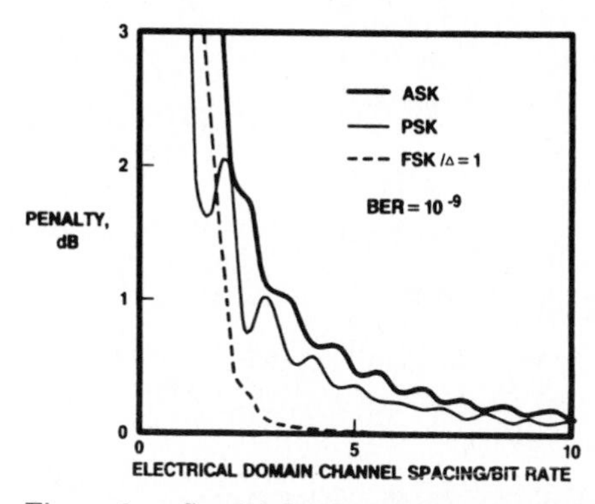

Fig. 6. Sensitivity penalty versus electrical–domain channel spacing for idealized ASK, FSK ($\Delta = 1$) and PSK multichannel systems: no phase noise; rectangular pulses; matched filter demodulators [4].

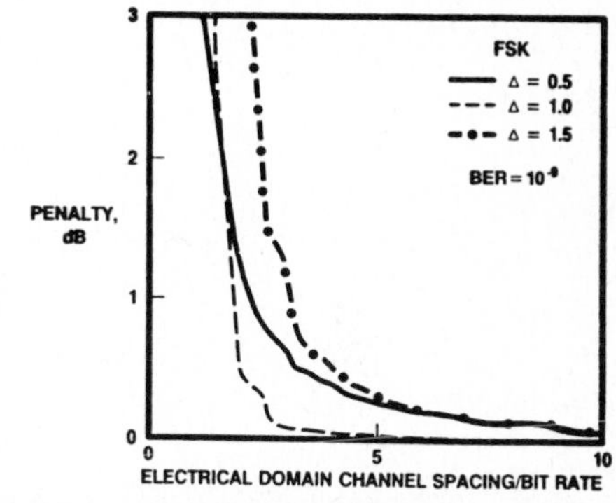

Fig. 7. Sensitivity penalty versus electrical–domain channel spacing for idealized FSK multichannel systems with different modulation indices: no phase noise; rectangular pulses; matched filter demodulators [4].

modulation formats: ASK, FSK and PSK. This, however, requires unrealistic infinitely–long *sinc*–shaped pulses and equally unrealistic brick–wall filters. More realistic nearly rectangular pulses lead to non–zero penalties, as shown in Fig. 6 for matched–filter demodulators [4]. Note that for FSK the modulation index Δ is assumed to be equal to 1 in Fig. 6; other values of Δ lead to larger penalties, as shown in Fig. 7. Inspection of Fig. 6 reveals that a channel spacing of $(2.0\text{–}3.6)\cdot R_b$ is sufficient for all three modulation formats if 1 dB penalty can be tolerated. The spacing required for 0.1 dB penalty ranges from $9.9R_b$ for ASK to modest $2.9R_b$ for FSK. Experimentally, D_{el} of $2.2R_b$ has been reported for FSK systems with 1 dB penalty [14], in a nice agreement with theoretical results.

5. IMPACT OF PHASE NOISE ON CHANNEL SPACING

We discuss below the impact of phase noise on two large classes of coherent systems. In Subsection 5.1, narrow–linewidth systems are discussed, including PSK and small–deviation FSK; in Subsection 5.2, wide–linewidth systems are discussed, including ASK and large–deviation FSK.

5.1. NARROW–LINEWIDTH SYSTEMS

PSK and small–deviation FSK systems encode information into phase changes and, therefore, cannot tolerate large linewidth: to keep sensitivity penalty below 0.5 dB, $\Delta\nu/R_b$ must be smaller than 0.1% for PSK [5] and smaller than 0.25% for FSK with unity modulation index [6]; here $\Delta\nu$ is the lasers' linewidth (the corresponding IF linewidth is $2\cdot\Delta\nu$). Small linewidth has a small (but noticeable) impact on crosstalk sensitivity penalty. As an example, Fig. 8 shows sensitivity penalty (measured with

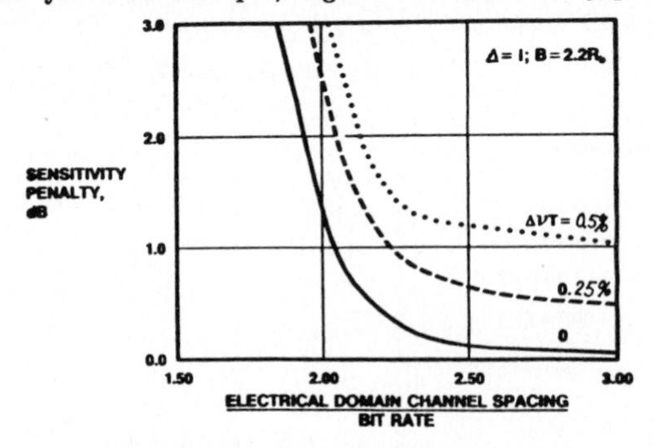

Fig. 8. Sensitivity penalty versus electrical–domain channel spacing for small–deviation FSK delay–and–multiply coherent receivers with non–zero linewidth [6].

respect to the ideal system with no phase noise) versus electrical domain channel spacing for FSK delay–and–multiply receivers [6]. Inspection of Fig. 8 reveals that as the linewidth increases from zero to 0.25% of the bit rate, the 1 dB channel spacing increases from $2.05R_b$ to $2.1R_b$. The reason for the increase is a non–linear interaction between phase–noise and crosstalk penalties [4].

5.2. WIDE–LINEWIDTH SYSTEMS

ASK and large–deviation FSK systems can tolerate large linewidth [7,8]: even when the linewidth is equal to the bit rate, the resulting sensitivity penalty is less than 2 dB. As the linewidth increases in such systems, the channel bandwidth and the crosstalk increase, too. To keep the sensitivity penalty small, one must increase the channel spacing substantially in such systems.
Table 1 gives approximate values of the IF bandwidth, intermediate frequency, and frequency deviation for single–filter FSK systems with Δ=4; applicable parameters for ASK systems are also shown. The table includes two cases: NRZ (no line coding) and AMI coding (discussed in Section 6).

TABLE 1. SYSTEM PARAMETERS FOR WIDE–LINEWIDTH SYSTEMS

Parameter	ASK–NRZ	FSK–NRZ	FSK–AMI
Intermediate frequency	$2R_b+14\Delta\nu$	$4R_b+30\Delta\nu$	$4R_b+32\Delta\nu$
IF bandwidth	$3R_b+20\Delta\nu$	$3R_b+20\Delta\nu$	$3R_b+20\Delta\nu$
Frequency deviation	N/A	$\pm(2R_b+16\Delta\nu)$	$\pm(4R_b+32\Delta\nu)$

Using the data of Table 1, one can find the optical–domain channel spacing (corresponding to 1 dB penalty) using computer simulations [9]; the result is shown in Fig. 9. Inspection of Fig. 9 reveals that as the lasers' linewidth increases from zero to

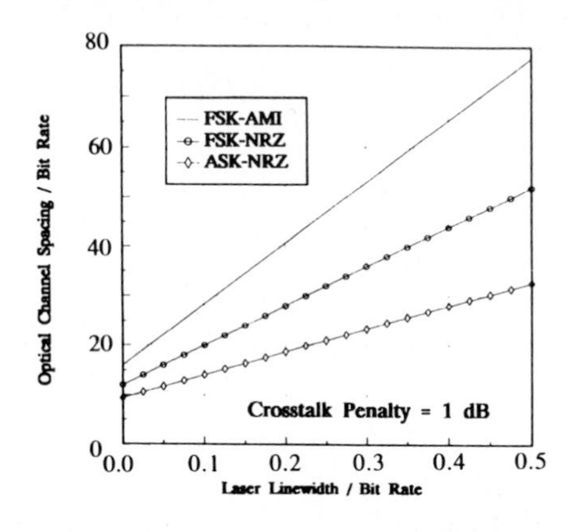

Fig. 9. Optical–domain channel spacing (corresponding to 1 dB penalty) versus linewidth for ASK and wide–deviation single–filter FSK systems [9].

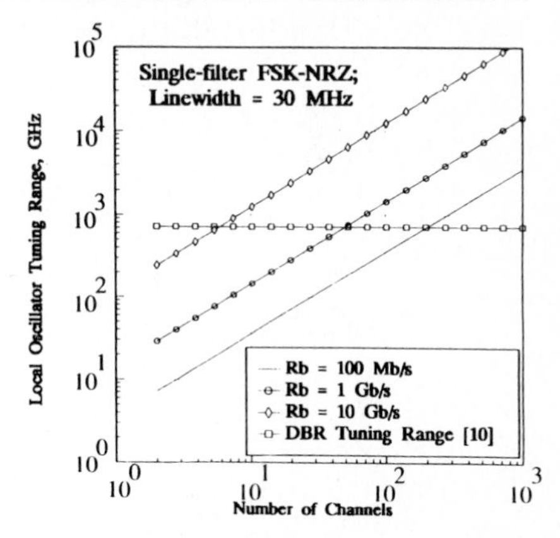

Fig. 10. Local oscillator required tuning range versus number of channels for single–filter NRZ–FSK systems with $\Delta\nu$=30 MHz.

$0.5R_b$, the optical–domain spacing increases from $10R_b$ to $32R_b$ for ASK, and from $12R_b$ to $52R_b$ for FSK. Thus, the lasers' linewidth has a serious impact on channel spacing and, therefore, on the tuning range of local oscillators. Fig. 10 shows the required tuning range versus the number of channels for several systems with NRZ FSK modulation and $\Delta\nu$ = 30 MHz. Inspection of Fig. 10 reveals that if the local oscillator has a 5.8 nm tunability range [10], the system can accommodate no more than 50 channels operating at 1 Gb/s each.

6. IMPACT OF LINE CODING ON CHANNEL SPACING

Line codes are frequently used in communication systems to overcome a non–uniformity of the channel frequency response [11]. For example, in FSK coherent systems, AMI coding has been used to overcome a non–uniformity of the frequency response of DFB lasers [12]. Since AMI effectively doubles the bandwidth of optical signals, it leads to substantially larger channel spacings than NRZ. Inspection of Fig.

9 reveals that AMI increases D_{op} to $77R_b$, compared to $52R_b$ for NRZ systems with $\Delta\nu = 0.5R_b$. Thus, AMI is a wasteful line coding technique; better results can be expected with more efficient line codes.

7. RECENT EXPERIMENTAL RESULTS

Table 2 summarizes several experimental measurements of channel spacing.

TABLE 2. EXPERIMENTAL MEASUREMENTS OF CHANNEL SPACING

System	ASK–NRZ	FSK–NRZ	FSK–NRZ	FSK–AMI
Modulation index	1	1	2	10
$\Delta\nu/R_b$	0.25	0	0.02	0.2
Bit rate, Mb/s	45	45	200	155
D_{op}, MHz	585	300	2,200	9,000
D_{op}/R_b	13	6.6	11	58
D_{el}, MHz	585	150	700	4,600
D_{el}/R_b	13	3.3	3.5	30
Ref.	13	14	15	16

Inspection of Table 2 reveals that while D_{el} is as large as $30R_b$ in some experiments [16], careful system design leads to D_{el} as small as $3.3R_b$ [14], close to the theoretically predicted value of $2.9R_b$ for 0.1 dB penalty [4].

8. CONCLUSIONS

Theoretically, the optical channel spacing D_{op} in binary coherent systems can be as small as one bit rate R_b. Experimental coherent systems operate with D_{op} ranging from 6.6 to $58R_b$. The major factors increasing D_{op} are: spectrum folding around the local oscillator frequency; sub optimum pulse shape; laser phase noise; and inefficient line coding. Large channel spacing, in conjunction with a limited tuning range of local oscillator lasers, limits the maximum number of channels in a coherent system. At the moment, this limit is about 50 channels for 1 Gb/s single–filter FSK–NRZ systems. Smaller channel spacing can be achieved using small intermediate frequency (with zero IF leading to closest channel spacing), narrow–linewidth lasers, FSK with a unity modulation index, and efficient line codes.

9. REFERENCES

1. L. G. Kazovsky, "Multichannel coherent optical communications systems", J. of Lightwave Technology, vol. LT–5, no. 8, August 1987, pp. 1095–1102.
2. B. S. Glance, "An optical heterodyne mixer providing image–frequency rejection", J. of Lightwave Technology, vol. LT–4, no. 11, November 1986, pp. 1722–1725.
3. W. R. Bennet and J. R. Davey, "Data transmission", New York, McGraw Hill, 1965, pp. 70–82 and 226–227.
4. L. G. Kazovsky and J. L. Gimlett, "Sensitivity penalty in multichannel coherent optical communications", J. of Lightwave Technology, vol. LT–6, no. 9, September 1988, pp. 1353–1365.
5. L. G. Kazovsky, "Performance analysis and laser linewidth requirements for optical PSK heterodyne communications systems", J. of Lightwave technology, vol. LT–4, no. 4, April 1986.
6. L. G. Kazovsky and G. Jacobsen, "Multichannel CPFSK coherent optical communications systems", J. of Lightwave Technology, vol. LT–7, no. 6, June 1989, pp. 972–982.
7. L. G. Kazovsky, P. Meissner and E. Patzak, "ASK multiport optical homodyne receivers", J. of Lightwave Technology, vol. LT–5,no. 6, June 1987, pp. 770–791.
8. G. J. Foschini, L. J. Greenstein and G. Vannucci, "Noncoherent detection of coherent lightwave signals corrupted by phase noise", IEEE Transactions on Communications, vol. 36, no. 3, March 1988, pp. 306–314.
9. A. Elrefaie, M. W. Maeda and R. Guru, "Impact of laser linewidth on optical channel spacing requirements for multichannel FSK and ASK systems", Photonics Technology Letters, vol. PTL–1, no. 4, April 1989, pp. 88–90.
10. S. Murata, I. Mito and K. Kobayashi, "Over 720 GHz (5.8 nm) frequency tuning by a 1.5 μm DBR laser with phase and Bragg wavelength control regions", Electronics Letters, vol. 23, no. 8, 9 April 1987, pp. 403–404.
11. K. S. Shanmugam, "Digital and analog communication systems", John Wiley and Sons, New York, 1979.
12. R. Noe, M. W. Maeda, S. G. Menocal and C. E. Zah, "AMI signal format for pattern–independent FSK heterodyne transmission", ECOC–88, Brighton, United Kingdom, September 1988, p. 175.
13. Y. K. Park et. al., "Crosstalk penalty in a two–channel ASK heterodyne detection system with non–negligible laser linewidth", Electronics Letters, vol. 23, no. 24, 19 November 1987, pp. 1291–1293.
14. B. S. Glance et. al., "Densely spaced FDM coherent star network with optical signals confined to equally spaced frequencies", J. of Lightwave Technology, vol. LT–6, no. 11, November 1988, pp. 1770–1781.
15. B. Glance, T. L. Koch, O. Scaramucci, K. C. Reichmann, U. Koren and C. A. Burrus, "Densely spaced FDM coherent optical star network using monolithic widely frequency–tunable lasers", Electronic Letters, vol. 25, no. 10, 11 May 1989, pp. 672–673.
16. R. Welter et. al., "16–Channel coherent broadcast network at 155 Mb/s", J. of Lightwave Technology, accepted for publication.

ANALYSIS OF INTERMODULATION DISTORTION IN TRAVELLING WAVE SEMICONDUCTOR LASER AMPLIFIERS

T. G. Hodgkinson and R. P. Webb

British Telecom Research Laboratories, Martlesham Heath, Ipswich, Suffolk, IP5 7RE, United Kingdom.

Using a modified form of a previously reported laser amplifier analysis technique, a communications theory model is derived which can be used to analyse carrier density intermodulation distortion effects in a travelling wave semiconductor laser amplifier. This model is then applied to study the effect of four-wave mixing on the amplification of an optical multiplex, and it is shown that for worst case input conditions the individual carriers can experience different gains depending on their position within the multiplex, the channel separation being used and the size of the multiplex. It is also shown that for an amplifier operating 10 dB below saturation with a power gain of 30 dB, worst case four-wave mixing effects should be becoming negligible when channel separations in excess of a few GHz are used.

1. INTRODUCTION

It is anticipated that future multichannel optical communication systems will incorporate optical amplification to either maximise transmission distance or overcome distribution losses, and that they will also use heterodyne detection in order to minimise the channel separation and hence allow the enormous bandwidth of singlemode fibre to be efficiently utilised. However, when relatively narrow channel separations are used and the optical amplification is provided by semiconductor laser amplifiers, the low frequency interchannel beat powers cause carrier density modulation and hence four-wave mixing [1-6]. Consequently, there is now an interest in whether the resulting intermodulation will significantly degrade system performance when an optical multiplex of closely spaced channels is amplified, and it is the aim of this paper to assess this degradation as a function of multiplex size and channel separation for worst case input conditions. Initially, the average cavity power laser analysis technique reported in [7] is modified to take account of input power variations, and it is then shown that with the amplifier operating 10 dB or more below saturation it can be modelled as a fixed gain process followed by amplitude and phase modulation stages. This model, which can be analysed using standard communications modulation theory, is then used to assess what minimum channel separation should be used if worst case four-wave mixing effects are to have negligible effect on system performance.

2. AMPLIFIER EQUATIONS: STATIC INPUT CONDITIONS

2.1. Average Cavity Power, Field Gain and Phase Delay

For an ideal travelling wave semiconductor laser amplifier (facet reflectivity equal to zero), it can be shown, using the results from [7], that the average cavity power normalised with respect to the saturation power (P_{av}/P_s) is

$$\frac{P_{av}}{P_s} = \frac{G_E^2 - 1}{2\log_e\{G_E\}}\frac{P_{in}}{P_s} \qquad .....(1)$$

and it can also be shown that the optical field gain (G_E) is

$$G_E = \exp\left\{ \frac{\frac{\Gamma g_o L}{2}}{\left(1 + \frac{P_{av}}{P_s}\right)} - \frac{\alpha_e L}{2} \right\} \qquad .....(2)$$

and that the phase delay (ϕ_E) through the amplifier is

$$\phi_E = \phi_o + \frac{\Gamma g_o L\alpha}{2}\left(\frac{\frac{P_{av}}{P_s}}{1 + \frac{P_{av}}{P_s}}\right) \qquad .....(3)$$

P_{in} is the input power, Γ is the radiation confinement factor, g_o is the unsaturated gain coefficient, L is cavity length, α is the linewidth enhancement factor and α_e is the effective loss coefficient (NB α was used to represent this parameter in [7]). ϕ_o, the unsaturated device phase delay, is a static phase offset which can be ignored for the rest of this analysis.

If the laser amplifier is taken to be operating 10 dB or more below saturation, the condition $P_{av}/P_S \leq 0.1$ is imposed and (1), (2) and (3) can be simplified by using the binomial expansion to a first order approximation. If account is taken of the fact that the effective loss coefficient is usually small enough to be ignored and that gain G_E is usually >> 1, then the normalised average cavity power, gain and phase delay expressions approximate to

$$\frac{P_{av}}{P_s} = \frac{G_E^2}{2\log_e\{G_E\}}\frac{P_{in}}{P_s} \qquad .....(4)$$

$$G_E = G_o \exp\left\{-\log_e\{G_o\}\frac{P_{av}}{P_s}\right\} \quad .....(5)$$

$$\phi_E = \alpha \log_e\{G_o\}\frac{P_{av}}{P_s} \quad .....(6)$$

where G_o, the unsaturated field gain, is equal to

$$G_o = \exp\left\{\frac{\Gamma g_o L}{2}\right\} \quad .....(7)$$

1.2. Gain and Phase Delay Characteristics versus Input Power

To derive the gain and phase delay characteristics as functions of input power it is simply a case of substituting (4) into (5) and (6), which gives

$$G_E = G_o \exp\left\{-\frac{\log_e\{G_o\}G_E^2}{2\log_e\{G_E\}}\frac{P_{in}}{P_s}\right\} \quad .....(8)$$

and

$$\phi_E = \frac{\alpha\log_e\{G_o\}G_E^2}{2\log_e\{G_E\}}\frac{P_{in}}{P_s} \quad .....(9)$$

It is immediately obvious from (8) that the amplifier gain characteristic can only be plotted as a function of input power by using iteration analysis techniques. However, as this is a relatively simple exercise in this case, the gain characteristic and the associated phase characteristic have been derived for the extreme operating conditions of $P_{av}/P_s = 0.1$ and G_o = 32 times (30 dB is a typical practical maximum power gain), these are the dashed line plots in Fig 1. These plots show that the static transfer characteristics are slightly curved, so it follows that they are both non-linear functions of input power. Although this is not a problem for static analysis conditions, it does make an analytic analysis of dynamic input conditions impossible. One way of overcoming this problem is to linearise the static transfer characteristics by replacing them with straight line approximations. Alternatively, numerical techniques could be used to avoid the need for linearisation, but this loses valuable insight which the linearised analysis retains at the expense of slightly reduced accuracy.

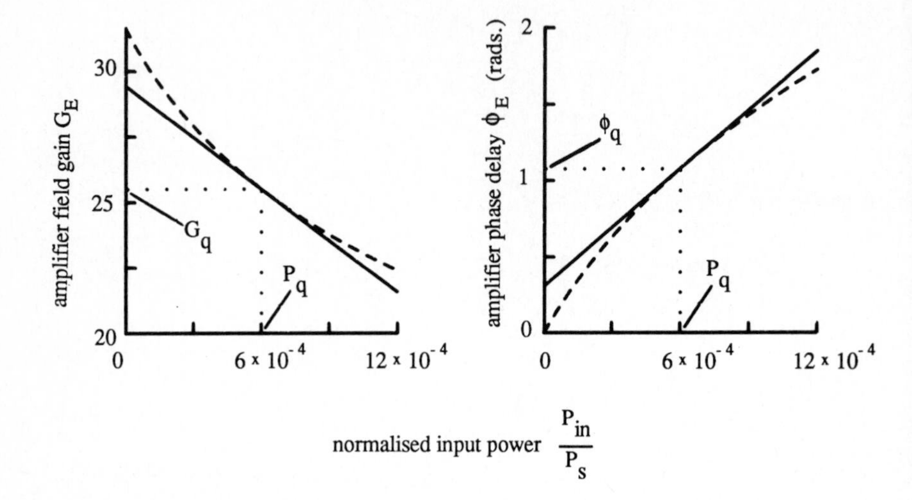

Fig. 1: *Non-linearised (dashed plots) and linearised (solid plots) gain and phase characteristics as a function of normalised input power (P_{in}/P_s), plotted for the extreme conditions $P_{pav}/P_s = 0.1$ and $G_o = 32$.*

2. LINEARISED GAIN AND PHASE CHARACTERISTICS

To linearise the gain and phase characteristics, it is first necessary to linearise the average cavity power so that it takes the form

$$\frac{P_{av}}{P_s} = K_s \frac{P_{in} - P_q}{P_s} + \frac{P_{avq}}{P_s} \qquad .....(10)$$

P_q is equal to half the peak input power and it is defined to be the quiescent input power, P_{avq} is the average cavity power associated with quiescent input conditions and K_s is the slope of the average cavity power characteristic at the quiescent input power point.

To derive slope K_s, (4) is differentiated with respect to P_{in}/P_s after substituting (5) for G_E, and if second and higher order powers of P_{av}/P_s are ignored, $\Gamma g_o L P_{av}/P_s \leq 0.3$ is assumed, and first order binomial and exponential expansions used, it can be shown after substituting P_q for P_{in} that slope K_s approximates to

$$K_s = \frac{K_{av} G_q^2}{2\log_e\{G_q\}} \qquad .....(11)$$

where

$$K_{av} = \frac{1}{\left\{1 + \frac{\log_e\{G_o\}(2\log_e\{G_o\} - 1)}{\log_e\{G_q\}} \frac{P_{avq}}{P_s}\right\}} \qquad .....(12)$$

Using (11) in (10) the linearised expressions for average cavity power can be expressed as

$$\frac{P_{av}}{P_s} = \frac{K_{av}G_q^2}{2\log_e\{G_q\}} \frac{P_{in} - P_q}{P_s} + \frac{P_{avq}}{P_s} \qquad .....(13)$$

and the associated linearised gain and phase expressions become

$$G_E = G_q - G_q \log_e\{G_o\} \frac{P_{av} - P_{avq}}{P_s} \qquad .....(14)$$

$$\phi_E = \phi_q + \alpha \log_e\{G_o\} \frac{P_{av} - P_{avq}}{P_s} \qquad .....(15)$$

where

$$\frac{P_{avq}}{P_s} = \frac{G_q^2}{2\log_e\{G_q\}} \frac{P_q}{P_s} \qquad .....(16)$$

and

$$\phi_q = \alpha \log_e\{G_o\} \frac{P_{avq}}{P_s} \qquad .....(17)$$

G_q and ϕ_q are, respectively, the gain and phase delay values associated with the quiescent input power.

For comparison purposes, these linearised characteristics have been plotted in Fig 1 (solid line plots) using the same P_{av}/P_s and G_o values as used for the non-linearised characteristics (dashed line plots). It is worth noting that the discrepancy between the solid and dashed line plots represents the worst case discrepancy because if an equivalent set of curves were derived for lower values of unsaturated field gain and/or average cavity power, it would be found that the linearisation accuracy improves.

3. LINEARISED AMPLIFIER EQUATIONS: DYNAMIC INPUT CONDITIONS

3.1. Static Analysis Modifications

Although it is clear from the static characteristics that the amplifier gain and the phase delay are both input power dependent, they fail to indicate that carrier lifetime limits the rate at which these two parameters can change. Therefore, to analyse dynamic conditions using the static results, the input power, average cavity power, gain and phase delay must all be treated as functions of time and P_{av} in (14) and (15) should be replaced by

$$P_{av} = P_{av(t)} * \left[\frac{1}{\tau_e}\exp\left\{\frac{-t}{\tau_e}\right\}\right] \qquad .....(18)$$

where [·] is the outcome of introducing carrier lifetime into the analysis, * indicates convolution and τ_e, the effective carrier lifetime [8], is equal to

$$\tau_e = \frac{\tau}{1+\frac{P_{avq}}{P_s}} \qquad .....(19)$$

τ being the carrier lifetime.

When the average cavity power analysis technique is used for dynamic conditions it is also being assumed that the maximum rate of change of input power will be negligible over a period compared to the cavity transit time, and that the $P_{av}/P_S \leq 0.1$ condition has been replaced by $P_{pav}/P_S \leq 0.1$, P_{pav} being the peak value of $P_{av(t)}$.

If the input power is expressed as a power change ($\Delta P_{in(t)}$) it can be shown using (13) through (15) that the associated average cavity power change is

$$\frac{\Delta P_{av(t)}}{P_s} = \frac{K_{av}G_q^2}{2\log_e\{G_q\}}\frac{\Delta P_{in(t)}}{P_s} \qquad .....(20)$$

and that for sinusoidal input conditions the amplifier gain and phase delay changes are

$$\Delta G_{E(t)} = -G_q\frac{\log_e\{G_o\}}{1+j\omega\tau_e}\frac{\Delta P_{av(t)}}{P_s} \qquad .....(21)$$

$$\Delta\phi_{E(t)} = \alpha\frac{\log_e\{G_o\}}{1+j\omega\tau_e}\frac{\Delta P_{av(t)}}{P_s} \qquad .....(22)$$

The only difference between these equations and the ones that would have been derived had the static results been used directly, is that (21) and (22) exhibit a lowpass frequency response as a result of introducing carrier lifetime into the analysis.

3.2. Dynamic Analysis Model

From equations (20) through (22) the equivalent communications theory laser amplifier model given in Fig 2 can easily be derived (NB all phase terms from the analysis appear as negative values to indicate that they they are phase delays). The lower arm of this model converts the input field into a power waveform, removes its d.c. term and then low pass filters its frequency components. The upper path amplifies the input field by an amount equal to the quiescent gain, and prior to this appearing at the amplifier output it is amplitude and phase modulated by the output from the low pass filter. This 'self' modulation effect, which is caused by the input power variation modulating the carrier density, is the mechanism which generates four-wave mixing in the semiconductor laser amplifier.

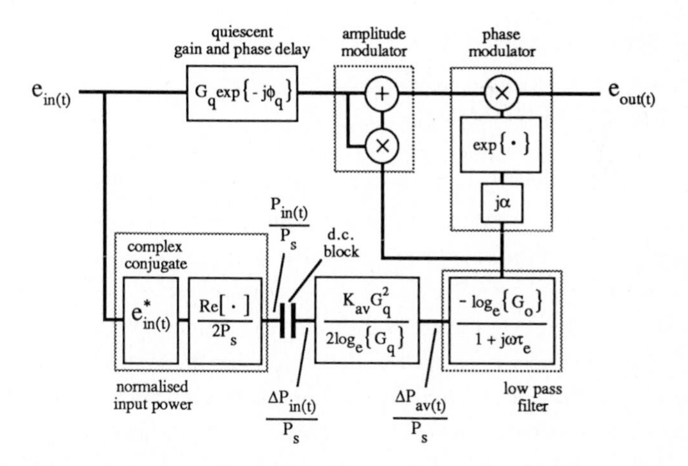

Fig. 2: *Equivalent communications theory laser amplifier model. In this model ⊕ and ⊗ represent addition and multiplication, respectively, Re[·] is the real part of [·] and * indicates complex conjugation.*

If the amplifier is operated at least 20 dB below saturation, which is equivalent to reducing the $P_{pav}/P_s \leq 0.1$ limit by an order of magnitude, it is reasonable for K_{av}, G_q and τ_e to be approximated as unity, G_o and τ, respectively, and for the phase modulation to be represented as an amplitude modulation process by using the narrow band frequency modulation approximation [9]. When this is the case and any frequency components produced by sidebands being modulated are negligible, the equivalent model simplifies to the one given in Fig 3, which is identical to the small signal model reported in [10].

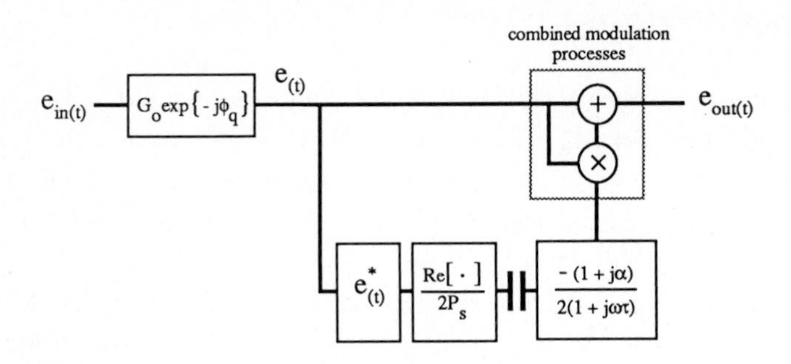

Fig. 3: *Simplified equivalent communications theory laser amplifier model. This model is used when P_{pav}/P_s is small enough for K_{av}, G_q and τ_e to be approximated as unity, G_o and τ, respectively.*

4. AMPLIFICATION OF AN OPTICAL MULTIPLEX

4.1. Worst Case Input Considerations

Ideally, a general analysis would take into account the statistical nature of the modulated channels, the random phase and polarisation relationships between the various carriers, their lack of time synchronism and any frequency instability, but this would make it intractable. One way of overcoming this difficulty is to analyse the worst case which corresponds to assuming that the optical carriers are unmodulated and that the interchannel beat power terms take their maximum values. Although this approach does not take account of any modulation format dependent effects or predict the expected average four-wave mixing effect, it will give an upper bound, the magnitude of which will indicate whether a problem is likely to exist or not. Another reason for performing a worst case analysis is that all these parameters apart from the modulation are likely to be very slow functions of time, so although the probability of worst case conditions occurring may appear to be negligible, they may last for an appreciable length of time compared to the data-rate. Should this prove to be the case the system must then be capable of giving satisfactory operation during worst case conditions.

4.2. Input Power Waveform

For an N channel multiplex comprising equal amplitude optical fields (E_{in}) separated by angular frequency $\Delta\omega$, the field input to the amplifier (e_{in}) is given by

$$e_{in} = E_{in} \sum_{k=1}^{k=N} \exp\{j(\omega_k t - \theta_k)\} \qquad .....(23)$$

where

$$\omega_k = \omega_1 + (k-1)\Delta\omega \qquad .....(24)$$

k is the channel identifier and θ_k is the field phase at $t = 0$.

When the input field is multiplied by its complex conjugate to derive the input power waveform, the beat power sinusoids generated by the cross multiplication terms take their maximum values when $\theta_k = 0$, and this gives an input power waveform which is equal to

$$P_{in(t)} = NP_{ch} \left\{ 1 + \sum_{n=1}^{n=N-1} \frac{2(N-n)}{N} \text{Cos}(n\Delta\omega t) \right\} \qquad .....(25)$$

where n is the beat frequency identifier and P_{ch} is the single channel input power.

This equation shows that for an optical multiplex the worst case input power waveform always comprises N-1 sinusoidal power beats, which are $\Delta\omega$ apart and in the angular frequency range $\Delta\omega$ to $N\Delta\omega$. It is also clear that the power beat amplitudes reduce in direct proportion to their frequency.

4.3. Worst Case Output Field

If the multichannel input power waveform ($P_{in(t)}$) is now transferred through the amplifier model given in Fig 2, it can be shown that the output field can be expressed as

$$e_{out} = E_{in} G_q \sum_{k=1}^{k=N} \left\{ 1 + \frac{\Delta G_{E(t)}}{G_q} \right\} \text{Cos}\left\{ \omega_k t - \phi_q - \Delta\phi_{E(t)} \right\} \qquad .....(26)$$

where the gain and phase variations are given by

$$\Delta G_{E(t)} = -G_q \sum_{n=1}^{n=N-1} \frac{\beta_n}{\alpha} \text{Cos}(n\Delta\omega t - \phi_n) \qquad .....(27)$$

$$\Delta\phi_{E(t)} = \sum_{n=1}^{n=N-1} \beta_n \mathrm{Cos}(n\Delta\omega t - \phi_n) \qquad .....(28)$$

and

$$\beta_n = \frac{(N-n)\alpha K_{av} G_q^2 \log_e\{G_o\}}{\log_e\{G_q\}\sqrt{1+(n\Delta\omega\tau_e)^2}} \frac{P_{ch}}{P_s} \qquad .....(29)$$

$$\phi_n = \mathrm{Tan}^{-1}\{n\Delta\omega\tau_e\} \qquad .....(30)$$

These equations show that the output field is amplitude and phase modulated by the N - 1 power sinusoids, and that the modulation depths are dependent on the single channel input power. It is also clear that the phase modulation depth (β_n) is always larger than that for the amplitude modulation (β_n/α) by a factor equal to the linewidth enhancement factor. It can also be observed that carrier lifetime effects result in the modulating power beats being attenuated and phase delayed as a function of beat frequency.

4.4. Output Field Amplitude for an Arbitrary Channel within the Amplified Multiplex

To avoid the analysis becoming increasingly cumbersome and unwieldy as a result of the cross modulation terms generated by multifrequency phase modulation, it will be assumed from here that the terms involving second and higher order Bessel functions and products of first order Bessel functions are negligible (the implications of this are discussed later). In addition to this it will also be assumed that to a first order approximation the amplitude modulation terms can be ignored as they are a factor of five smaller than those produced by the phase modulation. Expanding (26), using standard communications phase modulation theory [11,12], applying the Bessel function approximations and collecting like frequency terms, the resultant output field amplitude for the kth channel carrier can be shown to be given by

$$e_{out_k} = E_{in} G_q \left\{\prod_{x=1}^{x=N-1} J_{o(\beta_x)}\right\}\left[\mathrm{Cos}(\omega_k t) + \sum_{y=1}^{y=k-1}\left(\frac{J_{1(\beta_y)}}{J_{o(\beta_y)}} \mathrm{Sin}(\omega_k t - \phi_y)\right)\right.$$

$$\left. + \sum_{y=1}^{y=N-k}\left(\frac{J_{1(\beta_y)}}{J_{o(\beta_y)}} \mathrm{Sin}(\omega_k t + \phi_y)\right)\right] \qquad .....(31)$$

$J_{n(\beta)}$ terms are Bessel functions of the first kind (NB if conditions are such that $\beta \leq 0.3$ the approximations $J_{0(\beta n)} \approx 1 - (\beta_n/2)^2$ and $J_{1(\beta n)} \approx (\beta_n/2)$ can be used), x and y are dummy variable substitutes for n, and Π represents repeated multiplication. The first term in the square brackets is the selected carrier, the second represents interfering upper modulation sidebands and the final one represents interfering lower modulation sidebands (NB the first Σ term is ignored when k = 1, and the second is ignored when k = N).

The important fact to emerge from this equation is that the individual output fields are vector sums of the original fields and interfering modulation sidebands generated by the interchannel power beats. It can also be seen that the phase relationships between the interfering sidebands and a particular carrier are determined by both the carrier lifetime and whether the interfering signal is an upper or a lower modulation sideband.

4.5. Implications of Analysis Simplifications on the Upper Limit Phase Modulation Depth

In principle, to analyse the resultant output field it is simply a matter of specifying the phase modulation depth for the $\Delta\omega$ beat term (β_1), and then using

$$\beta_n = \beta_1 \left\{ \frac{N - n}{N - 1} \right\} \sqrt{\frac{1 + (\Delta\omega\tau_e)^2}{1 + (n\Delta\omega\tau_e)^2}} \qquad .....(32)$$

to derive the modulation depths associated with the higher frequency beat terms, and finally using a Bessel function generating algorithm [9] to derive the Bessel function values for use in (31). However, given that the $P_{pav}/P_S \le 0.1$ condition imposed earlier is equivalent to specifying that

$$P_{ch} \le \frac{2\log_e\{G_{Emin}\}}{G_{Emin}^2\left\{N + \sum_{n=1}^{n=N-1} 2(N - n)\right\}} \frac{P_{pav}}{P_S} \qquad .....(33)$$

where G_{Emin} is the amplifier field gain associated with the peak input power, it follows, from (29), that when choosing β_1 care must be taken not to select values which violate the single channel power restriction, otherwise reduced accuracy will be incurred. To derive the upper limit phase modulation depth associated with the $\Delta\omega$ beat term, it is simply a case of substituting (33) and n = 1 into (29) and then solving for the particular N, P_{pav}/P_S, G_o and $\Delta\omega\tau_e$ parameter values of interest. If P_{pav}/P_S and G_o take their maximum values (0.1 and 32, respectively), the upper limit modulation depth is also at a maximum (β_{1max}), its actual value being dependent on channel separation and multiplex size, Fig. 4.

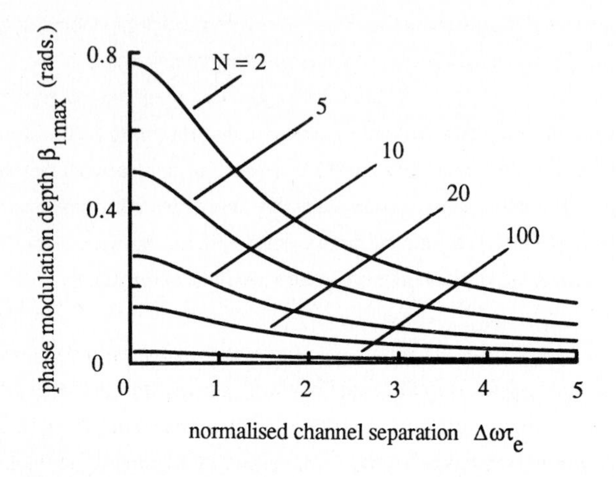

Fig. 4: *Upper limit phase modulation depth (β_{1max}) as a function of normalised channel separation frequency ($\Delta\omega\tau_e$) and multiplex size (N) for $\alpha = 5$, $P_{pav}/P_s = 0.1$ and $G_o = 32$.*

Having derived the maximum phase modulation depth values it is necessary to verify whether they are small enough for second and higher order Bessel terms and products of first order terms to have been assumed negligible when deriving (31). Second and higher order Bessel terms may be ignored when the second order terms are an order of magnitude smaller than the first. The corresponding maximum phase modulation depth must be ≤ 0.6 radians, whereas the limit set by $P_{pav}/P_S = 0.1$ is $\beta_{1max} \leq 0.78$ radians, Fig. 4. However, the 0.6 radian limit is only violated by the two and three channel multiplex, and for such small channel numbers little error is introduced if modulation depths upto 0.78 radians are used in (31): this has recently been verified by a two channel experiment which showed that good agreement is obtained between theory and practice for a β_1 value of 0.76 [8]. As far as the other assumption is concerned, it can be shown that the first order Bessel function products associated with multifrequency phase modulation increase approximately as 3^N, so even if the individual products are small it does not follow that their cumulative effect will be negligible. However, these terms diminish with channel separation and for the modulation depths plotted in Fig. 4 they are negligible provided $\Delta\omega\tau_e$ is ≥ 0.5. It should be noted that this constraint occurs only because of the approximations used when deriving (31), it is not caused by any limitations associated with the equivalent laser amplifier model. Consequently, a more accurate analysis of multifrequency phase modulation could be used to remove the lower limit placed on the channel separation, but such an analysis very quickly becomes intractable as N increases.

5. DISCUSSION

Using the appropriate β_{1max} values, the normalised output carrier powers (P_{NK}) have been derived for a ten channel multiplex for channel separations equal to $0.5\Delta\omega\tau_e$ and $10\Delta\omega\tau_e$, Fig. 5. It is clear from these results that for worst case input conditions the interfering modulation sidebands generated by four-wave mixing can cause the individual carriers to experience different gains depending on channel separation and carrier position within the multiplex (Fig. 5a). Although it is not clear from these plots, this gain variation is also dependent on multiplex size, unsaturated field gain (G_0) and how far below saturation the amplifier is being operated. The individual carriers experience different effective gains because the carrier lifetime generated phase delay (ϕ_n) is positive (negative) for lower (upper) interfering modulation sidebands. Therefore, the higher the carrier frequency with respect to the centre frequency of the multiplex the greater the number of interfering upper sideband terms and hence the smaller the resultant carrier and vice-versa. However, increasing the channel separation reduces the amplitude of the interfering modulation sidebands and a point is eventually reached where all carriers experience a similar gain (Fig. 5b).

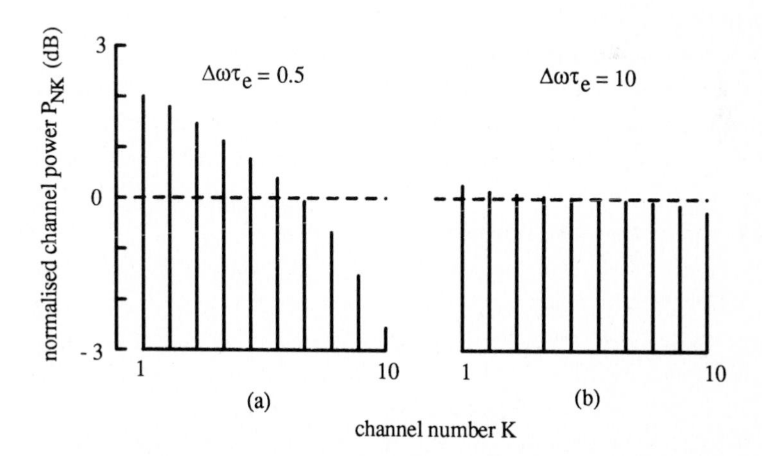

Fig. 5: *Normalised channel output powers (P_{NK}) for a ten channel multiplex with normalised channel separation frequencies of $\Delta\omega\tau_e = 0.5$ (a) and $\Delta\omega\tau_e = 10$ (b), for $\alpha = 5$, $P_{pav}/P_s = 0.1$ and $G_o = 32$.*

In Fig. 6 the normalised output power (P_{NN}) for the highest frequency carrier has been plotted as a function of channel separation for a range of multiplex sizes: the highest frequency carrier was chosen because this also shows the minimum possible gain values that can occur. The plotted results show that the effective carrier gain is dependent on both multiplex size and channel separation, and that the position of the minimum gain, which appears to takes its smallest possible value when the multiplex comprises five channels, never exceeds $\Delta\omega\tau_e > 1$. From the trend of these results it would appear reasonable to conclude that four-wave mixing effects should have become negligible for channel separations $\geq 10\Delta\omega\tau_e$ (typically of the order of $\geq$ 3GHz). Consequently, because it is likely that practical frequency multiplexed coherent systems will require larger channel separations than the ideal value of twice the data-rate in hertz, systems operating at Gbit/s data-rates are expected to automatically satisfy the minimum channel separation requirement.

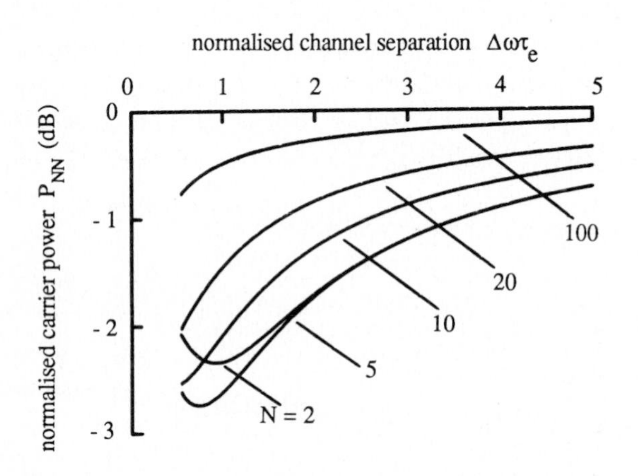

Fig. 6. *Normalised output power (P_{NN}) for the highest frequency carrier (channel N) as a function of normalised channel separation frequency ($\Delta\omega\tau_e$) and multiplex size (N) for $\alpha = 5$, $P_{pav}/P_s = 0.1$ and G_o = 32.*

6. CONCLUSIONS

Using a communications theory laser amplifier model it has been shown that for worst case input conditions, four-waving mixing effects in a travelling wave semiconductor laser amplifier can result in the individual carriers experiencing different gains, the actual gain being dependent on channel separation, multiplex size, channel position, unsaturated gain and how far below saturation the amplifier is operated. It has also been shown that the highest frequency carrier experiences least gain and that the position of the minimum gain never exceeds $\Delta\omega\tau_e > 1$. For worst case multichannel systems design, it is anticipated that it will be necessary to use channel separations $\geq$ 3 GHz for four-wave mixing effects to be negligible when using amplifiers giving 30 dB unsaturated gain and operating 10 dB below saturation. It is envisaged that the channel separations needed for practical Gbit/s frequency multiplexed coherent systems will automatically satisfy this requirement.

ACKNOWLEDGEMENT

Acknowledgement is made to the Research and Technology Board of British Telecom for permission to publish this paper.

REFERENCES

[1] INOUE K, MUKAI T, SAITOH T: 'Nearly degenerate four-wave mixing in a travelling-wave semiconductor laser amplifier', Appl. Phys. Lett., 1987, 51, pp.1051-1053.
[2] AGRAWAL G P: 'Amplifier-induced cross talk in multichannel coherent lightwave systems', Electron. Lett., 1987, 23, pp.1175-1177.
[3] DARCIE T E, JOPSON R M: 'Nonlinear interactions in optical amplifiers for multifrequency lightwave systems', Electron. Lett., 1988, 24, pp.638 - 640.
[4] JOPSON R M, DARCIE T E: 'Calculation of multicarrier intermodulation distortion in semiconductor optical amplifiers', Electron. Lett., 1988, 24, pp. 1372-1374.
[5] HODGKINSON T G, WEBB R P: 'Application of communications theory to analyse carrier density modulation effects in travelling wave semiconductor laser amplifiers', Electron. Lett., 1988, 24, pp. 1550-1552.
[6] FAVRE F, LE GUEN D: 'Contradirectional four-wave mixing in 1.51 μm near travelling wave semiconductor laser amplifier', Electron. Lett., 1989, 25, pp. 1053-1055.
[7] ADAMS M J, WESTLAKE H J, O'MAHONY M J, HENNING I D: 'A comparison of active and passive optical bistability in semiconductors', IEEE J. Quantum. Electron., 1985, QE-21, pp.1498-1504.
[8] WEBB R P, HODGKINSON T G: 'Experimental confirmation of laser amplifier intermodulation model', Electron. Lett., 1989, 25, pp. 491-493.
[9] HAYKIN S: 'Communication Systems: Second Edition', Pub. John Wiley and Sons, 1983.
[10] SALEH A A M: 'Nonlinear models of travelling-wave optical amplifiers', Electron. Lett., 1988, 24, pp.835 - 837.
[11] TAUB H, SCHILLING D L: 'Principles of communications systems', Pub. McGraw-Hill Kogakusha, 1971.
[12] WILLIAMS R A: 'Communication system analysis and design: a systems approach', Pub. Prentice-Hall International, 1987.

Quantum state control and non-demolition detection of photons

Nobuyuki IMOTO
NTT Basic Research Laboratories
Musashino-shi, Tokyo 180, Japan

Abstract

We discuss quantum state control of light and non-demolition detection of photons in connection with applications such as optical communications and optical measurement. The recent research activities on quantum optics including squeezed state generation and quantum non-demolition (QND) detection are reviewed.

1 Introduction

Recent research activities on quantum optics including squeezed state generation and QND detection mark first steps toward new-type optical detection, signal processing, and communications. The possibility of overcoming the *standard limit* of the shot noise in optical detection is becoming reality by the recent progress in theoretical and experimental investigations.

Squeezing and quantum nondemolition detection (QND detection) of light are often referred to as nonclassical effects as they cannot be explained by classical optics but require quantum optics. Of course, there are other topics that are explained only by quantum theory of the electromagnetic field such as spontaneous emission of light, Lamb shift, [1] and sub-Poisson photon statistics [2], and higher order correlation of photons. [3] In this sense, quantum optics have been already established experimentally. Squeezing and QND detection of light are applications of quantum optics because they are included in the present framework of quantum optics.

There is, however, a different feature of squeezing and QND detection from the other topics. The other topics are more and less phenomena that are not easily controllable, and therefore, we should just observe the phenomena but cannot utilize them. In squeezing or QND detection of light, however, we are controlling the quantum property of optical noise based on the law of quantum mechanics.

The first theoretical consideration on squeezed state of light appeared (although the terminology "squeezed state" was not used then) in middle of 1960's, [4][5] and the first proposal of QND detection appeared in middle of 1970's.[6] The theories have been also enriched both for squeezed states [7]-[9] and for QND detection.[10][11]. Theory of quantum communications [12] and quantum estimation theory [13] have been also developed.

The first experimental work on squeezed state appeared in 1985 [14] followed by several experiments on squeezed state generation.[15]-[19] The key to generate a squeezed state is use of nonlinear optical effect. The recent progress in experiment is supported by technological progress in nonlinear optics. Since a QND scheme with the optical Kerr effect has been proposed, [20] experimental efforts on QND detection have been also made extensively.[21]-[23] The key for QND detection is again a nonlinear optical effect. However, QND detection has not been demonstrated yet. [1]

On the whole, theory precedes experiment in this area. It seems, however, that some theoretical problems remain unconsidered. One of them is quantum treatment of a propagating beam. In a usual quantum theory for the electromagnetic field, the basic idea is to describe *time* evolution of spatial modes defined in a cavity. In many quantum optical experiments, however, we are interested in *spatial* propagation of a quantum state in a linear or nonlinear medium. If we use a monochromatic, stationary beam, it is hard to describe the beam by the usual time evolution picture. There are several approaches for this problem. In this paper, however, a recent approach will be introduced. This approach is thought to be a direct extension of usual frequency analysis in classical optics, microwave theory, or electronic circuit theory.

The organization of this paper is as follows. In Sec. 2, the quantum nature of light is described in terms of the shot noise. In Sec. 3, the experiments on squeezing of light are reviewed. In Sec. 4, the experimental efforts for QND detection of light are reviewed. In Sec. 5, a recent theory on QND detection with loss and error is reviewed. In Sec. 6, a recent theory on quantum mechanical treatment of propagating beam based on the frequency analysis is reviewed.

[1] Although the definition of squeezing is clear, there is some confusion in the definition of QND detection, as will be mentioned about later. According to the definition of the present paper, QND detection has not been demonstrated yet.

2 Shot noise as the quantum mechanical uncertainty

Let us consider an optical direct detection system, which contains a laser having no excess noise and a detector whose quantum efficiency is unity. The signal to noise ratio (SNR) of an optical communication system is determined by the noise power per unit bandwidth $P_{\text{noise}} = 2eIR$, where e is the electron charge, I is the dc photocurrent, and R is the receiver impedance. This noise level is known as the shot noise level. In practical, there are several other noise sources such as intensity fluctuation of the laser and receiver thermal noise. These noise terms are, however, called classical excess noise terms, which are in principle removable.

Fig. 1 shows the comparison between the classical and quantum theories on the shot noise. In the classical theory, an ideal optical beam is assumed to have no noise. The shot noise is regarded to be generated at the detector when the beam is randomly converted into photoelectrons. From a quantum mechanical viewpoint, however, the shot noise is explained as the *quantum optical noise of the light source.* An ideal detector is assumed to generate no noise within the bandwidth of interest.

CLASSICAL PICTURE :

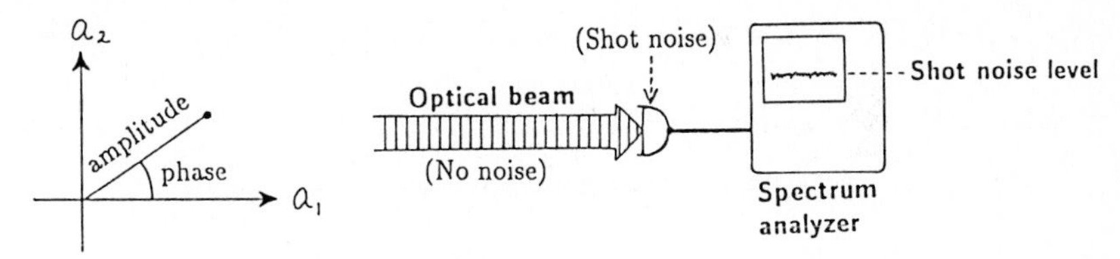

QUANTUM MECHANICAL PICTURE :

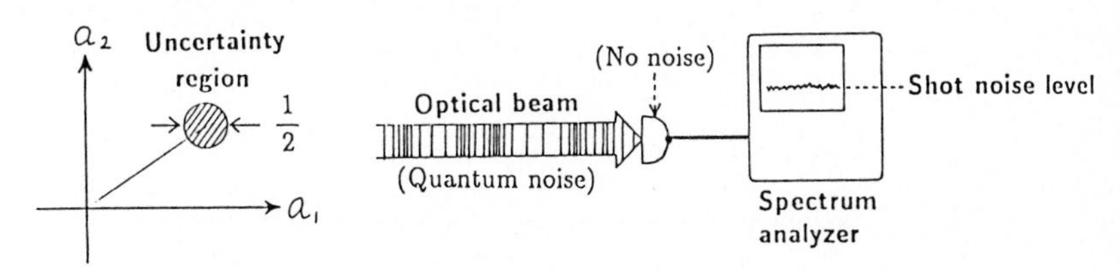

Fig. 1. Classical picture and quantum mechanical picture for the shot noise.

The quantum noise is due to Heisenberg's uncertainty principle, which cannot be violated, but the uncertainty balance can be controlled in ways described below. The origin of the shot noise is considered to be the equally distributed minimum uncertainty noise int eh in-phase amplitude (cosine-component) and the quadrature phase amplitude (sine-component) of the electric field, which obeys Heisenberg's uncertainty principle. This state is called coherent state of light.[24] The uncertainty principle, however, does not allow an imbalance in the uncertainty of the two amplitudes. This means that phase amplitude noise and the quadrature amplitude noise cannot be reduced below the shot noise level at the same time but that *one of them can be reduced below the shot noise level* and the other increased above the shot noise level. The aim of squeezing is to reduce one of the quantum noise to attain a better performance than that restricted by the shot noise.

An optical beam having a quantum noise imbalance in the two amplitudes is generally called a "squeezed state of light" (the noise is squeezed into one of the two amplitudes). How to squeeze an optical beam will be reviewed in Sec. 3.

Squeezing of light is to control quantum noise that is already contained in the light source. There is, however, another kind of quantum noise that is newly *added* by processing an optical beam. For example, amplification of an optical beam adds the spontaneous emission noise of the amplifier. If a signal beam is branched by a beamsplitter, as is shown in Fig. 2(a), signal to noise ratio in both two output ports are degraded from the original SNR of the input port.

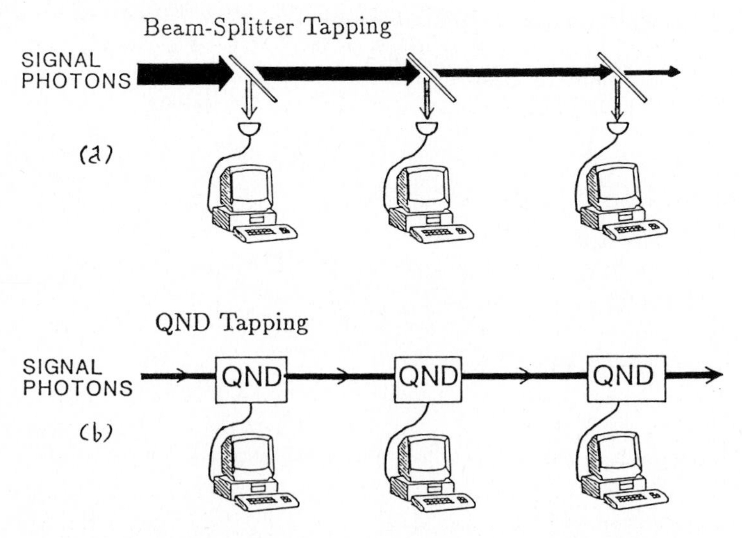

Fig. 2. (a) Classical signal branching and (b) QND branching.

A simple explanation of this SNR degradation is that a photon cannot be divided into parts but is transmitted or reflected by the beamsplitter in a stochastic way. This causes photon number uncertainty in the transmitted or reflected beam. A quantum analysis tells that a beamsplitter causes SNR degradation both in intensity and phase of the optical beam.

If one use a QND detector, however, as is shown in Fig. 2(b), then the SNR degradation in intensity can be reduced, for example. In this case, SNR degradation in phase is increased. However, if we use only the intensity as the information carrier, a signal branching free from the added quantum noise is possible. How to realize a QND detection of intensity will be reviewed in Sec. 4.

3 Squeezed state of light

The study of squeezed states of light began with a quantum mechanical analysis of parametric amplification [4] and a generalization of coherent states of light. [5] In 1976, Yuen developed the theory of two photon coherent states, [7] today referred to as squeezed states, which have noise imbalance in sine and cosine component of the electric field amplitude. He proposed the use of four-wave mixing as a generation scheme for squeezed states of light, [25] and also developed a theory of squeezed state optical communication, [12] and showed the advantage of using squeezed light, rather than coherent states, in communication. More detailed theories for generating squeezed states have been developed by D. F. Walls et al.[8][9]

Squeezed states were first observed in 1985 by Slusher and his co-workers using four wave mixing in sodium vapor. Several experiments followed using different materials and configurations. [15]-[19] Squeezed state is detected by a homodyne receiver by measuring the noise power spectrum in sine or cosine component of the optical beam. If one of the noise power spectrum goes down below the shot noise level, the beam is verified to be a squeezed state. This means that squeezing of light is to make a quantum correlation between two vacuum states in frequency modes around the frequency of the local oscillator, because the noise power spectrum is a result of beat between the local oscillator and two side-bands vacuum noise. The largest squeezing that has been obtained so far is by Kimble and his co-workers with the optical parametric oscillation (OPO) using $LiNbO_3$ crystal.[15]

A typical experimental configuration is shown in Fig. 3 (After L. Wu et al.). A single mode degenerate optical parametric oscillator (OPO) pumped by a second harmonic generator of a YAG laser emits a squeezed vacuum at the same wavelength as the fundamental beam of the YAG laser. The squeezed vacuum is detected by homodyne detection using a balanced mixer detector. The fundamental beam of the YAG laser is used as the local oscillator. A noise power spectrum below the shot noise level is obtained on an electric spectrum analyzer by adjusting the phase of the local oscillator in phase with the squeezed light.

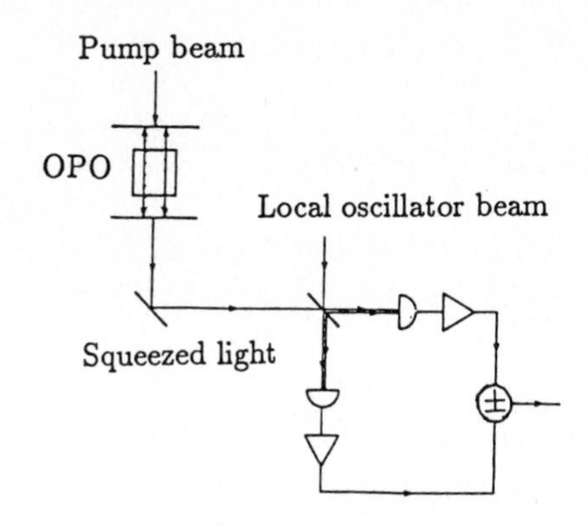

Fig. 3. Configuration for generation and detection of squeezed light using an optical parametric oscillator

An application of this squeezed vacuum is improvement of SNR of an interferometer. The SNR of an interferometer is usually determined by the natural vacuum field introduced from the other port of the beamsplitter that is used for dividing the input signal beam into two paths. If a squeezed vacuum is irradiated instead of the natural vacuum field with a proper relative phase to the signal beam, the SNR of the interferometer is improved. [26][27]

There exist nonclassical photon states other than squeezed states. For example number-phase squeezed states can have smaller intensity fluctuations and larger phase fluctuations than the shot-noise-limited light. The photon number state is the limiting case of a number -phase squeezed state for which the intensity is constant and the phase is completely uncertain. It can be shown that maximum entropy is obtained for a constant average optical power when number states are used, [28] and the use of these states or number-phase squeezed states in optical communications are therefore advantageous.[29] Generation of a number-phase squeezed state has been reported by NTT using a pump-noise suppressed diode laser by means of high-impedance suppression of injection current noise.[18] Figure 4 shows the experimental result in Ref. [18].

Another kind of squeezing is intensity noise correlation between two different frequency beams. This squeezed state was observed by Giacobino's group [19] by parametric down conversion.

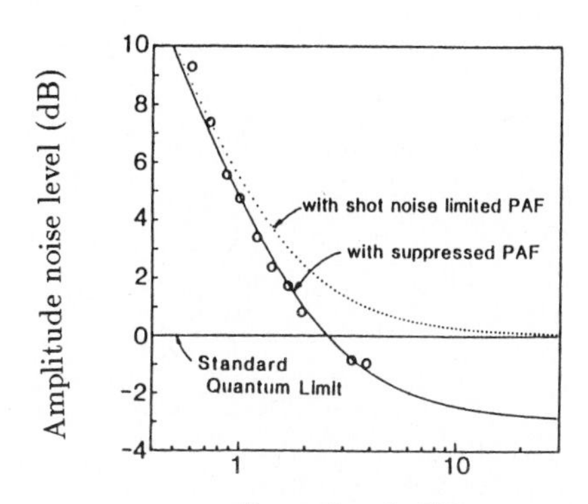

Fig. 4. Amplitude noise level as a function of pump level. Circle : experimental data. Solid line : Theoretical curve for pump noise suppressed laser. Dotted line is a theoretical curve for a conventional laser.

4 Quantum nondemolition detection of light

Quantum nondemolition (QND) detection is first proposed by Braginsky [6][30] and was aimed at improving the mechanical measurement accuracy of a gravitational wave detector. The general theory is developed by Unruh [10], and the theory of optical QND detection was developed by Caves et al. [11] and Walls et al. [31] An optical QND detection scheme using the optical Kerr effect was proposed, [20] and several experimental demonstration have been attempted.[21][22]

Figure 5 shows the schematic of a QND detection scheme using the optical Kerr effect. A signal beam passes through dichroic mirrors M1 and M2 and through a Kerr medium with no attenuation. The probe beam is transmitted through a Mach-Zehnder interferometer with the Kerr medium as one of the optical paths. The signal beam intensity modulates the refractive index of the medium via the Kerr effect, which, in turn, modulates the probe beam phase. A balanced mixer detector detects the probe beam phase, yielding a nondemolition-detection current for the signal beam intensity. The signal optical phase is modulated by quantum noise in the probe intensity, but the intensity of the signal is not changed by the Kerr effect. This means the signal phase is unpredictably changed in compensation for the nondemolition detection of the intensity. Basically, any phase sensitive detection can be used in the above scheme. Levenson and co-workers adopted a Fabry-Perot cavity as the phase sensitive interferometer, [21] and obtained quantum correlation between the signal and probe beam.

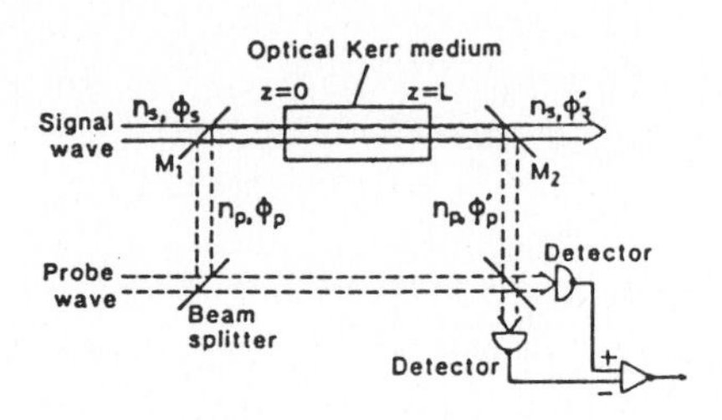

Fig. 5. Configuration for quantum non-demolition (QND) detection of optical intensity using the optical Kerr effect.

Here, a comment on the definition of QND detection should be made. There are several versions of definition for QND detection. The original definition is "such a measurement in which the repeated measurement gives the same measured value as the first measurement." This definition implies a non-destructive, complete measurement (measurement with no unpredictable disturbance and no measurement error). Such a complete QND detection is, however, not realizable in practical.

A more practical definition is "a measurement using an interaction which does not disturb the quantity to be measured." This definition is based on an idea of a nondemolitional detection via system-probe coupling. If the magnitude of interaction is infinite, i. e., if the quantum correlation formed between the system and the probe is complete, this scheme acts as a complete QND detection with no measurement error. Therefore, this definition allows a finite measurement error because the correlation between the system and the probe is not complete in usual cases. This definition has been widely adopted. This definition, however, assumes no optical loss in the detection scheme. In this case, the outgoing beam intensity is exactly the same as the input beam intensity, that is, the measurement error is precisely related to the quantum correlation between the outgoing signal and the readout signal. If we remind that squeezing of light is also to make a quantum correlation between two quantities, this definition of QND detection having no loss makes no difference between QND detection and squeezing. Sometimes, QND detection is confused as making a quantum correlation even though there is an optical loss.

Any existing nonlinear medium has a finite transparency. This makes difference between squeezing and QND detection. The optical loss of the medium plays two roles, one is absorption for the input beam and the other is reduction of quantum correlation between two outgoing beams. In squeezing of light, only the latter is important because the absorption of the input pump beam is not relevant. In QND detection, however, the loss not only degrades the measurement error, but also destroys the signal beam. The problem is, then, the definition of QND detection with an optical loss.

A theory on lossy QND detection has been developed [32] in which a lossy QND detection of photons is defined as a detection having less measurement error than a random deletion detection having the same insertion loss. This means that whether it is better or not to use random deletion of photons by a beamsplitter can be a criterion for a lossy QND detection scheme of photons. Based on this definition of lossy QND detection, a condition is obtained for the QND detection scheme using the optical Kerr effect. [32] According to the definition, QND detection is not yet demonstrated.

A numerical example shows that about 50 mW optical beam is required both for the signal and probe beams assuming an optical silica fiber as the Kerr medium having the third order nonlinear susceptibility of 3×10^{-33} MKS and loss value of 0.2 dB/km. This result shows that QND detection of photons is in principle possible using existing material and lasers.

5 Quantum treatment of propagating beam

5.1 General description

In a conventional quantum theory for the electromagnetic field, the basic idea is to describe *time* evolution of spatial modes defined in a cavity volume. The generator of time evolution is the Hamiltonian of the system. There are, however, a number of situations in which we are interested in *spatial* evolution of a quantum state. In squeezing of an electromagnetic beam, for example, an input coherent state beam is gradually squeezed as it propagates along a nonlinear optical medium. In amplification of the beam, the input beam is amplified and spontaneous emission is gradually added as it propagates along a gain medium. If the beam is stationary, the beam itself does not evolve in the time domain. Here, we develop a theory of quantum mechanics to treat spatial evolution of the field. There are several approaches for this problem.[33]-[38] In this section a recent approach, which is a direct extension of usual frequency analysis in classical optics, is introduced below. [39][40]

The present idea is to expand the field $\hat{\Psi}(t,x,y,z)$ with mode functions $\phi_j(t,x,y)$ as $\hat{\Psi}(t,x,y,z) = \sum_j \hat{a}_j(z)\phi_j(t,x,y)$ instead of the conventional expansion, $\sum_k \hat{a}_k(t)\psi_k(x,y,z)$. For example, $\phi_j(t,x,y)$ is further factorized into a frequency-mode function, $e^{-i\omega_j t}$, and a Gaussian-beam transverse-mode function. This approach, i.e., propagation of a Fourier component $e^{-i\omega t}$, is quite normal method in classical optics, microwave theory, and electronic circuit theory. In the usual quantum mechanics, however, this approach has been out of consideration from the first time, because quantum mechanics are constructed on classical Hamilton formula, which is already based on the time evolution picture. The differences between the present theory and the usual theory are compared in Table 1.

Table 1. Comparison of usual quantum theory and the quantum theory in this section

	Usual theory	Present theory
	Time evolution of spatial modes	Spatial evolution of time-domain modes
Picture	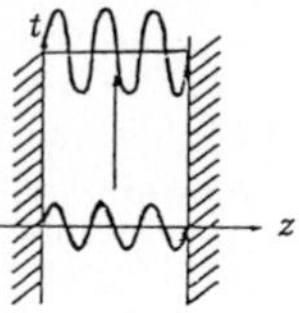 t, z	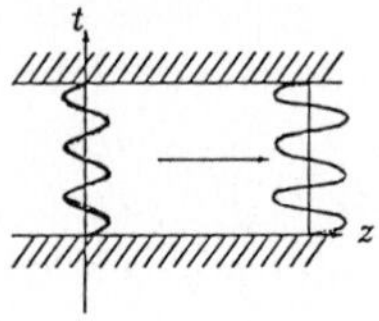 t, z
Mode suffix (Example)	k	ω
Field expansion	$E = \sum_k \hat{a}_k(t)\psi_k(x,y,z)$	$E = \sum_\omega \hat{a}_\omega(z)\phi_\omega(t,x,y)$
Equation of evolution	$\frac{d\hat{A}}{dt} = \frac{1}{i\hbar}[\hat{A},\hat{H}]$	$\frac{d\hat{A}}{dz} = \frac{1}{i\hbar}[\hat{A},\boxed{?}]$

The corresponding Heisenberg-type equation of *propagation* for $\hat{a}_j$ should be of the form

$$\frac{d}{dz}\hat{a}(z) = \frac{1}{i\hbar}[\hat{a}(z), \hat{I}_z(z)], \tag{1}$$

where $\hat{I}_z$ is a spatial evolution generator, which plays the role of the Hamiltonian in the usual picture. We will not go to the detailed theory, but it is derived that the spatial evolution generator $\hat{I}_z$ is defined, for the electromagnetic field, as

$$\hat{I}_z = \int\int_A \int_T (-T_{zz})dtdxdy, \tag{2}$$

where T_{zz} is the (z,z) component of the Maxwell's energy-momentum tensor, A is the cross-sectional area of the beam, and T is the time duration of observing a frequency mode. Thus the frequency mode separation is $\Delta\omega \equiv 2\pi/T$. The derivation of Eq. (2) can be find in Refs. [40] and [39]

It should be noted here that the usual Hamiltonian is equal to $\int\int\int_V (T_{tt})dxdydz$, where T_{tt}(= Hamiltonian density) is the (t,t) component of the Maxwell's energy-momentum tensor.

In solving practical problems, we need to express the spatial evolution generator by creation and annihilation operators, $\hat{a}^\dagger(z)$ and $\hat{a}(z)$. This can be done by substituting the quantized electromagnetic field into T_{zz}, which is expressed by

$$-T_{zz} = \left[E_z D_z + H_z B_z - \frac{1}{2}(\vec{E}\cdot\vec{D} + \vec{H}\cdot\vec{B})\right]. \tag{3}$$

The quantized electric and magnetic fields are generally expressed as

$$\vec{E} = \sum_j \vec{e}_j \hat{a}_j(z)\phi_j(t,x,y) + \text{H.c.} \tag{4}$$

and

$$\vec{H} = \sum_j \vec{h}_j \hat{a}_j(z)\psi_j(t,x,y) + \text{H.c.}, \tag{5}$$

where $\vec{e}_j$ is a vector having a magnitude of the electric field per photon and direction of the polarization of j-th mode, $\vec{h}_j$ is the orthogonal vector to $\vec{e}_j$ having a magnitude of the magnetic field per photon. The mode functions, ϕ and ψ, should be chosen so that $\vec{E}$ and $\vec{H}$ satisfy the Maxwell's equation.

The unperturbed part of the spatial evolution generator is obtained, using the linear material equations, $\vec{D} = \varepsilon\vec{E}$ and $\vec{B} = \mu\vec{H}$, as

$$\hat{I}_0 = -\sum_\omega \hbar k_\omega \left[\hat{n}_\omega(z) + \frac{1}{2}\right], \tag{6}$$

where $\hat{n}_\omega(z) \equiv \hat{a}^\dagger_\omega(z)\hat{a}_\omega(z)$ is the photon number operator. The meaning of the photon number in the present picture is the number of elementary quanta, $\hbar k_\omega$, which pass through point z with cross-sectional area A within time duration T. The equation of propagation leads to an unperturbed operator solution as $\hat{a}_\omega(z) = e^{ik_\omega z}\hat{a}_\omega(0)$.

When there is an interaction, the electric flux density becomes $\vec{D} = \varepsilon\vec{E} + \vec{P}_{\text{int}}$, where $\vec{P}_{\text{int}}$ is the polarization vector due to the interaction. The interaction part of the spatial evolution generator, $\hat{I}_{\text{int}}$, is obtained from $\vec{P}_{\text{int}}$.

5.2 Noise power spectrum

Quantum nature of the electromagnetic beam often appears in the noise power spectrum in direct, homodyne, or heterodyne detection. In the usual quantum theory, the noise power spectrum is expressed by Fourier transform of a temporally fluctuating quantity such as $\hat{a}_k(\Omega) \equiv \mathcal{F}[\hat{a}_k(t)]$ and $\hat{n}_k(\Omega) \equiv \mathcal{F}[\hat{n}_k(t)]$, where $\mathcal{F}$ means a Fourier transform. Here Ω is the frequency component of the noise, which is much smaller than the optical frequency ω. These quantities are integrated values over a cavity volume, and therefore include no spatial dependence of the propagating beam. In the present theory, however, the spatial dependence of the noise power spectrum is directly expressed by $\hat{a}_{\omega\pm\Omega}(z)$. Note that Ω is a parameter (because t is a parameter) in the usual theory, while Ω is a mode suffix and z is the parameter in the present theory.

In direct detection of an intense monochromatic beam, the electric field is expressed as

$$\hat{E}_y = \sqrt{\frac{\hbar k_\omega}{2\varepsilon AT}}\hat{a}_\omega(z)e^{-i\omega t} + \sum_{\omega'\neq\omega}\sqrt{\frac{\hbar k_{\omega'}}{2\varepsilon AT}}\hat{a}_{\omega'}(z)e^{-i\omega' t} + \text{H.c.}, \tag{7}$$

where ω is the center frequency of the intense beam. We assume that the photocurrent I emitted by the detector is proportional to the time averaged optical power as

$$I = \beta \cdot c\sqrt{\frac{\varepsilon_0}{\varepsilon}}\int\int_A \frac{1}{2}\left(\varepsilon\hat{E}_y^2 + \mu\hat{H}_x^2\right), \tag{8}$$

where β is the conversion efficiency from the optical power into the electric current. Substituting Eq. (7) into Eq. (8), and extracting the $\pm\Omega$ component, we obtain

$$I(\Omega) = \frac{\beta\hbar\omega}{T}\left\{\left[\left(1+\frac{\Omega}{\omega}\right)\hat{a}_\omega^\dagger\hat{a}_{\omega+\Omega} + \left(1-\frac{\Omega}{\omega}\right)\hat{a}_\omega\hat{a}_{\omega-\Omega}^\dagger\right]e^{-i\Omega t} + \text{H.c.}\right\}. \tag{9}$$

The electric power per unit frequency consumed at the resistor R of an electric spectrum analyzer is defined as $P(\Omega) \equiv \frac{2\pi}{\Delta\omega}R\overline{\langle[I(\Omega)]^2\rangle}$ $(= R\langle\int_0^T[I(\Omega)]^2dt\rangle)$. This is a single-sided power spectrum per unit frequency. The power spectrum $P(\Omega)$ is obtained as $P(\Omega) = \frac{2R}{T}(\beta\hbar\omega)^2\langle\hat{a}^\dagger\hat{a}\rangle$, assuming that modes other than ω are in the vacuum state, and using an approximation, $1 \pm \frac{\Omega}{\omega} \simeq 1$. If we assume that one photon is converted into one photo-electron with 100 % efficiency, we obtain $\beta = \frac{e}{\hbar\omega}$. In this case, the power spectrum becomes $P(\Omega) = 2eI_{dc}R$, where $I_{dc}(\equiv \langle\hat{a}^\dagger\hat{a}\rangle e/T)$ is the dc photo-current. This is known as the shot noise in the photodetection..

The noise power spectrum in homodyne or heterodyne detection is also obtained in a similar way. In this case, two power spectra, $P_{\cos}(\Omega)$ and $P_{\sin}(\Omega)$, are defined according to the relative phase difference between the signal and reference beams. It is shown that $P_{\cos}(\Omega) = P_{\sin}(\Omega) = 2eI_{dc}R$ if modes other than the center frequency are in the vacuum state.

In the squeezing of an electromagnetic beam, it is shown that $P_{\cos}(\Omega)$ and $P_{\sin}(\Omega)$ can be made imbalanced under the constraint of $P_{\cos}(\Omega) \cdot P_{\sin}(\Omega) =$

$(2eI_{dc}R)^2$. We will not go into the detailed calculation, but the position dependence of the power spectrum is obtained as follows. In the squeezing of the beam, the $\pm\Omega$ components, $\hat{a}_{\omega+\Omega}(z)$ and $\hat{a}_{\omega-\Omega}(z)$, are no more in the independent vacuum states but are correlated via nonlinear coupling in the medium. This makes it possible to lower the noise power spectrum $P_{\cos}(\Omega)$ or $P_{\sin}(\Omega)$ below the shot noise level. The degree of correlation increases as the beam propagates through the nonlinear medium, resulting in the z dependence of the spectrum.

The present theory provides a simple method of describing beam propagation problems in which the interaction is a function of only z and the quantum property of the beam evolves spatially along the z axis. By determining the time dependence of the field (frequency ω, for example), the rest of the dynamics are described by spatial evolution of the field. Classically, this method has been widely adopted especially in optics, microwave theory, and electronic circuit theory. The present method can give a quantum version of these well-developed methods. Basically, the present method is suited to systems which conserve I_z with respect to space, while the usual Hamilton formulation is suited to systems which conserve energy in a volume with respect to time.

6 Discussion

The experimental and theoretical research activities described here are first steps toward the application of quantum state control and non-demolition detection of photons. However, several problems remain for practical applications. Today, the degree of squeezing is not satisfactory, and an intense pump beam is also required. This is due to small nonlinearities of existing optical materials. Optical losses, including transmission losses, should also be considered, as optical loss destroys the quantum mechanical nature of nonclassical photonic applications. Better nonlinear optical materials are needed for $\chi^{(2)}$ process and $\chi^{(3)}$. Development of new nonlinear materials such as organic materials and MQW semiconductor materials will be interesting for nonclassical photonic applications.

The detection of the squeezed states is with homodyne detection, which requires a local oscillator beam derived from the pump beam for the squeezed state generation scheme. For communication or measurement in a distant place, it is not desirable to transmit both a noise-reduced signal and an intense local oscillator beam. Therefore, modulation and homodyne detection of squeezed state using phase-locked loop (PLL) should be investigated.

Squeezing and QND detection of an optical pulse will be also the future problem. There are theoretical approaches for squeezing and QND detection of optical solitons.[41]-[43] Not only squeezing or QND detection, but also cavity quantum electrodynamics [44] will be of importance in connection with the control of spontaneous emission [45][46] rate with a certain boundary condition.

Acknowledgments

The author would like to thank Dr. Y. Yamamoto, M. Ueda and K. Igeta for their useful discussions.

References

[1] W. E. Lamb, Jr. and R. C. Retherford, Phys. Rev. **72**(1947)241.

[2] R. Short and L. Mandel Phys. Rev. Lett. **51**(1983)384.

[3] R. Ghosh and L. Mandel Phys. Rev. Lett. **59**(1987)1903.

[4] H. Takahashi, *Advances in Communication Systems.* (Academic, New York, a. v. balahrishnan edition, 1965).

[5] D. Stoler, Phys. Rev. **D1**(1970)3217.

[6] V. B. Braginsky and Y. I. Vorontsov, Sov. Phys.-Usp. **17**(1975)644.

[7] H. P. Yuen, Phys. Rev. **A13**(1976)2226.

[8] P. D. Drummond and D. F. Walls, Phys. Rev. **A23**(1981)2563.

[9] M. D. Reid and D. F. Walls, Phys. Rev. **A31**(1985)1622.

[10] W. G. Unruh, Phys. Rev. **D18**(1978)1764.

[11] C. M. Caves, K. S. Thorne, R. W. P. Drever, V. D. Sandberg, and M. Zimmerman, Rev. Mod. Phys.**52**(1980)341.

[12] H. P. Yuen and J. H. Shapiro, IEEE trans. Inform. Theory **IT-26**(1980)78.

[13] C. W. Helstrom, *Quantum Detection and Estimation Theory*, (Academic Press, New york 1976).

[14] R. E. Slusher, L. W. Hollberg, B. Yurke, J. C. Mertz, and J. F. Valley, Phys. Rev. Lett. **55**(1985)2409.

[15] L. Wu, H. J. Kimble, J. L. Hall, and H. Wu, Phys.Rev.Lett. **57**(1986)2520.

[16] R. M. Shelby, M. D. Levenson, S. H. Perlmutter, R. G. DeVoe, and D. F. Walls, Phys. Rev. Lett. **57**(1986)691.

[17] M. W. Maeda, P. Kummar, and J. H. Shapiro, Opt. Lett. **12**(1987)161.

[18] S. Machida, Y. Yamamoto, and Y. Itaya, Phys. Rev. Lett. **58**(1987)1000.

[19] A. Heidmann, R. J. Horowicz, S. Reynaud, E. Giacobino, C. Fabre, and G. Camy, Phys. Rev. Lett. **59**(1987)2555.

[20] N. Imoto, H. A. Haus, and Y. Yamamoto, Phys. Rev. **A32**(1985)2287.

[21] M. D. Levenson, R. M. Shelby, M. Reid, and D. F. Walls, Phys. Rev. Lett. **57**(1986)2473.

[22] N. Imoto, S. Watkins, and Y. Sasaki, Opt. Commun. **61**(1987)159.

[23] A. La Porta, R. E. Slusher, and B. Yurke, Phys. Rev. Lett. **62**(1989)28.

[24] R. J. Glauber, Phys. Rev. **131**(1963)2766.

[25] H. P. Yuen and J. H. Shapiro, Opt. Lett. **4**(1979)334.

[26] Min Xiao, Ling-An Wu, and H. J. Kimble, Phys. Rev. Lett. **59**(1987)278.

[27] P. Grangier, R. E. Slusher, B. Yurke, and A. LaPorta, Phys. Rev. Lett. **59**(1987)2153.

[28] W. H. Louisell, *Quantum Statistical Properties of Radiation*, John Wiley & Sons, New York (1973).

[29] Y. Yamamoto and H. A. Haus, Rev. Mod. Phys. **56**(1986)1001.

[30] V. B. Braginsky, Y. I. Vorontsov, and K. S. Thorne, Science **209**(1980)547.

[31] G. J. Milburn and D. F. Walls, Phys. Rev. **A28**(1983)2065.

[32] N. Imoto and S. Saito, Phys. Rev. **A39**(1989)675.

[33] Y. R. Shen, in Proceedings of the International School of Physics ≪Enrico Fermi≫ *Quantum Optics* (Academic, New York, edited by R. J. Glauber), p.489(1969).

[34] I. Abram, Phys. Rev. **A35**, 4661(1987).

[35] H. A. Haus, Proc. IEEE **58**, 1599(1970).

[36] C. M. Caves and D. D. Crouch, J. Opt. Soc. Amer. **B4**, 1535(1987).

[37] M. J. Collet and R. Loudon, J. Opt. Soc. Amer. **B4**, 1525(1987).

[38] B. Yurke and J. S. Denker, Phys. Rev. **A29**, 1419(1984).

[39] N.Imoto, in Proceedings of the 3rd International Symposium of Foundations of Quantum Mechanics (ISQM-Tokyo'89), (Physical Society of Japan, 1990).

[40] N. Imoto, to be published in Phys. Rev. Lett.

[41] P. D. Drummond and S. J. Carter, J. Opt. Soc. Amer. **B4**(1987)1565.

[42] H. A. Haus, K. Watanabe, and Y. Yamamoto, to be published in J. Opt. Soc. Amer. B.

[43] K. Watanabe, H. Nakano, A. Honold, and Y. Yamamoto, Phys. Rev. Lett. **62**(1989)2257.

[44] S. Haroche and D. Kleppner, Physics Today **42**(1989)24.

[45] F. De Martini and G. R. Jacobovitz, Technical Digest of XVI International Conference on Quantum Electronics (IQEC'88 Tokyo), TuH-68(1988).

[46] Y. Yamamoto, S. Machida, K. Igeta, and Y. Horikoshi, in Proceeding of Sixth Rochester Conference on Coherence and Quantum Optics, (1989).

PART 3

OPTICAL MULTIACCESS NETWORKS

COHERENT OPTICAL FDM BROADCASTING SYSTEM WITH OPTICAL AMPLIFIER

Katsumi Emura, Makoto Shibutani, Shuntaro Yamazaki, Iljunn Cha, Ikuo Mito and Minoru Shikada

Opto-Electronics Research Laboratories
NEC Corporation, Japan

The capability of a coherent FDM technology, which could realize the largest scale systems, has been confirmed through a 10 channel coherent CATV experiment. To enlarge the system capacity, a traveling wave semiconductor optical amplifier was applied to the system. The experimental results indicate a system capability of more than 70 HDTV channel distribution to more than 10,000 subscribers, which may be realized only by the coherent FDM technology. High saturation level optical amplifier with cross polarized channel arrangement and multi local receiver would be effective for further improvement in the system capacity.

1. INTRODUCTION

Multichannel distribution systems, based on an optical WDM or FDM technology, are attractive for achieving large capacity, large subscriber networks [1]-[4]. Figure 1 shows a system example for CATV application. The target of these system could be more than 100 (High Definition TV) channel distribution to several tens of thousands of subscribers. However, several factors would limit these numbers. Optical tunable filters for channel selection are one

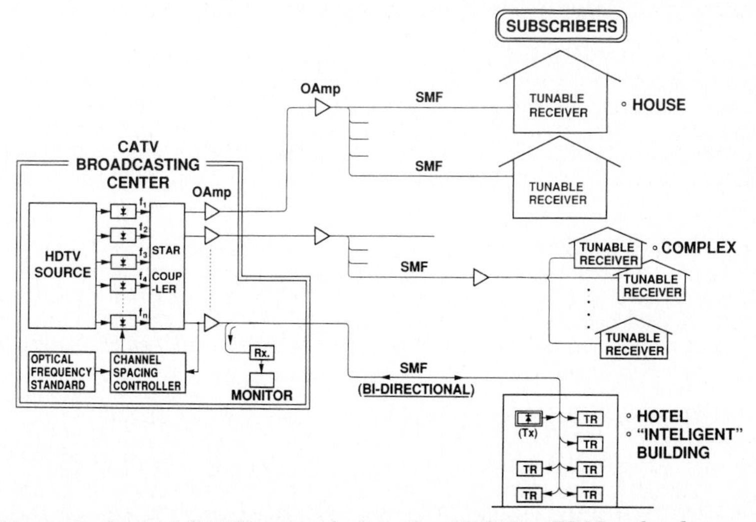

Figure 1. Optical CATV networks based on WDM or FDM technology

Table 1. Tunable optical filter performances

	Tuning Method (Speed)	Tuning Range	Filter Bandwidth	Loss	Crosstalk	Polarization Dependence	Temperature Dependence	Power Dissipation
Coherent	Injection current (ns)	~100Å	Arbitary ◎	◎	Very low ◎	△	1 Å / C	~20 mA
AO filter	RF (μs)	1.3~1.5 μm	~10 Å	~4 dB	15~20 dB	×(→○)	7 Å / C	~100 mW
EO filter	Voltage (ns)	~100 Å	~10 Å	~8 dB	15~20 dB	○	7 Å / C	±100 V
Tunable DFB filter	Injection current (ns)	~30 Å	~1 Å	20 dB Gain	20 dB	×	1 Å / C	~50 mA
Mach-zehnder	Thermal or PZT (ms)	Periodic △	Arbitary ○	~2 dB	20 dB	△	0.1 Å / C	~0.6 W
Fabry-Perot	PZT (ms)	Periodic ○	Artitary ◎	~1.5 dB	◎	◎	0.1 Å / C	~10 V
Brillouin amp.	Pump wave-length (ns)	~100 Å	~100 MHz	~10 dB Gain	10 dB	○	1 Å / C	~10 mW

of the most important components. Table 1 shows characteristics of the tunable filters [5]-[10]. The tunable range and resolution bandwidth will limit the available channel number. There would be two kinds of multichannel systems. One is the WDM with coarse (low density) frequency setting. The system with AO filters is a typical example for this case. The other is a dense packing system, with a typical example of coherent detection.

For subscriber number increase, utilization of an optical amplifier is extremely useful. However, the bandwidth of the optical amplifier is somewhat limited, compared with the optical fiber large bandwidth. Therefore, dense packing of signals is necessary to achieve large channel number system with an optical amplifier. Coherent detection is most attractive to achieve such a dense FDM distribution system, because of its high resolution filtering characteristics. Furthermore, coherent detection has high sensitivity and low crosstalk characteristics, both are indispensable for achieving high performance FDM systems.

Using this coherent technology, the authors have achieved a 10 channel optical broadcasting system [11] with a semiconductor optical amplifier, with an estimated capability of handling more than 100 channel distribution to more than 10,000 subscribers. This paper presents details regarding the system experiment and discusses the capability of such a densely packed FDM system.

2. A 10 CHANNEL COHERENT CATV SYSTEM EXPERIMENT

This section explains the basic configuration for the coherent FDM distribution system with optical amplifier. Figure 2 shows a blockdiagram of 10 channel coherent CATV system developed in the laboratory. At the transmitter end (central office), 10 optical sources are stabilized with 8 GHz channel spacings, so

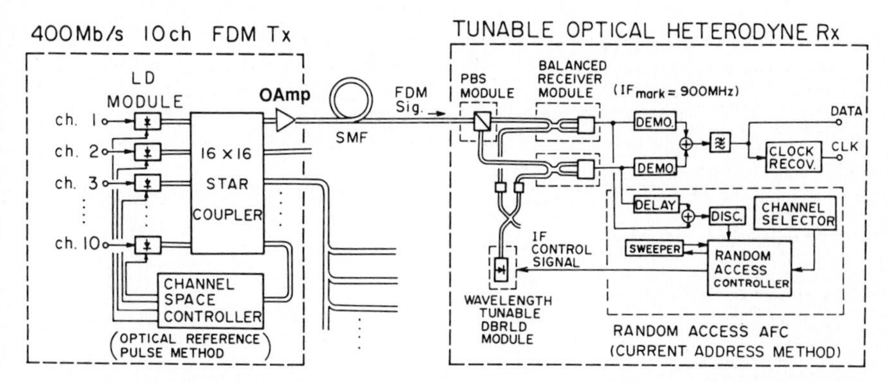

Figure 2. Blockdiagram of 10 channel coherent CATV experimental system

as to suppress crosstalk. Channel separation locking is accomplished by a "reference pulse method" [12]. Each light source (wavelength tunable DBR LD [13] in this case) is modulated in FSK format, either by 400 Mb/s NRZ PN pattern or by 100 Mb/s biphase code for a TV signal. All the 10 signal outputs are multiplexed with a 16*16 optical star coupler. Individual combined signals from the star coupler are distributed to subscribers. Optical amplifiers are attractive to obtain sufficient distributing margin. A semiconductor optical amplifier is implemented in the system.

Each subscriber has a tunable heterodyne receiver, which includes a balanced polarization diversity receiver [14] and a random access channel selection circuit [15]. A wavelength tunable DBR LD is used for the local oscillator for easy channel selection, because of its wide frequency tunability. A single filter detection [16] is used, since it has simple demodulation configuration and it tolerates a relatively large spectral spread and modulation depth instability. The polarization diversity receiver, employing an "active square-law combiner" [14], realizes an ideal square-law demodulation. Figure 3 shows the receiver configuration. The active square-law combiner improved the receiver sensitivity by 1 dB, compared with a conventional square-law receiver.

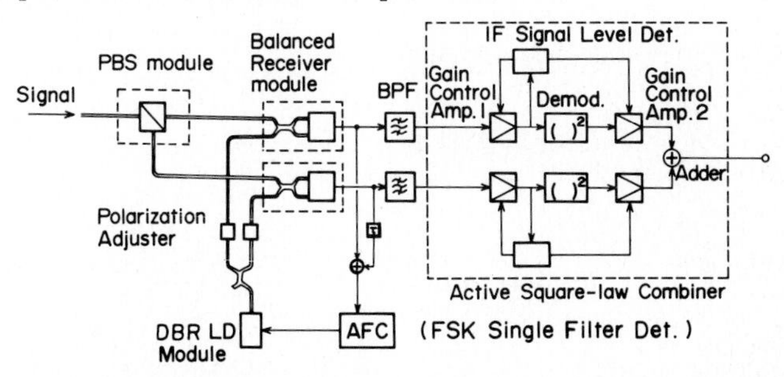

Figure 3. Balanced polarization diversity receiver configuration with active square-law combiners

The random access channel selection circuit realizes instantaneous channel selection. Figure 4 shows an operation procedure for the random access channel selection, based on a "current address method" [15]. The micro processor memorizes the local oscillator current value for individual channel reception at the initializing stage (when turn on the switch every morning or so). When a channel request signal is arrived, the controller directly accesses the memolized value. This sets the local oscillator frequency within an AFC capture range (114 GHz for the receiver used) for the desired channel. With this procedure, arbitrary channel selection has been achieved within 1 ms. This performance would be in the same level as that for conventional TV systems.

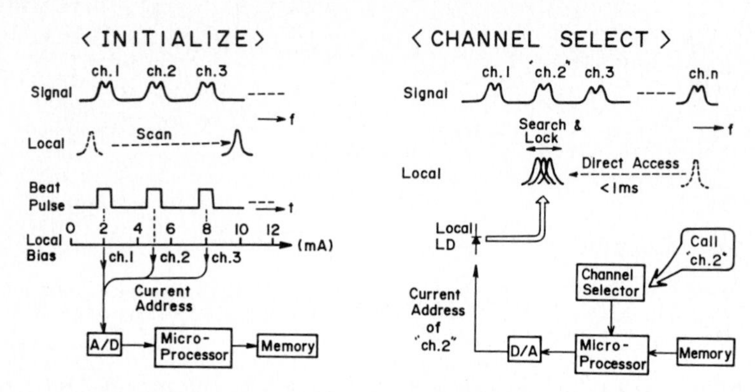

Figure 4. Random access channel selection procedure

To operate the system stably, the whole system was packaged into racks as a demonstration equipment, as shown in Fig. 5. The system worked fairly stably, with receiver sensitivities for individual channels of −45 dBm at 400 Mb/s. The sensitivity dependency on the polarization fluctuation was suppressed to less than 0.8 dB by the polarization diversity receiver. No crosstalk penalty was observed in 10 channel transmission with 8 GHz channel spacing. Table 2 shows the system loss budget. The total loss margin, including distribution loss, was 46.7 dB, which indicates more than 2000 subscriber accommodating capability for the system. Concerning the channel number capacity, the heterodyne receiver tunability would give the limit. Considering the reported 550 GHz continuous tunability of the wavelength tunable DBR LD, a 70 channel system would be possible with 8 GHz channel spacings.

Table 2.
Loss budgets for the coherent CATV experimental system

SIGNAL LIGHT POWER (MIN.)	+ 2.2 dBm
RECEIVER SENSITIVITY	- 44.5 dBm
TOTAL LOSS MARGIN	46.7 dB
FIBER LOSS (0.22dB/km x 10km)	2.2 dB
CONNECTOR LOSS	2.0 dB
DISTRIBUTION LOSS (3.5 dB/COUPLER 2048 SUBSCRIBERS)	38.5 dB
MARGIN	4.0 dB

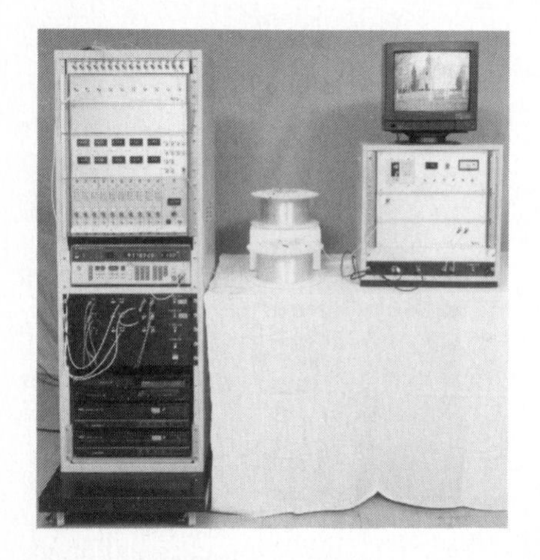

Figure 5. Coherent CATV experimental equipment

3. OPTICAL AMPLIFIER APPLICATION FOR THE COHERENT FDM DISTRIBUTION SYSTEM (SUBSCRIBER NUMBER INCREASE)

To increase the subscriber number in the coherent FDM distribution system, an optical amplifier application would be most attractive. To apply optical amplifiers most effectively, basic properties of the optical amplifier must be carefully investigated. The authors applied a semiconductor optical amplifier to the 10 channel coherent CATV experimental system and examined the basic characteristics. Figure 6 shows the structure and Table 3 shows characteristics for the semiconductor traveling wave amplifier (TWA), used in the experiment [17]. The TWA has DC-PBH structure with window regions for both facets. The window structure suppressed the spectral gain ripple as low as 0.6 dB. The TE/TM polarization dependency was also fairly low, only 1.3 dB at maximum. Both low ripple characteristics are suitable for multichannel amplification systems. The TWA was packaged into a module, as shown in Fig. 7, with 8 dB fiber-to-fiber gain.

Table 3. TWA characteristics

GAIN PEAK WAVELENGTH	1.56 μm
MEASURED WAVELENGTH	1.542 μm
FIBER TO FIBER GAIN	8.2 dB
INTERNAL GAIN	18.2 dB
SATURATION OUT-PUT POWER	+1.4 dBm
GAIN RIPPLE	0.6 dB
GAIN DEPENDENCY ON INPUT POLARIZATION STATE	1.3 dB

p-InP
n-InP
p-InP
WINDOW REGION 35 [μm]
AR FILM
AR FILM
n-InP
InGaAsP
WINDOW REGION 55 [μm]
ACTIVE REGION 600 [μm]

Figure 6. DC-PBH traveling wave semiconductor amplifier with window regions

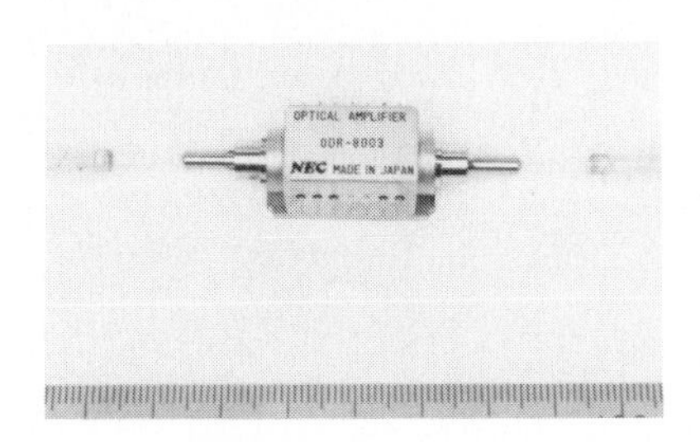

Figure 7. TWA module

The basic TWA noise characteristics were examined through one channel amplification experiment. Figure 8 shows bit error rate characteristics measured with a 400 Mb/s FSK heterodyne single filter detection scheme. The bit error rate characteristics strongly depend on the TWA fiber input power level, Pin, due to the existence of the TWA spontaneous emission noise. Figure 8 indicates that optical fiber input power to the TWA should be more than −26 dBm (more than −31 dBm for TWA chip input) to avoid an error rate floor.

A 10 channel simultaneous amplification experiment has been implemented by applying the TWA to the 10 channel coherent CATV experimental system. Figure 9 shows bit error rate characteristics for 400 Mb/s FSK single filter detection with 1.2 GHz FSK frequency deviation and 8 GHz channel spacings. In this case, the TWA was operated under saturation conditions. No power penalty was observed for each channel. More than −9 dBm total TWA input power only decrease the gain and does not affect the receiver sensitivity. The reason is that

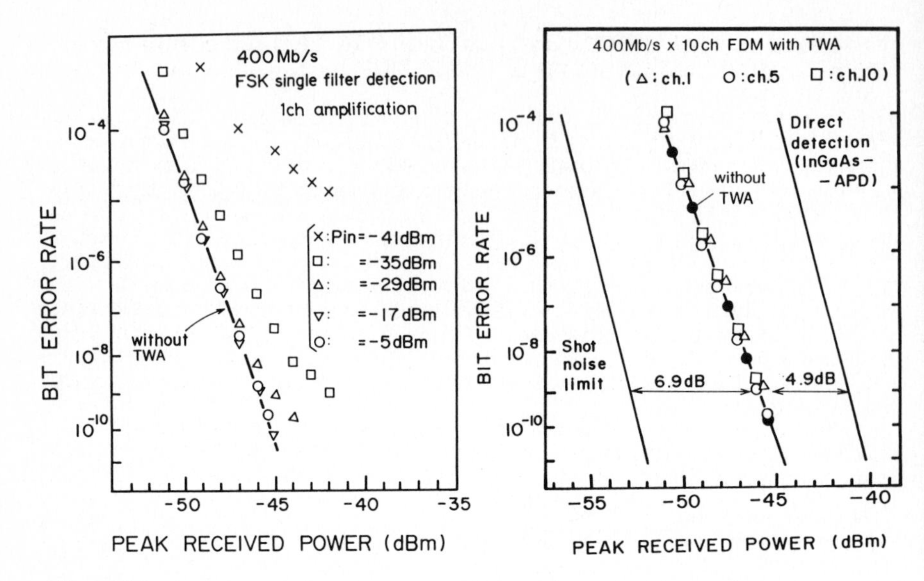

Figure 8.
Bit error rate characteristics for different TWA input level

Figure 9.
Bit error rate characteristics for 10 channel simultaneous amplification experiment

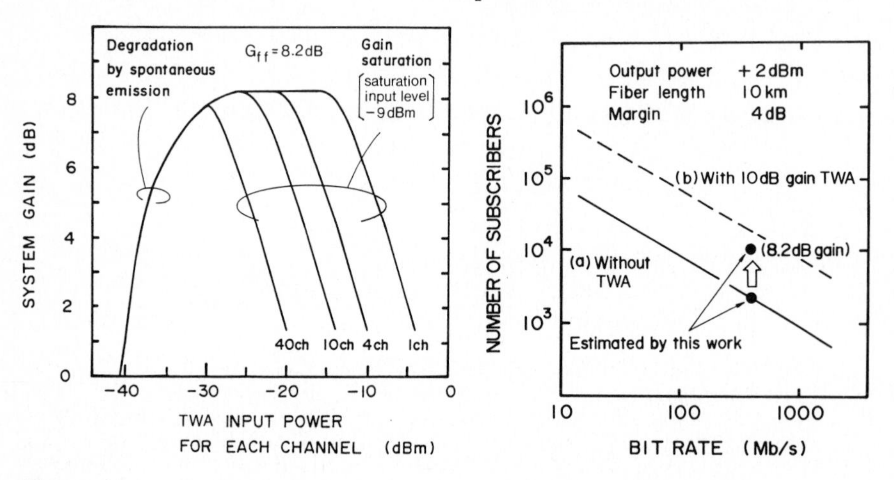

Figure 10.
System gain achieved by TWA

Figure 11. Subscriber number increase by 8.2 dB gain TWA

the FSK modulated signals with almost constant power level do not cause a signal power dependent gain fluctuation, even under TWA gain saturation conditions. Of course, no degradation was observed for individual channels under no saturation conditions, as far as the TWA input power for each channel exceeds −26 dBm,. The −9 dBm, limited by the TWA saturation and −26 dBm, determined by the TWA spontaneous emission noise would limit the dynamic range for the particular TWA used, as shown in Fig. 10. To enlarge the system, especially to increase the channel number, the TWA performance improvement is indispensable.

The 8.2 dB gain achieved with this 10 channel simultaneous amplification could increase the available subscriber number from 2000 to 10000, as shown in Fig. 11.

4. CHANNEL NUMBER INCREASE

With the 8 GHz channel spacings, no crosstalk penalty was observed. Therefore, it might be a good idea to design a more densely packed (narrower spacings) system, for channel number increase. To examine the feasibility of the dense FDM system, crosstalk characteristics were measured for a 2 channel system with TWA by changing the channel spacing, as shown in Fig. 12. These crosstalk characteristics greatly depend on the system configuration (400 MHz to 1.4 GHz single filter detection IF band for 1.2 GHz frequency deviation FSK signal) and the TWA input signal level. With less than 5 GHz channel spacing, a large power penalty was observed due to the nearly degenerate four wave mixing (NDFWM) induced interference [18][19].

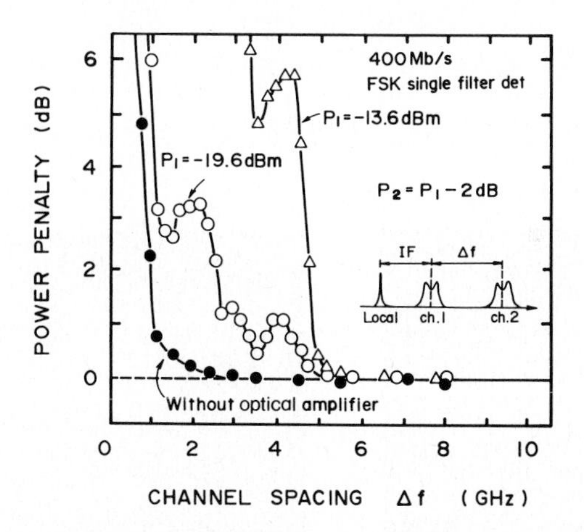

Figure 12. NDFWM induced crosstalk characteristics for 2 ch system
P_i: TWA fiber input power level

Since the NDFWM is generated by carrier density modulation in the optical amplifier, originated from interchannel beats of the incident signals to the TWA, a cross polarized channel arrangement (CPCA) [20][21] is effective. Figure 13 shows the channel polarization setting for the CPCA. The CPCA can easily be realized by optically multiplexing the signal using a polarization beam splitter. The CPCA suppresses the carrier density modulation, because the orthogonal polarization setting reduces the neighboring-channel beats. Figure 14 shows the CW operation spectrum for amplified FDM signals obtained by the polarization diversity optical heterodyne reception. The TWA input levels for the two signals were -14 dBm and -17 dBm and channel separation was 400 MHz. The optical amplification without the CPCA generated strong crosstalk lights, induced by the NDFWM, as shown in Fig. 14 (a). In this case, the signal to crosstalk light ratio was only 6 dB. On the other hand, in the case with the CPCA, the crosstalk lights were sufficiently suppressed to less than −27 dB below the signal level. Figure 15 shows NDFWM induced light level for several channel spacings for the 2 channel CW signal input condition. In the matched signal polarization case, even signal level itself reduced 7 dB at maximum, due to the NDFWM. On the contrary, in the CPCA case, the NDFWM induced light is suppressed more than 30 dB below the signal level, for any input polarization states. (As shown in Fig. 15, about 10 dB input polarization direction dependency was observed. This may be due to the gain difference for the TE and TM modes.) In Fig. 16, crosstalk characteristics are indicated for the CPCA case in the 2 channel 400 Mb/s FSK single filter detection experiment. With CPCA, hardly any power penalty due to the NDFWM was observed. The eye patterns for 3 GHz channel spacing cases are shown in Fig. 17. A clear eye opening was obtained with CPCA, in contrast to the closed eye obtained for the without CPCA case.

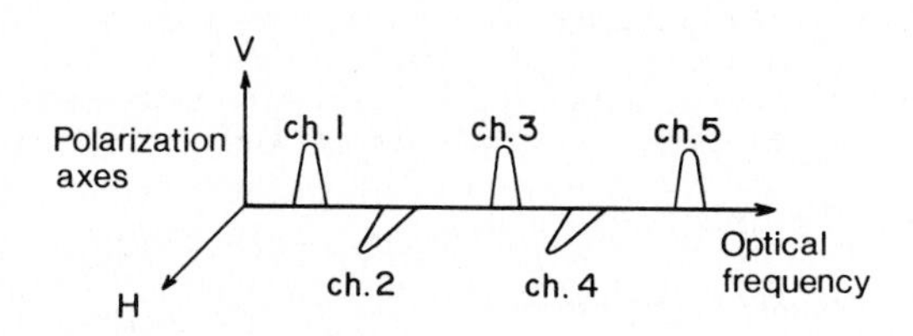

Figure 13. Cross polarized channel arrangement

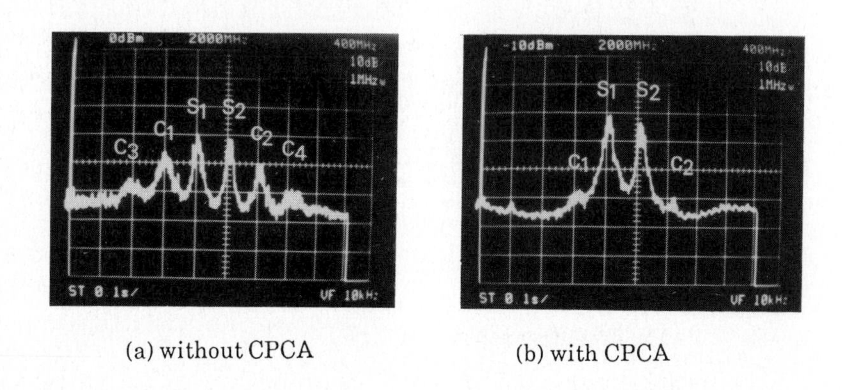

(a) without CPCA (b) with CPCA

Figure 14. TWA output spectrum measured by polarization diversity reception
(TWA input level S_1: −14dBm, S_2: −17dBm
Δf=400MHz H: 400MHz/div. V: 10dB/div.)

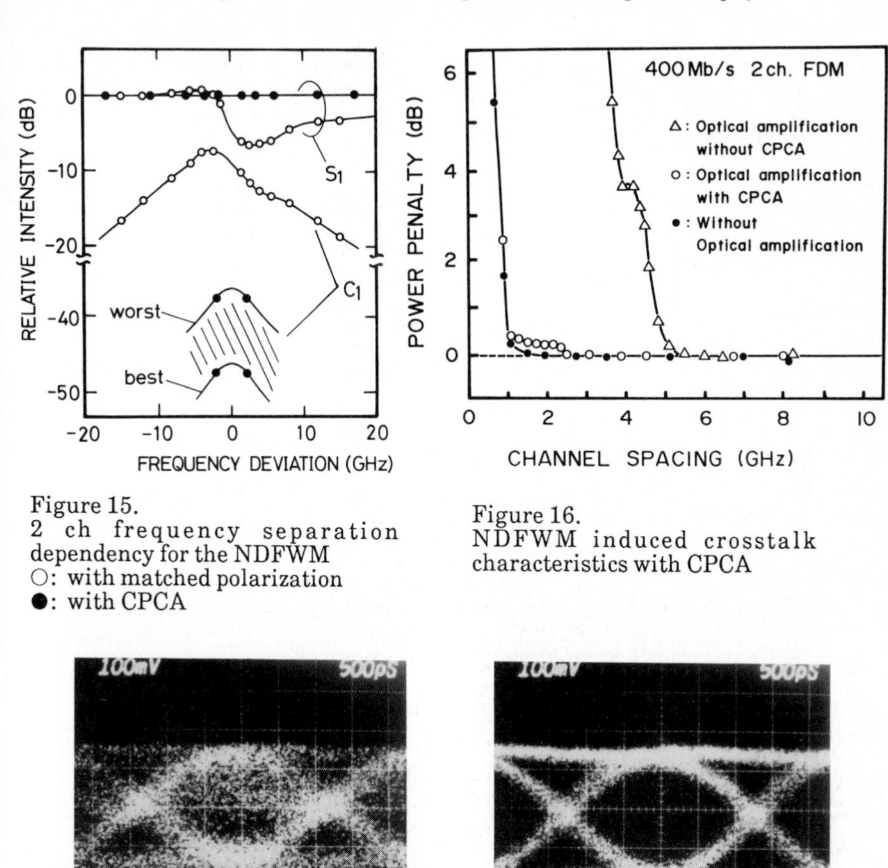

Figure 15.
2 ch frequency separation dependency for the NDFWM
○: with matched polarization
●: with CPCA

Figure 16.
NDFWM induced crosstalk characteristics with CPCA

(a) without CPCA

(b) with CPCA

Figure 17. Optically amplified signal demodulated eye patterns with 3 GHz channel spacing (500ps/div.)

These results indicate that the NDFWM between neighboring channels can be negligible with the CPCA. It is only necessary to consider NDFWM between every other channels, which have the same polarization states. This means that the CPCA expands the available channel number by closely setting the channel spacings. In the multichannel systems, the NDFWM effect accumulation should be considered, because the channel spacings are equally adjusted. Figure 18 shows the NDFWM level for 4 channel CW signal with CPCA compared with 2 channels case. The results indicate that the twice the channel number would double the NDFWM level. This should be considered for many channel systems.

The TWA saturation level is another factor which limits the channel number for the simultaneous amplification. Figure 19 shows available channel number for simultaneous amplification versus TWA saturation level characteristics.

The 200 channel system is feasible with a more than 10 dBm TWA saturation level and a 10 dB gain [22], which has already been reported. The high saturation level characteristics are one of the most important characteristics for optical amplifiers in FDM systems.

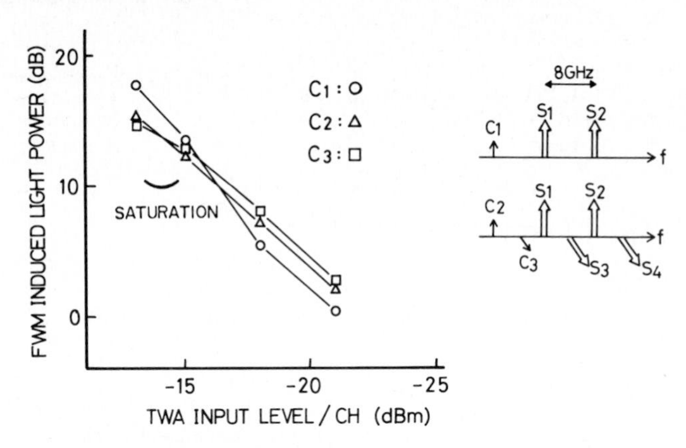

Figure 18. NDFWM induced light power for 4 ch amplification with CPCA

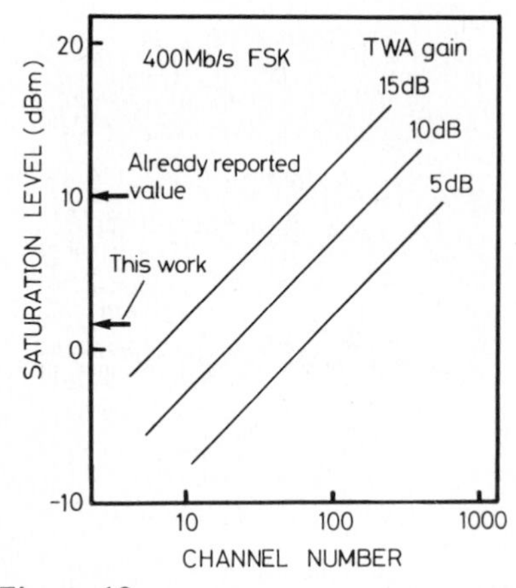

Figure 19.
Available channel number dependency on optical amplifier saturation level

Table 4.
Desired optical amplifier characteristics for large scale coherent FDM system

Gain (fiber - to - fiber)	> 15 dB
Saturation Output Level	> 10 dBm
Noise Figure	< 6 dB
Spectral Gain Ripple	< 1 dB
Polarization Dependency	< 1 dB

5. FUTURE TARGET FOR THE COHERENT FDM DISTRIBUTION SYSTEM

The previous estimations indicate that a high saturation level TWA and a wideband frequency tunable local LD are effective for realizing a large scale coherent FDM distribution system. Desired performance for the TWA for a large scale system is indicated in Table 4, which would realize a 10 dB dynamic range, even for the 200 channel system. The location of the TWAs should be carefully considered to operate them within their dynamic range. Cascading the TWAs in appropriate positions will realize more than 100,000 subscriber systems.

The channel number capacity is limited by the channel spacings and local oscillator tuning range, as shown in Fig. 20. The channel spacing is only limited by the crosstalk in the receiver, due to the fold over for the next channel, if the cross polarized channel arrangement (CPCA) is applied. Using high efficiency coding for (High Definition)TV signals would realize narrower channel spacings, that may increase the channel number to more than 100 with the 500 GHz local LD tuning range. To further increase the channel number, a multi local receiver, shown in Fig. 21, is effective. This receiver employs optical switch and several local oscillator lasers, whose frequency tuning ranges are shifted and overlapped with each other. This extends the local oscillator tuning range and accommodates several hundred HDTV channels.

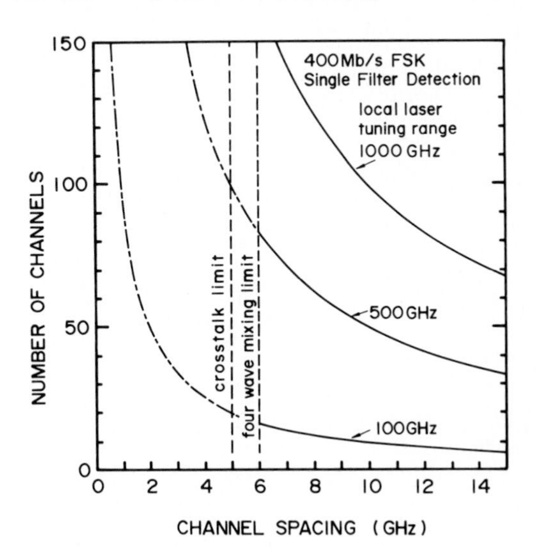

Figure 20. Available channel number dependency on local oscillator laser tuning range

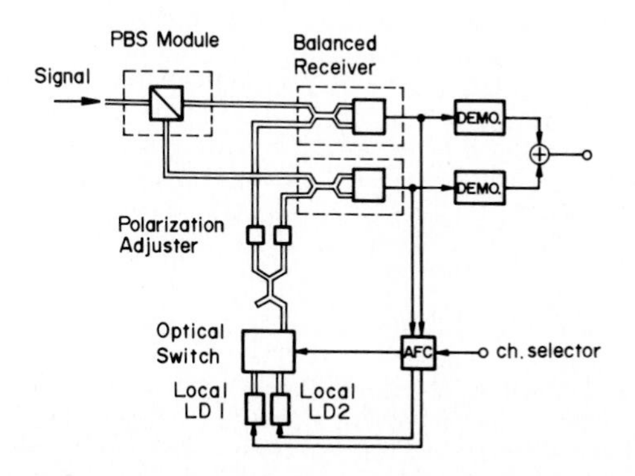

Figure 21. Blockdiagram of multi local optical heterodyne receiver

Considering the optical amplifier applicability with dense FDM packing, it is clear that the coherent FDM system is the only system that could achieve more than 100 channel, more than 10000 subscriber system, using technology available today, as shown in Fig. 22.

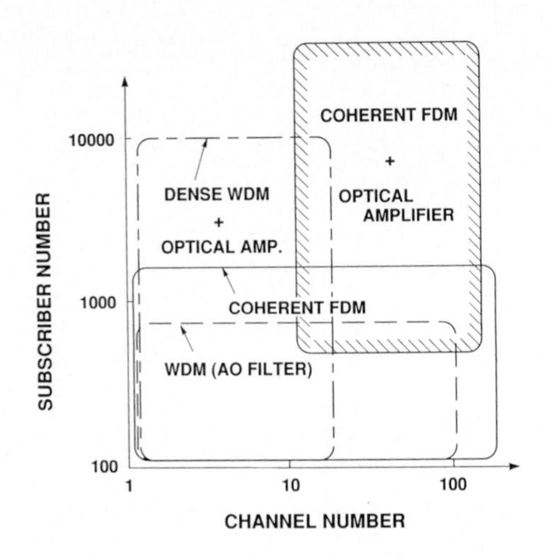

Figure 22.
Achievable system scale for coherent FDM distribution service networks with optical amplifiers

6. CONCLUSION

A 10 channel coherent CATV experimental system has been developed to confirm the coherent FDM system capability. A DC-PBH TWA was applied to this system to enlarge the system scale. With 8 GHz channel spacings, 10 channel simultaneous amplification with 8.2 dB gain has successfully been achieved without any receiver sensitivity degradation. This increases the system capacity from 2,000 subscribers to 10,000 subscribers. For channel number increase, still denser packing, with a cross polarized channel arrangement that suppresses the NDFWM effect, and multi local receiver for wider local oscillator tuning range, are effective. Using these technologies, the coherent FDM system could accommodate more than 100 channels and more than 10,000 subscribers, which is desirable for an actual CATV use. Since other technologies hardly achieve such a large scale systems, CATV distribution is one of the most attractive application regions for coherent FDM technology.

ACKNOWLEDGMENT

The authors would like to thank M.Sakaguchi, J.Namiki, T.Suzuki, and K.Minemura for their encouragement throughout this work. They also thank their colleagues for supplying optical devices and useful suggestions.

REFERENCES

[1] Wagner, R.E. et al., J. Lightwave Technol., LT-5 (1987) pp.429-438
[2] Stanley, I.W. et al., J. Lightwave Technol., LT-5 (1987) pp.439-451
[3] Nosu, K. et al., J. Lightwave Technol., LT-5 (1987) pp.1301-1308
[4] Kobrinski, H. et al., Electron. Lett., 23 (1987) pp.824-826
[5] Cheung, K.W. et al., Electron. Lett., 25 (1989) pp.375-376
[6] Warzanskyj, W. et al., Technical Digest on OFC'88 (1988) PD11
[7] Numai, T. et al., Electron. Lett., 24 (1988) pp.1526-1527
[8] Oda, K. et al., Technical Digest on IOOC'89 (1989) 20D2-4
[9] Kaminow, I.P. et al., J. Lightwave Technol., LT-6 (1988) pp.1406-1414
[10] Tkach, R.W. et al., Technical Digest on OFC'88 (1988) PD21
[11] Shibutani, M. et al., Technical Digest on OFC'89 (1989) ThC-2
[12] Shimosaka, N. et al., Technical Digest on OFC'88 (1988) ThG-3
[13] Murata, S., et al., Electron. Lett., 24 (1988) pp.577-579
[14] Shibutani, M. and Yamazaki, S. IEEE Photonics Technol. Lett., 1 (1989) pp.182-183
[15] Yamazaki, S. et al., Electron. Lett., 25 (1989) pp.507-508
[16] Emura, K. et al., Electron Lett., 20 (1984) pp.1022-1023
[17] Cha, I. et al., Technical Digest on IOOC'89 (1989) 20C2-2
[18] Agrawal, G.P., Electron. Lett., 23 (1987) pp.1175-1176
[19] Saito, T. and Mukai, T., J. Lightwave Technol., 6 (1988) pp.1656-1664
[20] Grosskopf, G. et al., Electron. Lett., 24 (1988) pp.31-32
[21] Shibutani, M. et al., Technical Digest on IOOC'89 (1989) 21B4-5
[22] Eisenstain, G. et al., Technical Digest on IOOC'89 (1989) 20PDB-13

COHERENT MULTICHANNEL EXPERIMENTS

Bernhard N.STREBEL

Heinrich-Hertz-Institut für Nachrichtentechnik Berlin GmbH
Einsteinufer 37
D-1000 Berlin 10, FRG

CMC experiments demonstrate a parallel and independent transmission of signals by optical carriers, the carrier frequency conversion in switching nodes and a first all optical signalling.

1. INTRODUCTION

The optical coherent multi carrier (CMC) technique uses the optical network for a parallel and independent transmission of broadband informations. Optical waves with narrowly spaced carrier frequencies and small linewidth are individually modulated, bundled in passive stars and transmitted to optical heterodyne receivers with variable access to the single carriers [1]. Typical applications are long distance multichannel connections, interoffice links or multichannel information distribution e.g. of TV or HDTV signals.
A second stage of development is the connection of several stations over a common passive optical star [2]. In this case the specific carriers of all transmitters are collected in a passive star and combined to a common optical frequency multiplex. This multiplex is transmitted to tunable heterodyne receivers of all stations resulting in bidirectional connections over the node.
The third stage of progress is a coherent optical multicarrier switching node [3] including carrier frequency conversion. This switching capability opens up a service integrated transparent coherent multichannel network.
At present the signalling information is expected to be separated from the user information signals. The signalling may be processed electronically and the routing may be performed by optoelectronic interfaces.
Just now first new experiments are started for a full optical network including optical computing of signalling informations and optical carrier routing. This would lead in the last consequence to the selfrouting information in a coherent optical network.
Most key experiments with coherent multicarrier networks are performed in the laboratory environment. This paper gives an overview on related experimental activities in the Heinrich-Hertz-Institut.

2. CUSTOMER ACCESS CONNECTION

A typical application of the CMC technique is the customer access connection (Fig. 1). Different services are integrated on one single mode fiber connection. The picture includes ISDN or B-ISDN and distribution services. Broadcasting programs like public TV, HDTV or audio services may be immediately modulated on laser waves within the local centre. The laser waves are guided to a combination and distribution star. Each fiber output of this star contains the multiplexed information channels. Each customer is connected to one of these outputs. Also a remote switching access to Pay TV and information banks is provided by a switching network.

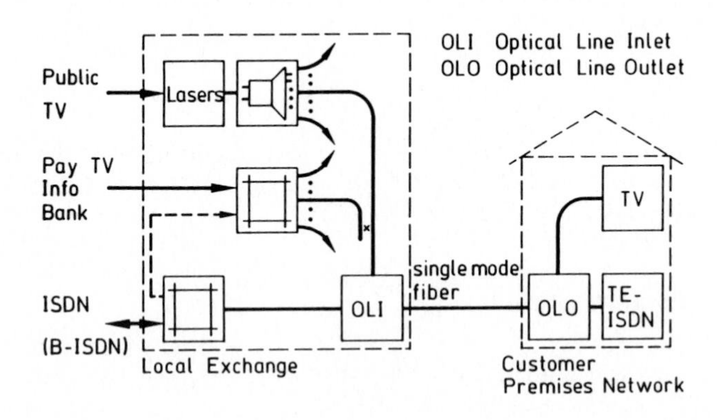

FIGURE 1
Customer access connection

The CMC method allocates one carrier to each broadband information e.g. to each TV channel. All carriers are combined in the optical line inlet and transmitted to the customer premises network (CPN). The CPN includes the terminals with coherent transmitters and tunable heterodyne receivers.

3. PRIMARY CMC FUNCTIONS

In order to allow more complex network experiments some basic CMC functions were experimentally verified.

3.1 Carrier frequency stabilization

For the frequency stabilization and monitoring of a carrier bundle a heterodyne spectrometer method (Fig. 2) may be used. It compares the bundle with the position of a reference frequency by superposition with the wave of a swept laser. The control unit analysed the positions of all output pulses and keeps the individual frequencies within a predefined window. This method is also used to control carriers in different waveguides.

3.2 Multichannel bidirectional carrier flow

The bidirectional operation of the multichannel network postulates a wave propagation without energy transfer between unidirectional or opposite travelling carriers. Measurements of four wave mixing in monomode fibers and optical amplifiers resulted in a proposal for a carrier frequency spacing of 5 GHz for unidirectional carriers. Opposite travelling carriers should have a spacing of more than 25 GHz due to Brillouin scattering.

3.3 Tunable channel access

The variable access to the single carrier of a bundle is performed by an optical heterodyne receiver or tunable frequency selective power splitters. The circuit of a heterodyne receiver (Fig. 3) with a computer controlled channel selection and two IF branches for different bit rates and modulation schemes was presented 1987.

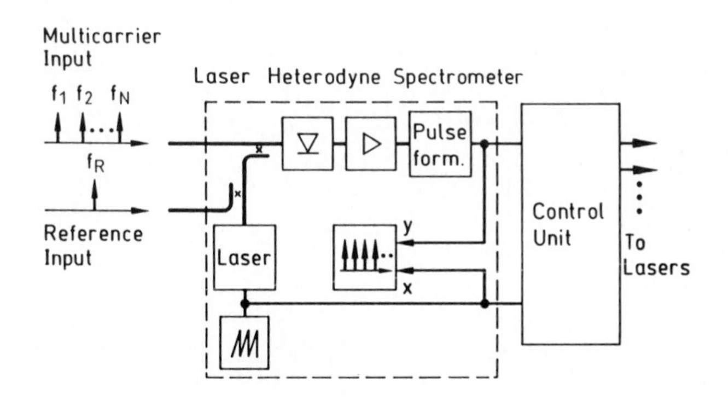

FIGURE 2
Carrier frequency stabilization

Tunable channel branching filters were manufactured as fiber microring or Mach-Zehnder filters.

3.4 Optic-optic frequency conversion

Signals can be converted from one optical carrier to another by the nonlinearity of an optical amplifier [4], if it is driven by two pump laser frequencies f_{p1} and f_{p2} (Fig. 4).

The pump frequency f_{p1} is allocated near the input frequency f_μ with two four wave mixing products. The signal is transferred to both sides of the pump frequency f_{p2}. One sideband is filtered out as the new modulated carrier frequency f_ν.

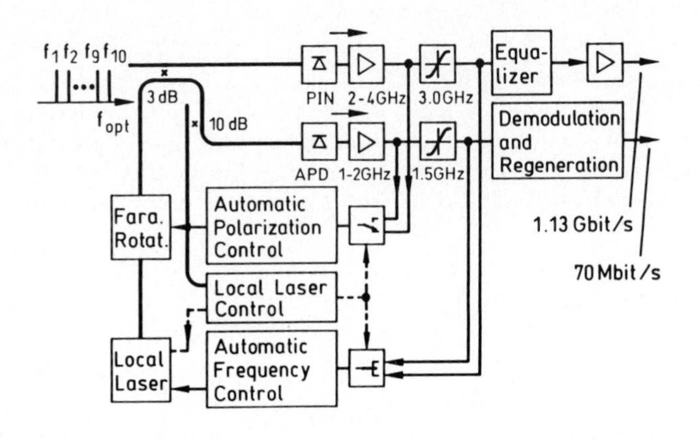

FIGURE 3
Channel access by heterodyning

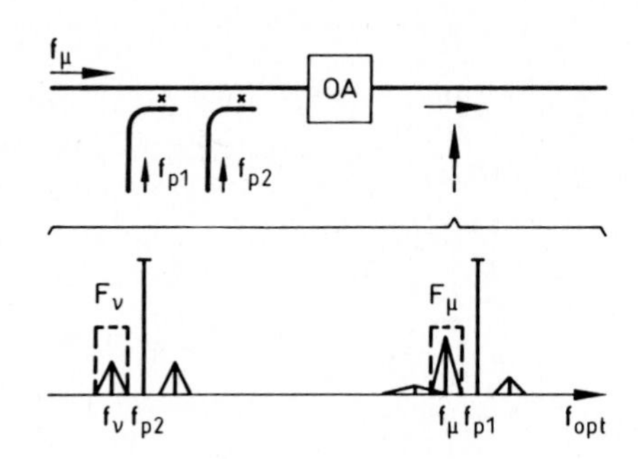

FIGURE 4
Frequency conversion in an optical amplifier

3.5 Phase noise cancelling transmission

In a coherent type optical fiber transmission with signals on fixed optical carriers the phase noise of the received signal results only form the transmitter and the local laser. The transmission over phase noise generating paths however deminishes the carrier coherence drastically. This problem can be overcome by a phase noise cancelling transmission method [5]. A novel transmission experiment uses two oppositely FSK-modulated phase coherent carriers, which were generated by optical filtering of the spectrum of a FSK/FM modulated laser (Fig. 5). After transmission the signal is detected in a nonlinear heterodyne receiver [6].

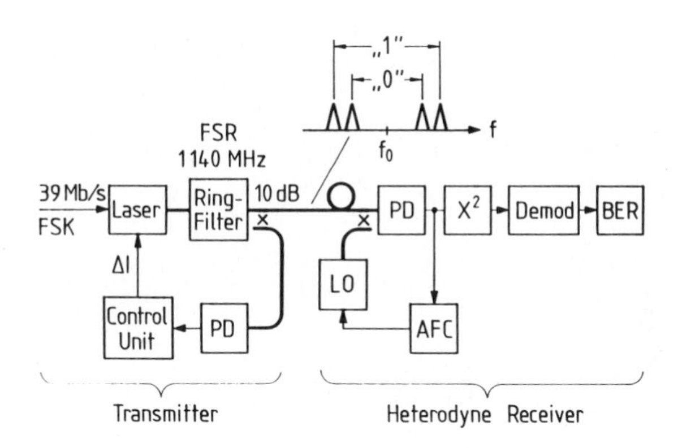

FIGURE 5
Phase noise cancelling FSK transmissing

4. SYSTEM EXPERIMENTS

A future integrated transparent coherent type optical network has to provide information distribution channels as well as optically switched bidirectional connections. To demonstrate the feasibility of a coherent network, a series of laboratory experiments simulating special services was performed.

4.1 TV/HDTV Distribution

One experiment demonstrated an optical coherent ten-channel distribution system with nine channels for 70 Mbit/s and one channel for 1.13 Gbit/s (Fig. 6). This corresponds to nine TV and one uncompressed HDTV channel. The setup comprised a carrier frequency stabilization unit (Fig. 2) and a computer controlled heterodyne receiver.

4.2 Switching via optical multiplex

Four stations were bidirectionally connected in an experimental setup (Fig. 7). The transmitters with fixed frequencies were combined in a passive combination star. The output of this star contains the complete frequency multiplex. The switching is performed by tuning the heterodyne receivers. A double channel heterodyne spectrometer was used to monitor transmitter and local lasers. Thus a computerized switching over the local laser currents was possible.

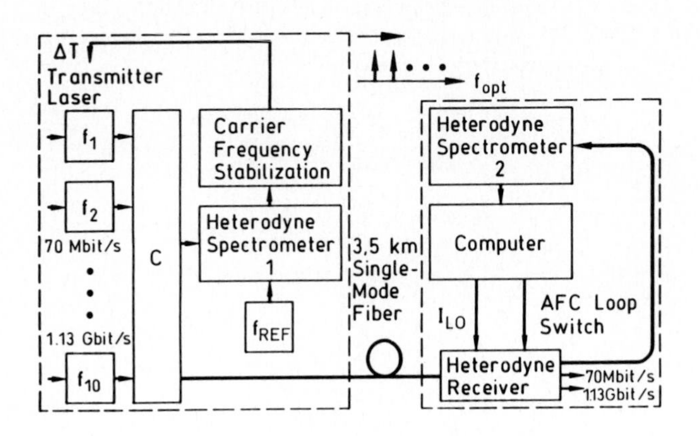

FIGURE 6
CMC distribution experiment

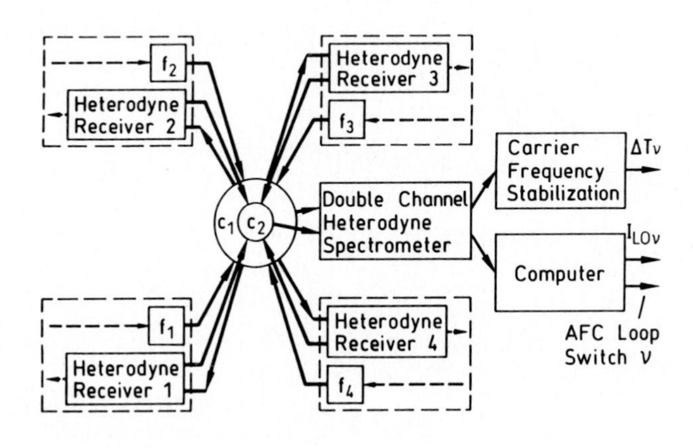

FIGURE 7
Carrier switching experiment

4.3 CMC switching node

The transmission part of a CMC switching node (Fig. 8)with input and output waveguides is one of the most important parts of the network. The input signal $S_{\mu q}$

corresponds to the carrier frequency f_μ and the waveguide q. The node is characterized by routing the input signals to an arbitrary carrier frequency position in an arbitrary output waveguide.
Features of this CMC node are:

- Use of high capacity of fibers by bundles of carriers,
- Switching of carriers with respect to frequency and space,
- Full transparency in true sense

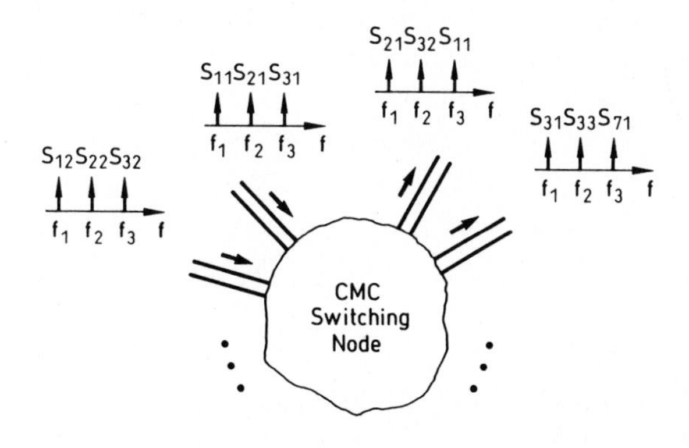

FIGURE 8
CMC switching node

First experiments were performed with semiconductor laser amplifiers and fiber components (Fig. 9). Two carriers with 5 GHz spacing were modulated with 34 Mbit/s NRZ coded signals in the PSK/FM scheme. A Mach-Zehnder interferometer with 10 GHz FSR served as a branching filter f_μ. Signals were converted from f_μ to f_ν in a DFB amplifier driven by two pump lasers f_{p1} and f_{p2} (Fig. 4.). Output filter F_ν consisted of two Mach-Zehnders in series with 25 and 5 GHz FSR respectively. Filters could be thermally tuned and all lasers were stabilized by a heterodyne spectrometer circuit.

4.4 Optically switched CMC node

A goal of these experiments is the demonstration of all optical signalling in a CMC network. A first example is the all optical program selector (Fig. 10) for remote switching [8]. Bistable DFB-lasers are switched by f_{on} and f_{off} pulses as optical gates for the pump frequencies of an optic-optic frequency converter.

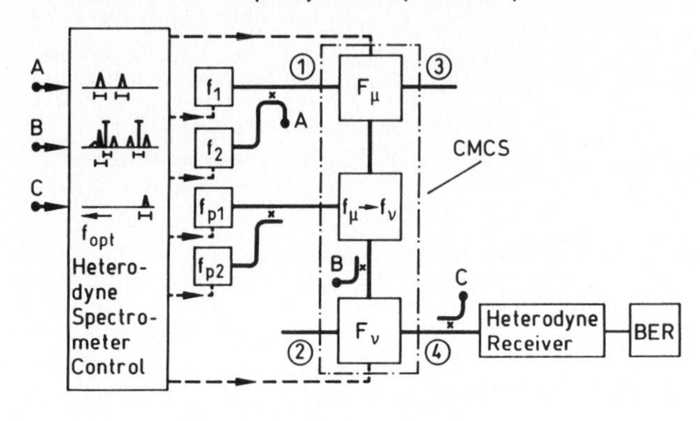

FIGURE 9
Carrier switching experiment

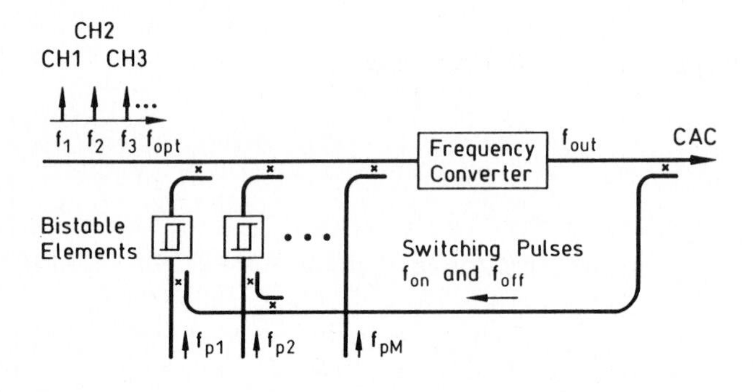

FIGURE 10
All optical program selector for remote switching

5. CONCLUSION

The research towards CMC networks is characterized mainly by laboratory experiments. After first setups for one channel coherent transmission the multichannel bidirectional function was demonstrated. The second level of experiments is service oriented for distribution and switching application. Now basic experiments appear for switching with optical signalling. The final goal may be the selfrouting of optical carriers.

ACKNOWLEDGEMENTS

The author would like to thank the Deutsche Bundespost for the support of the work.

REFERENCES

[1] B.Strebel, G.Heydt,Globecom'87, Tokyo, Conference Record Vol. 1, pp. 684-688

[2] C.Caspar, E.Großmann, B.Strebel, ECOC'87, Technical Digest, pp. 317-320

[3] E.-J.Bachus, R.P.Braun, C.Caspar, H.-M.Foisel, K.Heimes, N.Keil, B.Strebel, J.Vathke, M.Weickhmann, OFC'89, Houston, Postdeadline Papers, PD 13

[4] G.Großkopf, R.Ludwig, H.G.Weber, Electron.Lett., Vol. 24, 1988, pp. 1106-1107

[5] Y.H.Cheng, E.A.J.Marcatili, T.Okoshi, OFC'89, Houston, Techn. Digest, p. 41

[6] E.J.Bachus, R.-P.Braun, C.Caspar, H.-M.Foisel, B.Strebel, submitted as post-deadlinepaper to ECOC'89

[7] H.-M.Foisel, ECOC'87, Technical Digest, pp.287-290

[8] C.Baack, G.Heydt, ECOC'89, paper MoA1-1

DOES COHERENT OPTICAL COMMUNICATION HAVE A PLACE IN COMPUTER NETWORKS?

Paul E. Green and Rajiv Ramaswami

IBM T.J. Watson Research Center
Hawthorne, NY 10532

ABSTRACT: Coherent optical communication is becoming an accepted approach to designing single long-haul links, but computer networks require a different design philosophy with respect to number of separated stations, cost, ease of use, and other factors. Two basic architectural approaches to the realization of optical WDMA computer networks are compared qualitatively in this paper, the use of coherent reception and the combined use of incoherent reception and tunable Fabry-Perot receiver filters. It is concluded that the system gain and channel crowding advantages often attributed to the coherent approach are more easily realized by the incoherent-plus-filter approach combined with photonic amplification, (the latter often being required anyhow in coherent systems).

COMPUTER NETWORK REQUIREMENTS

This paper presents some speculations aimed at focussing on the possible role of coherent reception in future lightwave computer networks. It is generally accepted that for single point-to-point links, such as undersea or inter-city Telco trunking or inter-satellite space links, solutions of considerable complexity and cost are worth pursuing to increase inter-repeater distances. Coherent optics clearly has an important role for such systems. The picture for computer networks is not quite so clear.

Computer networks, i.e. interconnections of large and small processors to each other and to workstations and terminals, have requirements very different from those of long point-to-point links. For one thing, there are many nodes, not just two. In large local area networks and wide area computer networks, the number of nodes can be in the hundreds or even thousands. The size of nodes participating may range from large mainframes to desktop personal computers, all within the same network. At the high end of this range, the cost per port that can be economically supported may be several thousand dollars, but at the low end, manufacturing costs of several hundred dollars per port must be achievable. At the high end, good environmental conditions (temperature, vibration, etc.) are the rule, while the opposite is true at the low end. Ease of use and ease of network management and maintenance become more important toward the low end.

Computer networks require rapid re-addressing capability, so that those based on wavelength (frequency) tunability require at least sub-millisecond retuning speeds and will ultimately require sub-microsecond retuning in order to support fully dynamic packet switching. For circuit-switched local exchange or TV distribution networks, there is little demand for extremely fast switching times.

A wide range of received signal powers can be experienced in computer networks due to widely different transmitter-receiver path lengths. This requires that the receivers and other components have adequate dynamic range.

DESIGN POINT: THE WDMA METROPOLITAN AREA NETWORK

To fix ideas, let us pick an arbitrary design point and compare the advantages and disadvantages of coherent wavelength division networking relative to an alternative that uses only incoherent detection.

A metropolitan area network (MAN) of several tens of km. diameter is a reasonable choice, since this seems to be the distance limit over which it is likely that dark fiber will be available.

Perhaps the simplest model is one in which each station has a transmitter connected to the input arms of a passive optical star coupler located somewhere in the middle of the network, with a receiver connected to each of the star coupler outputs. In principle either the transmitters, the receivers, or both could be made tunable. However, because tunable receivers are a more immediately available technology, we arbitrarily assume fixed tuned transmitters and wavelength- (frequency-) tunable receivers. The question to be examined then is: should these receivers (and the corresponding modulation format) be of the coherent or incoherent type?

To offer a significant improvement over current all-electronic technology, we desire this optical network to support ultimately N = 1000 ports at per-port bit rates up to one gigabit per second. While some applications can be supported with retuning times as slow as milliseconds, the faster the tuning time the wider the range of applications that can be supported.

Of the many problems to be overcome in building such a MAN, two stand out, (1) the splitting loss of at least $10 \log_{10} N$ dB., and (2) the need to crowd these N channels into the available bandwidth without inter-channel crosstalk.

TWO ARGUMENTS IN FAVOR OF COHERENT RECEPTION

Coherent reception involves a tunable local oscillator (LO) that beats with the incoming signal, a photodetector, an IF (intermediate frequency) filter tuned to the beat frequency, followed by suitable electronic detector circuitry.

For highly dense multiple wavelength networks, it is often argued that coherent reception [1] provides an ideal solution to these two problems of link budget and channel crowding. Ideally some 15 dB. of loss can be recovered just by the better weak-signal detectability of coherent relative to incoherent [2]. As for tight channel spacing without crosstalk, coherent optics allows in principle a spacing as tight as several multiples of the modulation rate.

THE ALTERNATIVE: INCOHERENT DETECTION, FILTERS, AMPLIFIERS

In our studies of future computer networks, particularly MANs, we have looked for ways in which either coherent or incoherent reception could be made sufficiently simple, inexpensive, and easy to use that we could meet the expected future computer networking requirements listed earlier.

The incoherent approach we have chosen [3] uses on-off keying (OOK) for direct modulation of the laser and uses piezoelectrically tunable Fabry-Perot interferometers preceding the photodetectors at the receiver. The receiver decisions are made on the basis simply of the energy detected. As N increases to the point where the splitting loss is excessive, photonic amplification must be used. The choice between doped fiber amplifiers and travelling wave laser diode amplifiers was made, for the moment, in favor of the latter because of the smaller bandwidth and large optical pump power requirements of fiber amplifiers. However, the bandwidth performance of doped fiber amplifiers is improving rapidly [4]. Although travelling wave diode amplifiers require control of facet reflections (and external reflections) to very low values, recent angle-facet designs [5] may alleviate this requirement considerably.

As for channel spacing, the important quantity is the number of channels, not their spacing. With tunable Fabry-Perot filters of a given finesse, the number of channels that can be fitted into one free spectral range, while undergoing at most a 1 dB. link budget penalty is roughly 1.6 times the finesse [6, 7]. The free spectral range can be adjusted (by setting the cavity length) to be as large as the usable bandwidth. This usable bandwidth, as set by the attenuation minimum at 1.5 μ is about 100 nm.

If the limit is defined by the selection of today's commercially available DFB lasers, it is about half that figure, although as quantity needs arise, vendors are prepared to provide coverage of the entire 100 nm. by varying the semiconductor composition. If diode travelling wave amplifiers or doped fiber amplifiers are used, the bandwidth limit today is typically 50 and 35 nm., respectively.

For values of N that are small enough that amplification is not required (less than, say, 50), achieving the required value of finesse (80 or more) is not a significant problem; such filters are commercially available today [8] with finesse of several hundred. For higher values of finesse, tighter control on beam size and mirror coatings is required, and commercially available filters with finesse as high as 10,000 have been reported [9]. A particularly promising approach to economical high-finesse Fabry-Perot filters is to cascade two incommensurable cavities in a "Vernier" arrangement [3, 10, 11].

The spacing between channels in such a system is thus often much wider than with the coherent option. For example, with N = 100 and FSR = 50 nm., the spacing is 0.5 nm., or 60 GHz., much wider than any reasonable single-port modulation bandwidth. However, as N approaches 1000, the channel spacing of the incoherent approach will converge to that of the coherent option.

PROBLEMS IN APPLYING THE COHERENT APPROACH TO COMPUTER NETWORKS

Several unavoidable factors introduce considerable complexity in coherent systems.

Because of the high channel selectivity, both transmitter and LO lasers must be frequency-stabilized to the point that drift is small compared to the channel spacing. This involves using feedback loops of some sort [12, 13] to lock onto either the transmitter laser wavelength or to some common reference source that it shares with the transmitter. The stabilization scheme must also permit the laser to tune rapidly to a new wavelength, and then lock it at that wavelength.

Very narrow laser linewidths are needed to avoid serious degradation in detected signal-to-noise ratio [14], more so for lower modulation rates. This is exacerbated by the fact that even small reflections from connectors or other components reaching the laser cause its linewidth to increase, requiring a high degree of isolation, typically 50-60 dB. Alternatively, phase diversity techniques [15] that do not require knowledge of the phase of the incoming signal can be used at the receiver, at the cost of increased complexity and a few dB. of detectability loss.

In order to get the maximum receiver detectability, the local oscillator state of polarization (SOP) must match that of the incoming signal. In fact, if the two SOPs are orthogonal, the detected signal output will be zero. Unfortunately, the SOP of the received signal will generally be very different from that of the launched signal by a slowly varying random amount, because the fiber does not preserve the SOP. This can be alleviated either by polarization diversity techniques at the receiver [15], polarization scrambling either at the receiver or transmitter [16], polarization shift keying [17], polarization tracking and compensation [18], or by fitting out the entire network with polarization preserving fibers. In any event, this introduces additional complexity into the system. Also, except for the last two options, there is some detectability loss.

As remarked earlier, the local oscillator power must be high enough to overcome the receiver thermal noise. However, this introduces additional intensity noise in the detection process, arising from fluctuations in the local oscillator output. This may be overcome by using a balanced mixer at the receiver [19], but at the cost of additional complexity and tight requirements on the accuracy of the balance.

When link budget improvements are required beyond those achievable with coherent reception alone, amplifiers are going to be required anyhow. The internally generated spontaneous noise from the amplifier will beat with the local oscillator signal and the product is likely to be the dominant receiver noise component. Thus one loses the detectability advantage that coherent was expected to achieve (based on the expectation that the detectability limit is given by the LO shot noise, not by receiver thermal noise or APD multiplicative noise).

The use of amplifiers in the path also leads to loss of some of the tight channel spacing advantage of the coherent option. If the channels are spaced closer together than the reciprocal of the carrier recombination lifetime (about 1 nsec. for diode amplifiers), then there is intermodulation due to modulation of the carrier density at beat frequency rates [20]. This effect is not seen with wider channel spacing since the carrier density cannot change fast enough to produce this effect.

Finally, availability of tunable lasers is an important prerequisite for multi-channel coherent systems. While external cavity lasers are tunable in seconds over 100 nm. or more, as tuning speed increases, tuning range decreases [21]. For injection-tuned three-section DBR lasers, retuning time is on the order of tens of nanoseconds [22], but the tuning range is about 10 nm. (with jumps). These devices are still at the research stage. Direct detection systems can be built today without tunable lasers.

PROBLEMS IN APPLYING THE INCOHERENT APPROACH

Some of the same problems of stability of the local oscillator for a coherent receiver must be dealt with in a tunable filter receiver, although they are less severe. Specifically, at the beginning of a new connection the receiver passband peak must be moved rapidly to acquire the new wavelength, and then must either remain stably at that new wavelength or must track the transmitted signal (or some common reference shared with the transmitted signal). For modest numbers of channels the tolerance on these functions is very much in favor of the incoherent approach, since the filter passband is much wider than the modulation bandwidth. However, as N grows to greater than several hundred, this advantage disappears.

Also, as the number of channels gets very large, transmitter laser chirp will deliver part of the energy into adjacent channels, thus increasing the crosstalk.

The requirement for high optical isolation mentioned in connection with transmitter lasers and LO lasers for coherent reception are also present with laser amplifiers, although the angle-facet approach, which is an option for amplifiers is not an option for lasers.

One disadvantage of the incoherent approach for computer network service is considered to be that Fabry-Perot filter retuning and settling time will be orders of magnitude slower than the same functions with tunable lasers. It is unlikely that either piezo-tuned or electrostatically tuned silicon micromachined filters [23] will ever achieve sub-microsecond tuning speed, whereas that is a distinct future possibility for tunable lasers, even including acquisition and settling time. Eventually, acoustooptic tunable filters should alleviate this limitation.

The micromachined silicon device may or may not be capable of large tuning speed improvments over that of piezo-controlled units. Although the moving mass of the silicon device is much smaller than that of available piezo-tuned devices, the force applied is also much smaller. The principal attractions of the silicon device are the possibilities for cheap lithographic fabrication, perhaps including some of the electronics on-chip.

A particularly promising approach to submicrosecond tunable filters is the acoustooptic approach [24], but, as currently realized, the selectivity of these filters is insufficient.

The use of amplifiers instead of coherent receivers introduces spontaneous noise, crosstalk due to gain saturation, and intermodulation distortion . In the direct detection approach, crosstalk due to gain saturation is the most significant effect [25]. In a system using amplifiers, the bit error rate is extremely sensitive to the receiver decision threshold, and so some form of adaptation is required at the receiver as power level fluctuates when nodes enter and leave the network. Diode amplifiers have typically several dB. of polarization sensitivity and (unless extremely low facet reflectivities are achieved) of Fabry-Perot spectral ripple. Unless the receiver has adequate dynamic range and the link budget has ample margin to allow these variations, they could pose a problem.

CONCLUSIONS

Our studies aimed at early experimental prototype WDMA computer networks have convinced us that for the foreseeable future the answer to the question posed in the title of this paper is "No. Coherent optical communication does not have a place in computer networks".

This answer is true today for the N = 32 prototype we are building [3], and we expect it to remain so for some years. (We are currently doing analytical modelling and experimental verification to confirm more quantitatively many of the qualitative assertions about the incoherent approach that have been made in this paper).

The reasons for our pessimistic conclusion about the coherent approach come from two observations. First, looking at the list given earlier of shortcomings of today's coherent approach in the light of the requirements peculiar to computer networks convinces us that there is not a good match. The requirements on complexity, cost and need for fine-tuning of today's state-of-the art coherent systems, even taking into account some dramatic recent improvements (e.g. [26]), seems to us to argue that they are expensive and hard to use. Second, both the principal advantages of coherent (detectability gain and channel spacing) are achievable by the simpler, cheaper alternative.

We are very encouraged to believe that small-N, reliable, user-friendly, incoherent WDMA networks can be built today from custom commercial components that ought to be capable of considerable later cost reduction. The most technically aggressive component is the photonic amplifier, which becomes necessary with growth in network size (diameter and number of ports). They will be needed for coherent networks too, as the 15 dB. theoretical detectability advantage eventually proves insufficient or not fully realizable.

Experimental work remains to be done to show that tunable coherent systems can achieve the sub-microsecond tunability speed that seems unachievable for piezo- or electrostatically-tuned Fabry-Perot filters.

On the other hand, as pointed out at the beginning of this note, coherent detection has an important role in deep space, undersea and intercity point-to-point links. As these requirements drive the evolution of ever more reliable, economical and usable coherent subsystems and components, perhaps the fruits of this work will "trickle down" to the computer communication community.

REFERENCES

[1] T. Okoshi and K. Kikuchi, *Coherent Optical Fiber Communications*. Kluwer Academic Publishers, 1988.

[2] Y. Yamamoto, "Receiver performance evaluation of various digital optical modulation-demodulation systems in the 0.5-10 micron wavelength region," *IEEE J. Quant. Electron.*, vol. 16, pp. 1251-1259, 1980.

[3] N.R. Dono, P. E. Green, Jr., K. Liu, R. Ramaswami and and F.F. Tong, "Wavelength division multiple access networks for computer communication," *subm. to IEEE JSAC*, vol. 7, no. 7, August 1990.

[4] C.G. Atkins et al., "High-gain, broad spectral bandwidth Erbium doped fibre amplifier pumped near 1.5 μm," *Electron. Lett.*, vol. 25, no. 14, pp. 910-911, July 1989.

[5] C.E. Zah et al., "1.3 μm GaInAsP near-traveling wave laser amplifiers made by combination of angled-facets and anti-reflection coatings," *Electron. Lett.*, vol. 24, no. 20, pp. 1275-1276, 1988.

[6] S. R. Mallinson, "Crosstalk limits of Fabry-Perot demultiplexers," *Electron. Lett.*, vol. 21, no. 17, pp. 759-760, August 1985.

[7] W. M. Hamdy and P. A. Humblet, "Crosstalk analysis and filter optimization of single- and double-cavity Fabry-Perot filters," *submitted to IEEE JSAC*, vol. 8, no. 6, August 1990.

[8] K. Reay, P. Atherton and T. Hicks, "Multipass FPIs using roof prisms," *Private communication*, April 1988.

[9] Newport Corporation, ""Supercavity"," *Fountain Valley, CA*, 1988.

[10] J.E. Mack, D.P. McNutt, F.L. Roesler and R. Chabbal, "The PEPSIOS purely interferometric high-resolution scanning spectrometer," *Applied Optics*, vol. 2, pp. 873-885, 1963.

[11] A.A.M. Saleh and J. Stone,, "Two-stage Fabry-Perot filters as demultiplexers in optical FDMA LANs," *J. Lightwave Tech.*, vol. 7, no. 2, Feb. 1989.

[12] M.W. Maeda, R.E. Wagner, J.R. Barry, and T. Kumazawa, "Frequency identification and stabilization of packaged DFB lasers in the 1.5 micron region," *OFC'89 Tech. Digest*, vol. (Houston, TX), p. 38, Feb. 1989.

[13] B. Glance et. al., "Densely spaced WDM coherent star network with optical frequency stabilization," *OFC'88 Tech. Digest*, vol. (New Orleans, LA), pp. 93-94, Jan. 1988.

[14] L.G. Kazovsky, "Coherent optical receivers: performance analysis and laser linewidth requirements," *Opt. Engrg.*, vol. 25, no. 5, pp. 575-579, May 1986.

[15] L.G. Kazovsky, "Phase and polarization diversity coherent optical techniques," *J. Lightwave Tech.*, vol. 7, no. 2, pp. 279-291, Feb. 1989.

[16] I.M.I. Habbab and L.J. Cimini Jr., "Polarization switching techniques for coherent optical communications," *J. Lightwave Tech.*, vol. 6, no. 10, pp. 1537-1547, Oct. 1988.

[17] S. Betti et al., "State of polarisation and phase noise independent coherent optical transmission system based on Stokes parameter detection," *Electron. Lett.*, vol. 24, no. 23, pp. 1460-1461, Nov. 1988.

[18] N.G. Walker and G.R. Walker, "Polarisation control for coherent optical fibre systems," *Br. Telecom. Tech. J.*, vol. 5, no. 2, April 1987.

[19] G.L. Abbas, V.W.S. Chan, and T.K. Yee, "A dual-detector optical heterodyne receiver for local oscillator noise suppression," *J. Lightwave Tech.*, vol. LT-3, no. 5, pp. 1110-1122, Oct. 1985.

[20] R.M. Jopson and T.E. Darcie, "Calculation of multicarrier intermodulation distortion in semiconductor optical amplifiers," *Electron. Ltrs.*, vol. 24, no. 22, pp. 1372-3, 1988.

[21] C. A. Brackett, "Dense WDM Networks," *Proc. ECOC'88*, pp. 533-540, September 1988.

[22] S. Murata, I. Mito and K. Kobayashi, "Tuning ranges for 1.5 μm wavelength tunable DBR lasers," *Electron. Lett.*, vol. 24, no. 10, pp. 577-578, April 1988.

[23] S. R. Mallinson and J. H. Jerman, "Miniature micromachined Fabry-Perot interferometers in silicon," *Electron. Lett.*, vol. 23, no. 20, pp. 1041-1043, September 1987.

[24] K.W. Cheung, M.M. Choy, and H. Kobrinski, "Electronic wavelength tuning using acoustooptic tunable filter with broad continuous tuning range and narrow channel spacing," *IEEE Photonics Tech. Lett.*, vol. 1, no. 2, pp. 38-40, Feb. 1989.

[25] R. Ramaswami, "Amplifier induced crosstalk in multi-channel optical networks," *subm. to IEEE JSAC*, vol. 8, no. 6, August 1990.

[26] M.J. Creaner, et al., "Field demonstration of 565 Mbit/s DPSK coherent transmission system over 176 km of installed fibre," *Electron. Ltrs.*, vol. 24, no. 22, pp. 1354-6, 1988.

COMBINED SPECTRAL & SPATIAL MULTIPLEXING SCHEMES FOR SWITCHING NETWORKS

David W Smith

British Telecom Research Labs.,
Martlesham Heath, Ipswich, IP5-7RE
United Kingdom

Abstract
An optical switching fabric is described that has the capability of Tbit throughput. The design based on a bus structure combines both spatial and temporal multiplexing and has the desirable properties of non-blocking operation low crosspoint count and a linear growth characteristic. The architecture of these multidimensional networks and the required optical technology to realise them are discussed.

1. INTRODUCTION

Optical transmission has progressively impacted all levels of communications systems. First, in long haul transmission where range and capacity offered by single mode fibre is unsurpassed. Then, within the local access network to provide flexibility for future, but yet unspecified, services. Finally, optical interconnect is expected to overcome the communications bottleneck that is expected to develop within buildings, equipment racks, circuit boards and from chip to chip. In particular, optics is now poised to influence the design of switching machines for both broadband networks and advanced computer systems.

In principle, an optical switching fabric could be developed that emulates an existing electronic architecture, but this may not be the most appropriate evolution. Electronic switching systems take their form from the extensive capabilities of VLSI;i.e great complexity at low cost. Whereas, current optical communications technology has its roots set in long "thin" transmission applications not requiring extensive optoelectronic integration. The early application of optics to large switching machines is most likely to take on a hybrid approach,i.e optics for interconnect and electronics for routing and control.

For optical switching to be realistic for large systems either a two or more fold increase in optoelectronic circuit integration must be achieved, and device scaling rules may prevent this, or alternative architectures must be developed. A promising way ahead is to develop architectures that allow trading of the latent spatial and temporal bandwidth of optical systems against a reduction in component complexity.

This paper outlines the concept of a switching fabric that achieves the above objective by combining space, frequency and possibly time switching within a multidimensional optical network (MONET).

2. FUTURE SWITCH REQUIREMENTS

Switching system research appears to be at a crossroads. In addition to the range of technological routes open there are also an increasing spread of application directions, including; circuit switching for broadband networks ; ATM; network management (for SONET etc.); processor reconfiguration for large computer systems. It is not yet clear what combination of these or other applications will endure but what we can be sure is that future switches will handle ever increasing information throughput and networks will have far greater functionality and intelligence.

Current opinion suggests that optical switching starts to appear attractive when the aggregate data through a network node approaches Tbit data rates. Data flow of this order could be expected through the nodes of a future broadband network or within tomorrows super-computer. But, set against the potential throughput of an optical switching system are the control difficulties. Current optical technology is best suited to circuit switching under electronic control. This may be satisfactory for network management functions, like protection switching, where millisecond reconfiguration is acceptable, but for packet switched inter-computer communications nanosecond set-up times may be required. For some applications it will also be necessary to consider the need for multicast or broadcast operation and possibly the option of demand assignment or preassignment working. In addition, other desirable features required from switching networks may include, potential for graceful growth, ease of control and non-blocking operation.

The prospects of meeting some of these requirements by either optical space or optical frequency switching will now be discussed and the need for multidimensional networks outlined.

3. SPACE SWITCHING

In principle very large optical space switches could be produced by exploiting the spatial bandwidth of optical interconnections. Using free space optics it is possible to image more than 10,000 connections within an area of 1sqcm, with negligible crosstalk, and with each connection having the capability of transmitting Tbit signals. To make the connections between input and output arbitrarily rearrangable there are a number of possibilities including: programmable hologram (photorefractives)[1]; mechanical or acoustooptic beam steering[2]; digital light deflector using liquid crystal devices[3]; spatial light modulator (SLM) in a vector matrix multiplier configuration[4].

The alternative to freespace optics would be to use waveguide switches. A switch matrix can be made by cascading crosspoint elements which could be either fibre devices [5] or planar integrated optics devices. Lithium niobate is currently the most mature IO technology for waveguide switches, but even so matrices much beyond 16 X 16 become unwieldy (although a double pass can be used to provide some reduction in physical sizes [6]). Semiconductor integrated optics offers an alternative route but as yet there is no evidence to suggest that this will lead to significantly larger switch matrices.

The possible matrix size of selected freespace and waveguide optical switches are compared in the table below.

Switch type	Limitations	Possible size	Ref.
Vector Matrix multiplier (Liquid crystal SLM)	Crosstalk, Loss for singlemode I/O =20LogN	32 X32 (100 X100?)	4
Digital light deflector (Liquid crystal SLM & calcite crystals	Physical length Loss when used with passive distributor =10LogN	1 to 16 (64 X 64 would be > 100mm long)	3
Lithium Niobate waveguide IO	Physical length : loss & wafer size Crosspoint control complex	8 X 8 (16X16?)	6

Table 1 COMPARISON OF OPTICAL SPACE SWITCH SIZE

At present it does not seem clear that large space switches with many hundred of ports could be made without combining many smaller blocks into one of the well known wiring patterns, such as CLOS, perfect shuffle, rearrangable etc [see 7 for examples]. The need for complex interblock optical wiring and/or control complexity greatly reduces the attractiveness of these multistage networks. Moreover, the wiring patterns required may not lend themselves to graceful switch growth.

Alternatively, significant reduction of the number of space switch points is possible by time sharing (i.e TST) architectures. Unfortunately, with optics there is currently no attractive way of performing time switching on a large scale (although semiconductor laser bistables show some promise of performing a limited role for this in future[8]). At present the only practical route ahead for STS architectures is to combine optical space switching with electronic time switching.

4. FREQUENCY (wavelength) SWITCHING

Because of the vast temporal bandwidth of optical systems it is possible to build a high capacity switch directly around a shared access network. The network could be of a bus, ring or star topology, and the multiplexing could be by TDMA, FDMA or possibly spread spectrum techniques[9]. For a shared access network with Mbit or even Gbit throughput TDMA is probably the most likely approach. However, to achieve a total capacity approaching Tbits, optical FDMA currently looks particularly interesting and this is being pursued in a number of laboratories around the world.

In a shared access OFDM scheme frequency selection could be by tunable filter or by heterodyne detection. In either case large networks would require the ability to have many sources at different frequencies closely referenced together. With heterodyne channel selection at the terminal receivers it would also normally be necessary to have a tunable source to provide the local oscillator. The size of the network that can be realised will be determined by the tunability of the source, the allowable channel spacing for acceptable crosstalk penalties , and the power budget. The latter is a severe restraint on the size of passive shared media networks, such as bus (loss goes as 20logN where N =number of terminals) or star (10logN), but can be circumvented by the use of optical amplifiers.

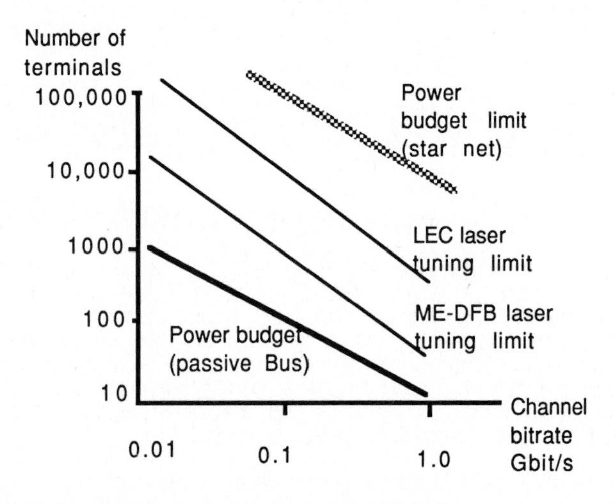

FIG1 Frequency switched networks :Size limitations

FIG1 shows the possible size of an optical frequency network for various constraints. Although it looks possible to make frequency switched networks of several hundred nodes the frequency referencing, registration and control problem are significant.

To increase the size of a frequency switched network it could be possible to build a multistage frequency switch by interconnecting several smaller networks through frequency translating devices (see FIG2). This arrangement would allow a limited number of optical frequencies to be reused many times. Frequency translation could be by intermediate conversion back to electronics (via heterodyne detection) and remodulation [10,11].

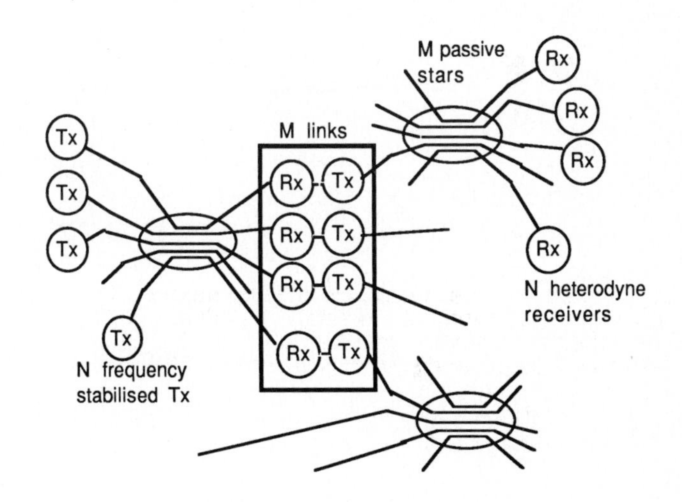

FIG 2 Outline showing how frequency switched sub-networks can be combined using a frequency translating link (i.e heterodyne detection & remodulation). For non-blocking operation more links would be needed.

To make a large multistage non-blocking network a great many such frequency translation links would be required. For example, a 3 stage CLOS network of 8192 X 8192 (8192 allows an optimal design to be considered) could be built that needs just 128 frequencies but would require a total of 40,704 tunable receivers and 40,704 frequency stabilised transmitters. To make such a scheme practical much integration of the functions performed by the frequency translating links would be necessary. In principle, in the long term, many optoelectronic conversions could be avoided by the use of non-linear optical signal processing. For this application a possible candidate technology is the non-linear laser amplifier [12].

However, for the time being a single stage of frequency switching looks to be a more realistic starting point, provided the problems of frequency referencing of terminals can be solved or avoided. One way round the problem is to provide a common set of reference frequencies that can be accessed by all terminals [13]. In this arrangement (see FIG3) reference frequencies both for the transmitters and receiver local oscillators are selected using optical space switches from a multiway reference bus. However, although this avoids the problem of laser tunability and registration the total number of crosspoints and spatial paths will now be much the same as for a conventional space switch.

It seems that by choosing either space multiplexing or frequency multiplexing alone it will be difficult to build large optical switched networks. A possible solution to this dilemma is the multidimensional optical network.

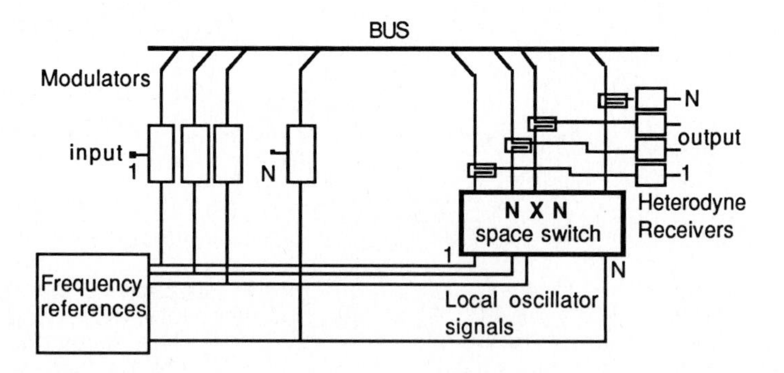

FIG3 Frequency switched network with local oscillator selection by optical space switch

5. MULTIDIMENSIONAL NETWORK (MONET)

By combining both optical frequency multiplexing and space division multiplexing it is possible to build switch structures which have a size equal to the product of the two individual multiplexing technologies. Therefore with, say, 100 input optical selector switches and 100 optical frequencies it could be possible to build a 10,000 input/output broadband switching machine. The key to combining the various degrees of switching freedom is by the use of a shared access multipath bus [14] to form the equipment back plane. This design provides a switch fabric that can easily be extended in small linear growth steps, just by plugging in additional circuit boards when necessary. Furthermore, the bus could also be extended to directly link additional equipment racks sited in different geographical locations.

To avoid the problems of channel referencing, between receivers and transmitters, the backplane would not only carry the signal multiplex but would also carry the common reference frequencies needed for modulation and demodulation. The operation of the generalised network, shown in FIG4, is described below.

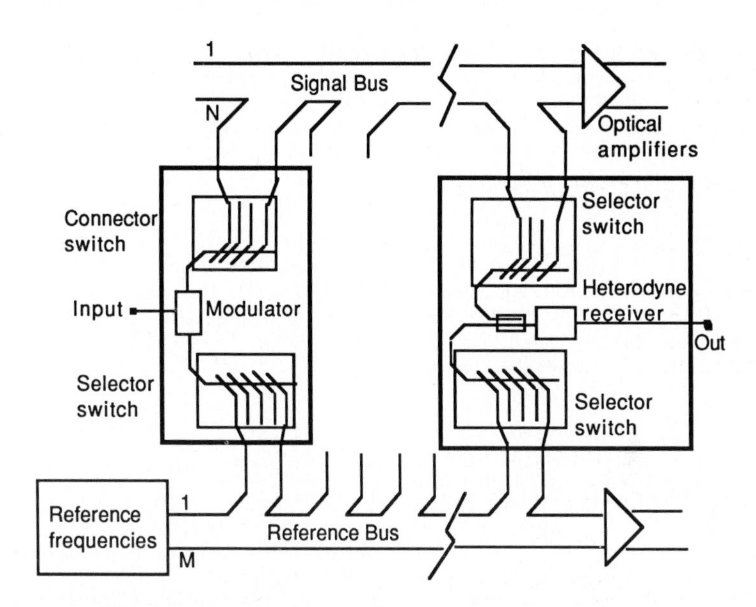

FIG4 Multidimensional optical network (MONET) switch (showing one transmitter and one receiver circuit out of N X M)

Each individual electrical data channel is applied to a separate transmitter that consists of an electrooptic modulator fed with a frequency selected from the reference bus. The selection of the reference frequency is performed by an M:1 optical space switch. The output of each modulator is then coupled to a selected position on the spatial bus by a N:1 optical connector switch. Thus each input channel out of N X M has its own unique location on the multidimensional signal bus.

To extract signals from the bus a small amount of power is tapped out of the appropriate spatial path and connected to a frequency selective optical receiver. The latter is achieved using self-heterodyne detection.

An optical space switch of size N:1 is used to select the appropriate signal from the signal bus and one of size M:1 to select the required local oscillator needed for demodulation. Not shown in FIG(4) are the frequency shifters needed in the local oscillator path for self-heterodyne detection. These could be located within each of the M X N receivers but a preferred option is to locate them within the reference bus, in this case only M frequency shifters are required.

By having dynamic channel selection at both the input and output of the network demand assignment operation is possible. This mode of operation could be useful to enable further crosspoint and equipment sharing by either (a) reducing availability by exploiting the calling rate statistics or (b) by time sharing crosspoints by either synchronous or packet working. However, for applications where preassignment of channels is allowable there could be a reduction in the number of crosspoints by limiting switching to either the transmitter or receiver.

6. ADDITIONAL DIMENSIONS

Although the network as previously shown uses just space and frequency multiplexing it is possible to add time multiplexing to increase the possible connectivity of the bus. In addition, further dimensions such as coarse grain WDM, CDMA, polarisation etc could be added. As the number of dimensions increases the number crosspoints and spatial highways is further reduced (as table 2 shows, [20])

With Log_2 N degrees of multiplexing freedom the information theory limit for crosspoint reduction is reached, however there may be little practical advantage gained by going beyond 3 degrees . As can be seen in the above table, the realisation of the 8192 terminal network by using just three multiplexing dimensions looks practical in terms of spatial highways, optical switch size, optical frequency referencing and time slot selection. The total data throughput of such a bus could be, say, 8192X140Mbit/s i.e >1Tbit/s and yet only require, <1.2Gbit/s time multiplexing, 32 optical frequencies and 32 spatial paths. Of course increasing the number of multiplexing dimensions beyond two (space & one temporal) does not increase the switch throughput but makes its more easily realisable.

Example for N=8192 inputs	
Single stage crossbar	67,108,864
Three stage CLOS	4,161,536
2 dimensional bus (64 signal paths 128 frequencies)	1,556,480
3 dimensional bus (32 spatial paths, 32 frequencies, 8 timeslots)	565,248
Multidimensional with Log_2N degrees	106,489

Table 2 Crosspoint number for various switch architectures

7. TECHNOLOGY

7.1 The Bus

The heart of the network is the bus. The bus forms the back plane that links all equipment cards. The approach we have been investigating is the use of connectors that tap only a small fraction of the power from the bus. To form the backplane, D fibre looks particularly attractive. In D fibre the core of the fibre is close to the flat, within 1μm, allowing power to be coupled easily from the interface. Connection is made by placing a second similar fibre across the bus.

In the tapped bus architecture it is only necessary to couple quite small values of power from the bus (say <1%) this relaxes the tolerances required for multiple fibre connection. The critical dimension in the multi-D-fibre connector is the relative hight of the flats of each fibre with respect to the mounting substrate. By use of a special pressing technique it is possible to achieve control to better than 1μm. Using this approach a 4 fibre bus and bus connector have been fabricated [15] which has better than 2dB uniformity between the 4 taps.

The number of equipment cards that can be connected to the passive bus will be determined by power budget considerations. In the basic scheme all optical power is obtained from the common reference sources that must provide power for all transmitters and local oscillators. If it is assumed that the local oscillator power must be at least 10dB larger than the signal power at each receiver this will be the determining factor.

For a system requiring -20dBm local oscillator power at each receiver card (although close to quantum limited detection has been achieved with -30dBm local oscillator power at 140Mbit/s [16]) up to 1000 receivers could be served by a single reference laser coupling 13dBm into the bus. Of course, in practice there would be excess losses in the selector switches and bus connectors. Also larger local oscillator power may be necessary to overcome heterodyne intermodulation products. Furthermore, the above calculation assumes that the power from the reference source is evenly distributed which requires graded tap coefficients along the length of the bus, this would complicate the bus design. A more realistic arrangement would be to use a common coupling coefficient but this would reduce the number of connections possible.

To make large bus switches it is expected that laser amplifiers will be essential to overcome distribution losses. Both semiconductor diode lasers and fibre lasers are candidates for this application. The fibre laser offers the attraction of better linearity, this feature may be critical for multichannel operation.

7.2 Frequency referencing

For a small number of frequencies it may not be necessary to have a complex referencing scheme. Indeed it may be allowable to use conventional DFB lasers free-running about selected central wavelengths. But for larger networks or where it is attractive to couple more than one bus switch together some form of centralised referencing scheme will be necessary. Several techniques have been considered including comb generation using microwave modulation [17], scanning spectrometers[18] and locking to atomic absorption lines[19].

7.3 Transmitters

The design of the transmitters is simplified by avoiding the need for stabilised frequency sources on each card, instead, electrooptic modulators are directly coupled to the reference bus. The modulators lend themselves to planar integration, possibly in arrays, and for this either Lithium Niobate or semiconductor based integrated optics could be used.

7.4 Space switches

The dominant component cost of the bus switch is likely to be selector/connector space switches. As previously stated, one of the main virtues of the architecture is the reduction of the complexity and number of space switches required. The N:1 selector/ connector switch is in itself of considerably simpler construction than the N X N crossbar since it has both fewer crosspoints and control wires. In fact, some optical realisations of the N:1 selector switch can have the theoretical minimum of control wires [3]

The selector switches themselves could be realised using any of the technologies described in the earlier section. Although waveguide structures, such as Lithium Niobate, are probably more appropriate, both at the transmitter, because the bus is singlemode, and at the receiver, because of the wavefront matching needed for heterodyne detection.

7.5 Receivers

If self-heterodyne detection is used the problems of frequency locking and acquisition of local oscillator lasers are avoided. The desired frequency offset required could be provided by an acoustooptic frequency shifter located in the reference bus. The use of self-heterodyne detection could also reduce the sensitivity of the system to laser phase noise, provided the local oscillator and signal bus paths are matched within the coherence length of the reference source.

The electrical output from the heterodyne detector will be at the difference frequency determined by the acoustooptic shifter. This signal must be further demodulated to produce the baseband data.

8. CONTROL

Control is a significant part of the switching problem. It has been assumed here that switch control will be electronic, this will remain until non-linear optical processing technology significantly matures.

Control is likely to be most difficult for multistage switches that use special routing algorithms to reduce the number of crosspoints needed , particularly those that are rearrangably non-blocking. In contrast , the control for the MONET switch can be localised to the selector switches unique to each terminal. This should lead to simple routing algorithms and the fast set up times necessary for packet working.

9. CONCLUSION

The concept of a switching fabric has been described that could find application both within future telecommunication systems and computer systems. The structure exploits the strengths of optical transmission , namely, spatial and spectral bandwidth and trades these against complexity, which in optical systems is the premium commodity.

The switch design, which is based on a bus structure, is simple to control, is easy to grow, is non-blocking, requires close to the minimum number of crosspoints and has the capability of Tbit throughput. These high throughputs are expected to be achieved by combining, space, frequency and time multiplexing methods. By using these in combination it is possible to avoid pushing any one or other of these degrees of switching freedom to unrealistic limits.

Key technologies needed to implement this switch architecture include: D fibre optical backplane; M:1 optical selector switches; optical amplifiers.

ACKNOWLEDGEMENTS

The author wishes to acknowledge P Healey and S Cassidy for their contributions to the development of the MONET concept and the Director for Research and Technology of British telecom for permission to publish this work.

REFERENCES

1. J P Herriau et al, "Optical beam steering for fibre array using dynamic holography", 100C/ECOC, Venice, conference digest, pp419, 1985
2. P C Huang et al, " Non-blocking photonic space switch architectures utilising acoustooptic deflectors", Topical meeting on photonic switching, Salt Lake, 1989.
3. P Healey, "Optical interconnection networks employing multiplexed crosspoints", i.b.i.d.
4. P Healey & D W Smith, "A broadband bi-directional parallel electrooptic space switch", OSA annual meeting, 1987.
5. S A Cassidy & P Yennadhiou, "Optimum switching architectures using D-fibre optical space switches", IEEE J/SAC Vol 6, 1988
6. P J Duthie et-al, "A new architecture for large integrated optical switch-arrays", Photonic Switching, Springer-Verlag, 1987
7. A A Sawchuk & B K Jenkins, "Dynamic optical interconnections for parallel processors", Proc. SPIE, vol 625, pp143, 1986

8. H Goto et al, "Photonic time division switching techniques", Photonic Switching, Springer-Verlag, pp151, 1987
9. D W Smith & P Healey, "Optical spread spectrum multiple access techniques", IOOC, Kobe, 1989.
10. E J Bachus et-al "Optical coherent multichannel switching experiment", ibid, paper20d2-2
11. M Fujiwara et-al, "Line capacity expansion schemes in photonic switching", ibid, paper21a2-2
12. D W Smith & G R Hill, "Optical processing in future coherent systems", Globecom, Tokyo, pp678, 1987
13. I W Stanley et-al, "The application of coherent optical techniques to wide-band networks", IEEE/JLT, vol LT-5, pp439, 1987.
14. D W Smith et al, "Extendible optical interconnection network", Topical meeting on photonic switching, Salt lake, 1989.
15. S Cassidy et al, "Extendible optical interconnection bus fabricated using D fibres", IOOC, Kobe, 1989.
16. T G Hodgkinson et-al, "Experimental assessment of a 140Mbit/s coherent optical receiver at 1.52μm", Electron. Lett., 17, pp523, 1982.
17. T G Hodgkinson & P Coppin, "Comparison of sinusoidal and pulse modulation techniques for generating optical frequency combs", Electron lett., 25, pp509, 1989.
18. E J Bachus et-al, "Coherent optical fibre subscriber line", post deadline paper IOOC/ECOC, Venice, 1985
19. Y C Chung, "Frequency locking of a 1.3μm DFB laser using a minature argon glow lamp", IOOC, Kobe, 1989
20. D W Smith, P Healey & S Cassidy, "Multidimensional optical switching networks", to be presented at GLOBECOM89.

RESEARCH ON COHERENT OPTICAL LANS

A. FIORETTI, E. NERI, S. FORCESI

ALCATEL FACE Research Centre
Via Nicaragua 10, Pomezia 00040, Italy
Tel. +39.6.912851 Fax +39.6.912851 ext. 255 Telex 613340

A.C. LABRUJERE, O.J. KONING AND J.P. BEKOOIJ

PTT-Research Neher Laboratories
P.O. Box 421, 2260 AK Leidschendam, The Netherlands
Tel. +3170436019 Fax +3170436477 Telex 312336 prnl nl

ABSTRACT

This paper illustrates the UCOL network concept which is currently under development as a laboratory prototype under the framework of the European programme ESPRIT; the acronym UCOL stands for Ultra-wideband Coherent Optical L.A.N.

The system architecture is reviewed first, then the constraints set by multichannel capability on network operation and capacity are addressed.

1. INTRODUCTION

The UCOL project follows a feasibility study investigating the promising potential of coherent optical techniques in the context of private area networks. The results is a plan for an ultra-wideband coherent optical local area network, that will demonstrate experimentally the performance and the flexibility of a coherent multichannel system.

The project adopts an all-round approach to the development of a system aiming to provide integrated support of narrowband and broadband services (data, voice and video) directed at the needs of specific market segments. This means that the project addresses both technology issues (specific of the physical layer) and the development of the necessary higher layers, for management and resource allocation for example, thereby insuring that the mutual interdependencies are taken into account.

The integration of any number of services and the flexibility in bandwidth allocation with agile mixing of high and low bandwidth users calls for an efficient multiplexing scheme which, as explained in [1], can be implemented by taking advantage of the fact that the only actual "broadcast" section of a star topology is the centre of the distribution hub. Frequency domain switching between channels is exploited essentially to obtain inexpensive adaptive network reconfiguration without hardware modifications[2].

In the following the optical part of the project will be discussed.

2. OPTICAL ARCHITECTURE

UCOL is a passive network interconnecting a number of network interfaces (i.e. transmitter/receiver pair) to a central passive star by means of single mode fibers operating at a wavelength of 1550 nm.

The available system gain allows to foresee a main configuration offering a connection over a star radius of up to 10 km for 512 Network Interfaces (NIs).

It is clear that other configurations could be obtained by varying the trade-off between number of NIs and maximum distance from the star centre: as an example a possible alternative configuration giving priority to the distance from the star centre at the expense of a reduced number of NIs can be dimensioned with longer connections (up to 25 km) and 128 NIs. Like other network based on a star topology, most of the losses are introduced by power splitting inside the central star, but also propagation, splices and aging loss have been considered in the power budget. Each NI is directly connected to the central star in order to avoid the additional loss due to local splitters/combiners.

All the interfaces are able to transmit over a set of FDM digital channels, frequency locked to a common reference source, each one with a TDM access mode. This concept allows the maximum flexibility as, on each channel, makes the transmission of different information rates possible.

A TDMA scheme on each optical frequency has been selected to insure efficient handling of real time packets. Frequency domain switching between channels is therefore exploited essentially to obtain inexpensive adaptive network reconfiguration without hardware modifications.

The modulation scheme is the Differential Phase Shift Keying (DPSK) with a transmission gross rate of 200 Mb/s; this scheme offers good detection sensitivity and makes TDMA implementation easier. In fact the modulation process does not affect the transmitter frequency making locking to the reference line less critical.

The UCOL optical hardware is therefore made up of a central star and a number of Laser Subsystems. Each Laser S/s is composed of:

- one or more NIs (the maximum number of NIs within each Laser S/s is 16);
- a Regeneration Block (RB), i.e., an entity able to lock the reference line received from the star and produce a set of equally spaced frequencies, used for the synthesis of the channel frequencies.

Several approaches [3] have been devised for the practical implementation of the RB: the one illustrated in this paper adopts a mode-locked laser as the frequency generating element. This approach for the comb line generation/synchronization consists in frequency locking all channels to one common reference line distributed from one main Master Laser throughout the UCOL star network to all Laser S/s in order to provide an optical frequency standard for all comb generators and subsequent transmitter/receiver stations.

Within each Laser S/s a RB has to provide the following functions:

- Recognition/identification of the reference line emerging from the network
- Frequency locking of the mode-locked laser for frequency comb generation
- Replacement of main ML by one of the others in case of fault.

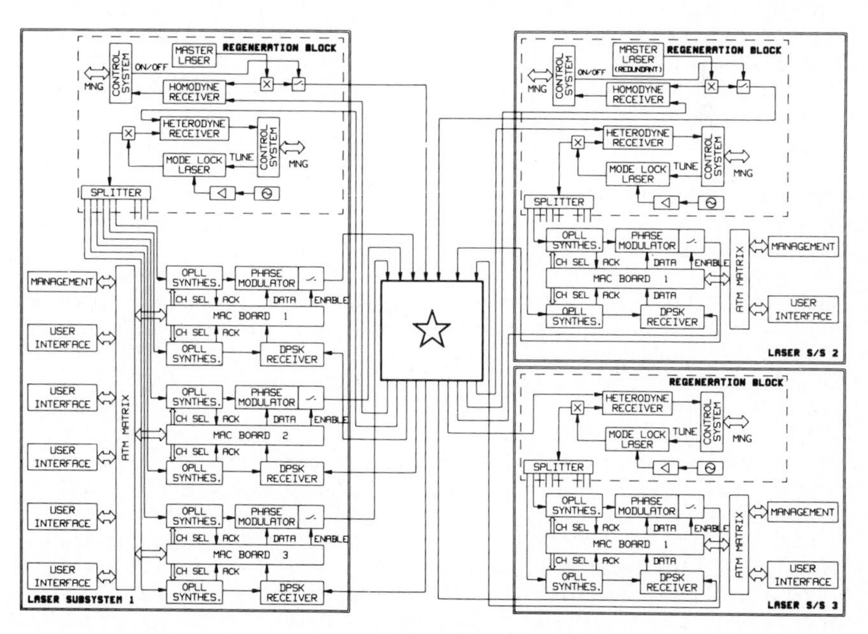

Fig. 1
UCOL block diagram

The basic elements of the scheme: Master Laser and homodyne receiver together with heterodyne receiver and mode-locked laser, are shown in Fig. 1. The two control systems shown interact with the network management to provide suitable operation of the optical hardware during normal operation and fault diagnosis. The mode-locked laser is driven by a microwave signal in order to generate a comb of optical frequencies; the output is shared, by means of a splitter, over the network interfaces. One of the MLs in the network acts as a common reference: in this case its output is also connected to the central star through a switch for distribution, and its homodyne receiver detects, for management purposes the presence of the reference in the network. In case of failure one of the other MLs becomes the master sending the reference frequency to the star. When the ML reference line is not distributed to the network its output is disconnected from the network by the switch.

All the mode-locked lasers generating the comb of carriers are locked to the reference through the heterodyne receiver. The UCOL reference line reaches the Laser S/s with power levels around -40 dBm. The regeneration scheme has first to detect and identify the weak reference signal in the center of the ~40 GHz broad frequency comb of DPSK modulated carriers, where the next signal carrier is spaced 4.85 GHz apart of the reference line. This is done by coherent heterodyne detection and filtering with low intermediate frequency (IF ≤2 MHz). The mode-locked laser is frequency locked to the reference line by using a frequency discriminator for low IF (≤2 MHz). This approach is capable to detect the low input power level of the reference line emerging from the UCOL star (receiver sensitivity ≅-50 dBm)[3]. The required time for frequency locking is an important parameter for the initialization procedure of the UCOL system and is mainly depending on the tuning characteristics of the LO laser: Laser S/s locking times of around one second seem to be feasible.

The central star acts as frequency multiplexer and on each output all carriers are present. The NIs structure is common to all the described schemes; the transmitter is composed of an OPLL synthesizer, a phase modulator and a switch, whereas the receiver uses the same subsystem, i.e. the OPLL synthesizer as LO, together with a polarization diversity hybrid and DPSK demodulator: obviously the phase modulator and switch are not needed. Each NI is fed with the set of harmonics deriving from the RB; the OPLL allows exploitation of the spectral range between each pair of carriers: its function is to select the required harmonic and to shift its output in order to obtain both the desired channel and local oscillator. Besides the frequency shifting function the OPLL is also required to narrow the line of the DFB laser operating as Current Controlled Oscillator (CCO).

The mode-locking based carrier generation produces a spacing between carriers of several GHz. For the present system, the comb-line spacing has been set at 7.6 GHz. In the OPLL, the optical fields of the CCO and a single comb-line are combined and the resulting intermediate frequency is then phase-locked to an electrical offset frequency using a broadband microwave mixer. Once the loop is locked, the resulting CCO frequency, which is transmitted into the network, will be shifted from the input reference by the offset frequency. Each offset frequency corresponds to a different channel. Similarly, at the receiver, an OPLL is used to generate the correct local oscillator frequencies, from the comb-lines, in order that a given channel can be selected.

The use of an OPLL as a frequency shifting element has some important implications on the frequency allocation plan which is shown in Fig. 2 in the case of a comb of 5 carriers. The central carrier is used only to provide the reference to the individual regeneration blocks. The need to avoid the interference terms due to the demodulation of individual channels has suggested an asymmetric distribution of the data channels around the reference carriers. Also the actual carriers cannot be used as data channels due to the need to operate the PLL with a predetermined minimum IF frequency.

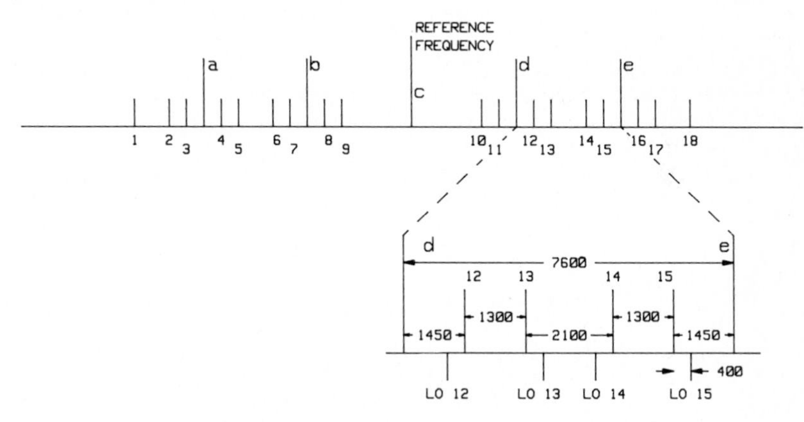

Fig. 2
Optical frequency allocation scheme

3. REFERENCE CARRIER GENERATION BY MODE-LOCKING

Internal gain modulation at a multiple of the roundtrip frequency of its cavity results in phase-locking of successive longitudinal modes. The frequency spectrum of such a mode-locked laser is a comb with lines separated by the free spectral range (FSR) of the cavity.

For use in a FDM coherent optical communication system, comb generation by mode-locking is a promising technique. First of all it offers a good comb stability; mode hopping, which is difficult to avoid in a single mode laser, does not occur in a mode-locked laser, because of the existing coupling between the longitudinal modes. Secondly it will be shown that mode-locking of an external cavity semiconductor laser can be obtained by current modulation at subharmonic frequencies, allowing relatively low frequencies, which are less difficult to generate, to be used. The most important feature of mode-locking concerns the wide spectrum. It will be shown that by mode-locking a very comprehensive spectrum (5÷100 lines) can be created. This implies that a FDM communication system based on a mode-locked laser, has potentially a large spectral bandwidth capacity.

In the UCOL-demonstrator, the frequency allocation scheme is based on a comb of just 5 lines. The control loop for carrier generation sets a demand on the power per comb line (for UCOL with an OPLL: -35 dBm). The transmission scheme also requires a minimal width of the individual lines of 450 kHz. The frequency spacing btween the comb lines is determined by the number of transmitters locked to a comb line and the tolerable intermodulation between the channels. The chosen frequency spacing is 7.6 GHz, corresponding to a mode-locked laser external cavity length of about 2 cm.

Hereafter some spectral aspects of the mode-locked laser are presented. Experiments, evaluating the performance in the frequency domain have been performed. For ease of operation, experiments were performed at low modulation frequencies (approximately 1 GHz).

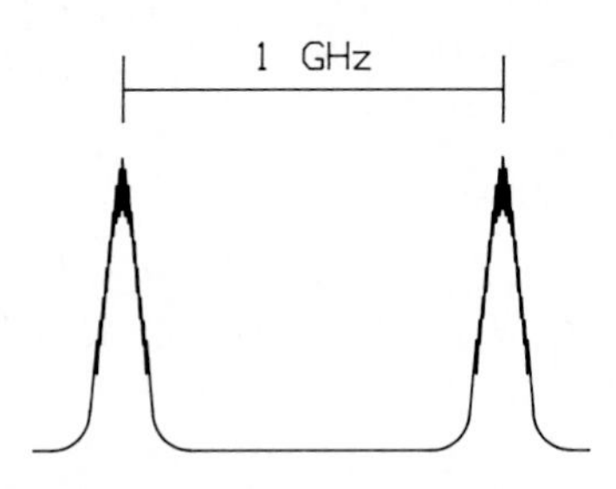

Fig. 3
Spectrum of mode-locked laser: modulation frequency 1 GHz

The laser used in the present experiments is composed of a Fabry-Perot laser chip (at one side AR coated), a collimating lens and an external diffraction grating. The distance between facet and grating was about 15 cm, corresponding to a FSR of the external cavity of about 1 GHz. The operating wavelength is 1.53 μm and can be tuned over 70 nm by rotating the external grating. In Fig. 3 a typical mode-locked spectrum on a Fabry-Perot Interferometer (FPI) is shown, obtained by gain modulation at the roundtrip frequency of about 1 GHz. The width of the comb lines could not be resolved by the FPI. The corresponding single mode laser operation is shown in Fig. 4. By rough intensity estimation, it follows that the cw laser intensity becomes distributed over all comb lines.

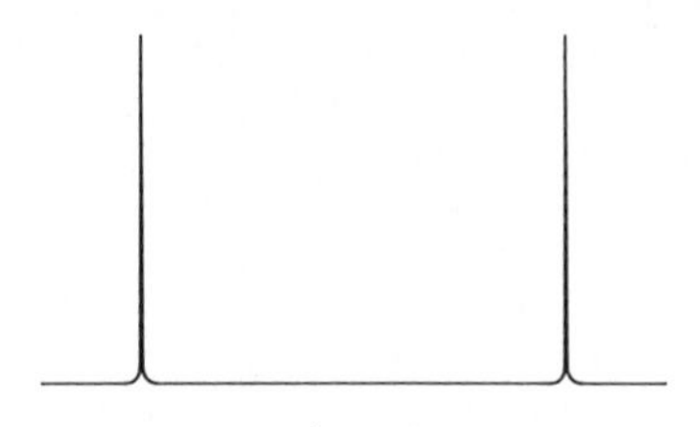

Fig. 4
Single mode spectrum: E.C.M. length ~15 cm

The number of locked modes depends on both the bias current and the modulation current through the laser chip. In Fig. 5 the number of modes with optical power greater than half the maximum value is plotted versus modulation power. The maximal number of locked modes is however determined by the spectral filtering of the diffraction grating. The spectral bandwidth of a grating is determined by its resolving power and hence by the diameter of the beam falling onto it. This effect is also visualized in Fig. 5, by two series of measurements, obtained with different intracavity aperture. By focusing the beam onto the grating, we have observed a very extended comb spectrum using an optical spectrum analyzer. From the spectral broadening of the laser line, the number of locked modes was estimated to be about 50.

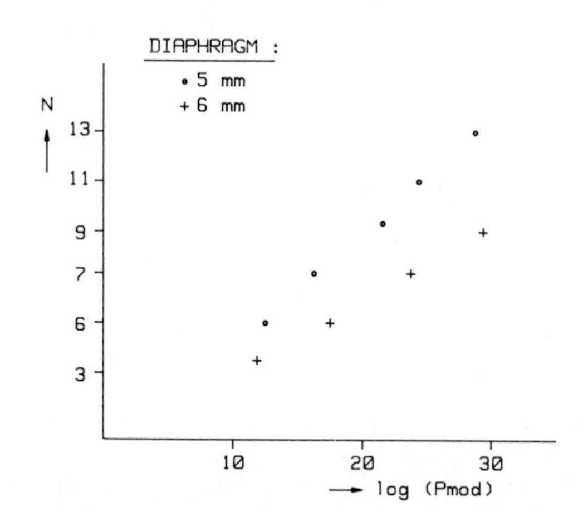

Fig. 5
Relation between modulation depth and number of comb lines

In fig. 6 the beat spectrum of the mode-locked laser and a cw single mode external cavity laser is plotted. This spectrum can be separated into two distinct parts. The lines indicated with M_i originate from mutual mixing of the various comb lines and therefore they are positioned at multiples of the modulation frequency. It must be noted that line M_1 partly results from the AM component of spontaneous emission. The lines 1÷6 originate from mixing with the single mode laser. The distance between two adjacent lines equals therefore exactly the modulation frequency and the linewidth could be determined to be less than 100 kHz.

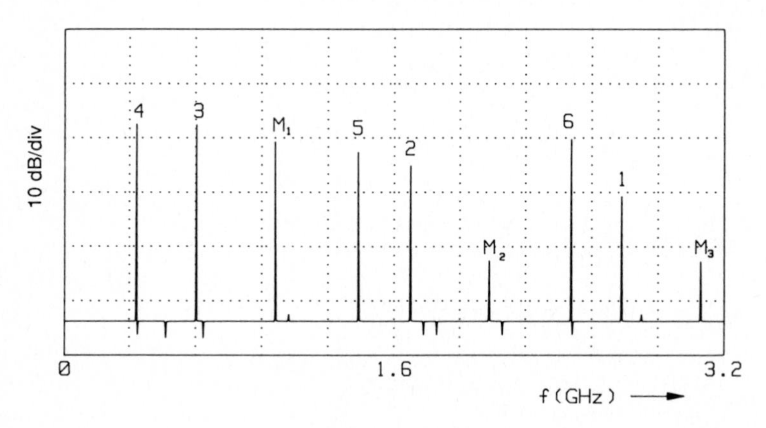

Fig. 6
Beat spectrum of a single mode laser and the mode-locked laser

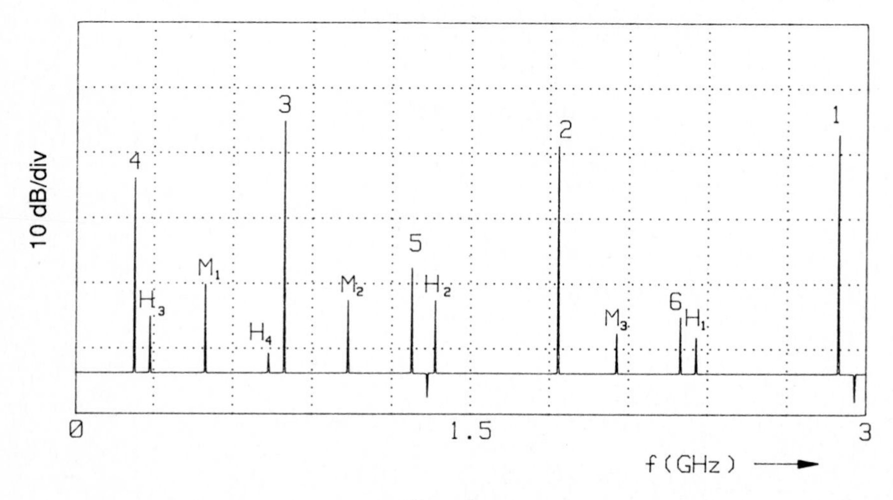

Fig. 7
Spectrum of mode-locked laser at subharmonic modulation

In Fig. 7 a spectrum obtained by subharmonic (here half the roundtrip frequency) modulation is shown. It can be seen that the spectrum consists of three parts. The lines M_i originate partly from AM components in the spectrum and partly from mutual beating. The strongest part is the beat spectrum of the lines of the mode-locked with the single mode laser (1÷6). Furthermore, frequency components at the modulation frequency (half the mode-locking frequency) are observed. The lines H_i are caused by mixing subharmonic comb lines with the cw single mode laser. However, the intensity of these spurious lines is more than 20 dB below those of the 'real' mode-locking spectrum.

The linewidth of the comb lines has been measured using self heterodyne techniques to be about 40 kHz. In Fig. 8 both the cw and the mode-locked laser spectra are shown. As was demonstrated by Rush et al. [4], the linewidth of a comb line equals the spectral width of the unmodulated laser. Therefore the UCOL linewidth demand for individual comb lines can be translated to a requirement on the cw laser.

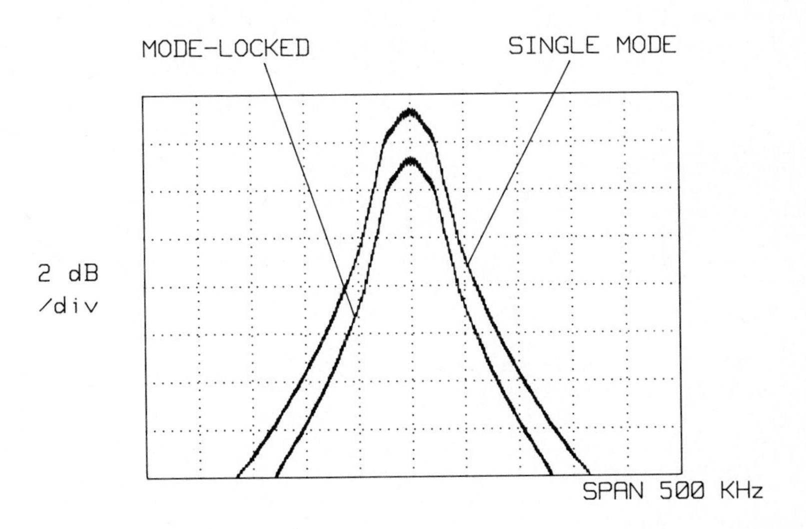

Fig. 8
Linewidth measurement for mode-locked and single mode laser

4. OPTICAL RECEIVER

The location of several optical carriers within the same spectral region, the bursty signal due to TDM access and the imperfections of laser sources have imposed a series of constraints upon receiver design. The UCOL receiver, based on a polarization diversity optical front-end in a balanced configuration followed by a double arm DPSK demodulator is shown in Fig. 9.

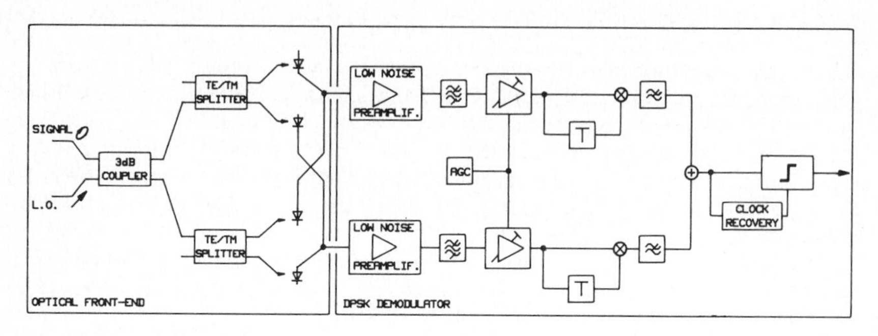

Fig. 9
Schematic of UCOL polarization diversity receiver

The received signal is combined with the LO signal, coming from the OPLL, in a 3 dB directional coupler; this mixed signal is then split into orthogonal polarization components by means of two TE/TM splitters and detected by four photodiodes. The two pairs of outputs corresponding to the same polarization are substracted to suppress intensity noise; the two output signals feed the two arms of the demodulator. Each arm is composed of a low-noise preamplifier (2.5 pA/√Hz) based on the current-current feedback scheme [5], an IF amplifier and relevant AGC, an IF filter and a delay line demodulator. The two demodulated signals, each one corresponding to its polarization component, are added and then sent to a threshold comparator. The clock recovery function is accomplished by a dedicated circuit expecially designed for fast acquisition.

A theoretical analysis of the receiver BER has been performed in the context of an ideal single channel system: the receiver sensitivity is about -61.0 dBm in order to obtain a BER of 10^{-6} with a phase noise standard deviation of σ_{ϕ}=0.168. This figure is reduced by the following impairments:

- excess loss in the optical hybrid:	1.3 dB
- 10% unbalance at the 3 dB coupler outputs:	0.1 dB
- 15 dB TE/TM x-talk at the TE/TM splitter o/p:	0.4 dB
- 20 dB conversion mode x-talk at the TE/TM splitter o/p:	0.3 dB
- 2.5 pA/√Hz preamplifier thermal noise:	0.8 dB
- LO RIN:	0.3 dB
- interference due to adjacent channels:	3.0 dB
- imperfections in data modulation:	0.2 dB
- ±1 MHz IF frequency offset:	0.3 dB

that leads the sensitivity to -54.3 dBm. Additional penalties due to demodulator imperfections (i.e. ISI, recovered clock jitter, phase and amplitude channel distortions) could further degrade the sensitivity to about -50.0 dBm.

In order to obtain an adequate system gain and therefore an adequate network size and geographical coverage it has been selected to operate the DPSK demodulator at a C/N value resulting in a BER of 10^{-6}, a relatively low value for MANs application. Therefore error correction techniques become necessary to meet the required channel quality. In order to estimate the performance at the physical layer of this network it is useful to notice that discontinuous mode of transmission on such a network implies that the BER is no longer a meaningful parameter to characterize performance.

The need to support wideband communication and to make the system able to connect to pre-existing networks has already been mentioned. As a consequence, this project has considered the result of the rapid progress made in the development of ATM (Asynchronous Transfer Mode) techniques [6][7], particularly in public broadband communication.

From the technical standpoint ATM emerges as an universal transfer technique with no special functions tailored to particular services, and is therefore relevant for a system such as UCOL which is aiming at providing a transparent communication world.

These considerations have led to the adoption of ATM techniques with the specific aim to insure that the objective of the project be met, i.e.:

- to define a network architecture independent of the specific services;
- easy interworking with public and local networks that will be based on ATM techniques, thereby minimizing the gateway functions.

The most apparent impact of the introduction of ATM techniques in UCOL is related to the way the information is carried through the network. The adoption of a TDMA scheme on each frequency matches the segmentation of the information flow into a number of fixed-length cells called for by ATM.

The asynchronous nature of ATM translates into a non deterministic ordering of different information streams on each optical channel. In UCOL the transport mechanism on each channel along the fiber is framed TDM, where the ATM cells are inserted in time slots within the frame structure.

The information on each individual optical channel is organized in periodic frames of constant length (1 ms). The frame format which is shown in Fig. 10 uses three major fields:

- Frame alignment field, consisting of 1472 bits;
- Queue status (QS) field, that basically carries the information related to packet priority and number of packets waiting for transmission in each individual node queue;
- Information field, where each slot can accomodate an ATM cell of the format 64+5 bytes.

It is important to identify within the frame the three fields allocated respectively to the Queue Status, the header of the ATM cell and the cell data field. An incorrect QS field may cause the loss of the entire frame. It is important to realize that the performance of the system is not determined by the basic bit error rate (BER) but by the loss rate of these three fields. The basic BER of the channel operating in continuous mode, 10^{-6}, forces the adoption of error correction techniques. The system requirement for the QS field call for a QS loss probability $<10^{-14}$.

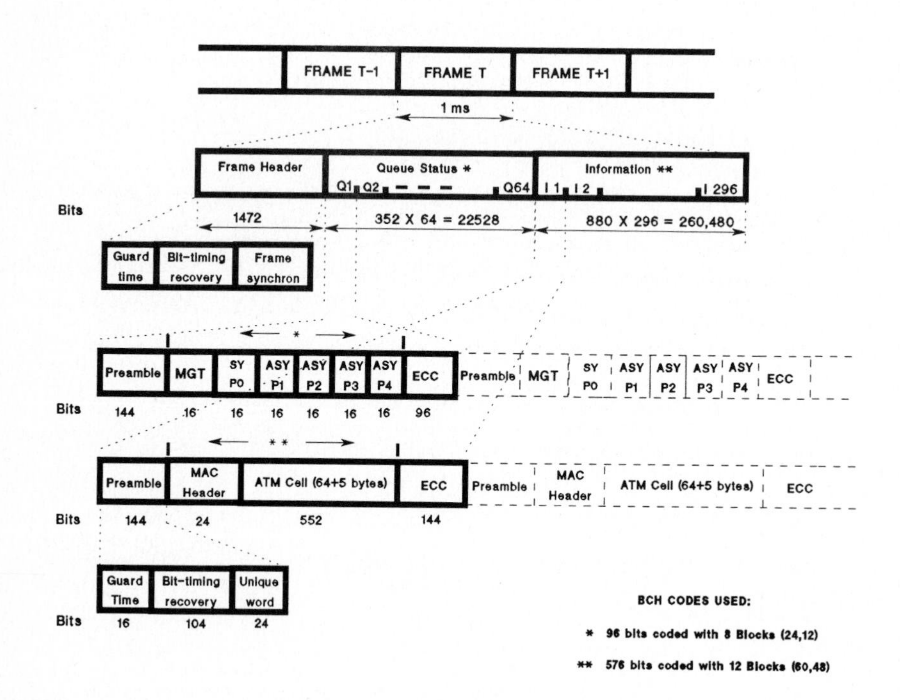

Fig. 10
Frame format

In evaluating the degree of protection needed for the ATM cell header, reference has been made to the ATM standard using 69 bytes, 5 of which are used for the header which contemplates allocation of one byte to Error Correction Coding. The code of the standard ATM cell has been dimensioned in such a way as to guarantee a cell loss rate of 1 every 1.5 Yrs, in a typical channel characterized by a random distribution of errors, a BER of 10^{-8} and a data rate of 150 Mb/s.

The decision taken in this project was to protect the ATM cell header in such a way as to obtain the same result in terms of ATM cells loss rate that is obtained with a standard ATM cell.

Considerations on the management of synchronous traffic in addition to the bursty transmission mode of the network and the bit rate lead to select a Forward Error Correction (FEC) tecnique based on block codes.

In DPSK demodulators characterized by white Gaussian noise an error model exhibits pairs of errors which represent approximately 10% of the total number of errors [8].

The UCOL receiver is affected by source phace noise and interchannel interference, in addition to white Gaussian noise, which aggravate the situation by increasing the percentage of double errors possibly introducing burst of errors of length equal to three.

The simultaneous presence of random errors and short burst of errors in addition to the frame format and the de-coding speed has led to select BCH codes [9], implemented with a suitable interleaving technique. The degree of interleaving is strictly dependent on error statistics in the channel and to the hardware complexity introduced; however a constraint is that the data field to be protected, which is defined by the frame structure, has to be segmented in a number of blocks with length multiple of the degree of interleaving. It is therefore necessary to balance code length and segmentation efficiency.

In order to account for the worst case, loss rate of fields has been calculated in the case of continuous transmission of ATM cells.

The required performance for the QS field can be obtained by adopting the Extended Golay Code (24,12) [9], capable of correcting all errors of weight ≤ 3 and detect patterns of errors of weight 4. Each QS field is made up of 12 bytes: by implementing the above code using 8 blocks (24,12) the probability to receive one such block with errors is 10^{-19}. With this choice the redundancy introduced in terms of transmission of the QS is therefore 2.

By coding the cell header with a BCH code (60,48) correcting two errors/block, the loss/day of ATM cells is 10^{-3} i.e. a cell every second year.

Block correction codes determine a concentration of errors in the incorregible blocks. As the error distribution in a cell with errors is such that there is no difference for the user to receive a cell with errors or to lose it, it was selected to adopt the same code used for the cell header also in the data field. The resulting rate of ATM cells data field with errors is approximately 3.7/Yr.

System evolution might exhibit requirements for higher reliability, but adoption of FEC techniques limits the possibility to further improve the quality of service. In fact, although a random error distribution would in principle allow use of a further correction code for quality applications, due to the differential modulation scheme used, errors are correlated: this makes it extremely impractical to implement any further correction stage in series to FEC used at the physical level.

The only way to overcome such a difficulty consists in a conservative selection of the FEC, as done here, allowing for the foreseeable evolution of the system.

5. CONCLUSIONS

In conclusion coherent optics make it possible to build a network with a large capacity and advanced performance in handling the information flow.

Despite stringent requirements on source linewidth, phase modulation in conjunction with carrier generation from a single source would appear to be an attractive solution for the optical configuration of the network.

FEC techniques can be used to relax technological constraints on the network subsystems and/or to extend both the network geographical coverage and the size of the central star. The main objective of future work is to show experimentally the increased performance and the flexibility built into the proposed system concept.

ACKNOWLEDGEMENT

This work was partially supported by the Commission of the European Communities in the context of the ESPRIT programme.

REFERENCES

[1] A. Fioretti, C.A. Rocchini, P.J. Wilkinson, A.J. Haylett, *Design Issues of O.S.I. Layers in Integrated Services Multichannel Metropolitan and Regional Area Networks,*

Proc. Int. Workshop on LAN Management, Berlin, 1987.
[2] A. Fioretti, C.A. Rocchini, P.J. Wilkinson, A.J. Haylett, *A New Protocol for Multiservice Integration over a High Speed Fibre Optic L.A.N. based on Star Topology,* IFIP WG 6.4 Workshop, Aachen, 1987.
[3] A. Fioretti, E. Neri, S. Forcesi, A.E. Green, P.N. Fernando, A.C. Labrujere, O.J. Koning, J.P. Bekooij, G. Veith, H. Schmuck, *Technology Aspects of a Coherent Optical M.A.N.,* Spie's OE/Fibers '89, Coherent Lightwave Communications, Boston, 5-8 September 1989.
[4] D.W. Rush, G.L. Burdge, P.T. Ho, *The Linewidth of Mode-Locked Semiconductor Lasers Caused by Spontaneous Emission: Experimental comparison to Single Mode Operation,* IEEE J. Quantum Electron. 22, 1986, p. 2088.
[5] R.F.M. Van den Brink, *Optical Receiver with Third-Order Capacitive Current-Current Feedback,* Electron. Lett. 24, 1988, pp. 1024-1025.
[6] W.R. Byrne, A. Papanicolau, M.N. Ransom, *World-wide Standardization of Broadband ISDN,* Intern. Journ. of Digital and Analog Cabled Systems, Vol. 1, 177 (1988).
[7] J.P. Coudreuse, *ATM: a Contribution to the Debate of Broadband ISDN,* Intern. Journ. of Digital and Analog Cabled Systems, Vol. 1, 177 (1988).
[8] J. Salz, B.R. Saltzberg, *Double Error Rates in Differentially Coherent Phase Systems,* IEEE Transactions on Communications Systems, June 1964, pp. 202-206.
[9] W. Peterson, E.J. Weldon, *Error Correcting Codes,* MIT Press, Cambridge, Massachussets, and London, England, 1972, pp. 120, pp. 269-307.

Optically-Processed Control of Photonic Switches

Paul R. Prucnal
Philippe A. Perrier

Department of Electrical Engineering
Princeton University
Princeton, New Jersey 08544
USA

As transmission rates in fiber-optic networks increase, routing optical signals using electronic processing will become increasingly difficult. This limitation can be overcome by performing the required processing optically. Self-routing of optical signals through a switching node using optically-processed control is demonstrated. Packet headers are encoded with packet destination addresses using either optical code-division or time-division encoding schemes. An optical routing controller reads the destination addresses and appropriately sets the optical switch using an optical look-up table. The results of several experiments demonstrating optical control of a photonic switch are described.

1. INTRODUCTION

Today's communication networks, whether of the packet- or circuit-switched variety, are characterized by hybrid architectures. In these architectures, electronic nodes are connected by optical links. Switching is usually performed electronically because of the advanced state of electronic switching technology.

An obstacle to fully utilizing the large bandwidth-distance product of the optical fiber is the relatively low bandwidth of the opto-electronic interfaces, the electronic switches, and the electronic signal processing. This obstacle could be overcome if signals remained in optical form during switching and signal processing.

The first step to completing the optical transmission path is to replace electronic switches with photonic switches. Photonic switches have several advantages over their electronic counterparts. First, photonic switches have a large transmission bandwidth (in excess of 1 THz), and faster switching time (sub-picosecond) than electronic switches [1]. Second, photonic switches eliminate the delays associated with opto-electronic conversion. Third, avoiding the opto-electronic interface decreases the hardware complexity and failure rate of the node. These features will be especially important in future multiple-services communication systems.

The second step in completing the optical transmission path is to replace the electronic signal processing associated with switching by optical signal processing. This can avoid a data flow bottleneck at the input to the photonic switch and eliminate the need for flow control. Unfortunately, the capabilities of present-day optical signal processing technology are rather limited. In order to control the photonic switch using optical processing, the network must be designed to simplify the processing at each node as much as possible. In this paper, we will first discuss the types of network architectures and routing strategies which are best-suited to optical processing. Then, we will present an optical routing processor and address-encoding schemes for optically controlling a photonic switch. Finally, we will describe the results of several experiments that have been performed demonstrating optical control of a photonic switch.

2. OPTICAL ROUTING PROCESSOR ARCHITECTURE

Many functions must be performed to control a switching node, including retiming, regeneration, amplification, and routing control. Performing these functions in real time may require extensive parallel processing hardware. To implement these functions with our present limited optical processing capability, we need to minimize the complexity of the functions that must be performed at each switching node. Network architectures have already been developed in the context of fast packet switching which separate low-level functions, such as switching and routing, from high-level functions, such as session setup. In these architectures, low-level functions are performed in the core of the network, whereas high-level functions are performed in the periphery of the network (a core and edge logical network structure [2]). The high-level functions require a large amount of processing, but only low-bandwidth, and can easily be performed electronically. The low-level functions divide the processing burden among all the nodes. Low-level functions must be performed at high speed, which may not be possible with electronics. We will consider an optical network architecture that includes electronically-processed, low-bandwidth, high-level functions at the periphery of the network, and optically-processed high-bandwidth low-level functions inside the network.

2.1. Optical Fixed-directory Self-routing

One important low-level function is routing. Routing strategies that direct messages to the desired path with simple algorithms are most suitable for optical processing. Since "deterministic" routing specifies the routes for all source-destination pairs in advance, the processing required at each node in this case is minimal. As a specific example, "fixed-directory" routing maintains a routing table containing an outgoing link for each destination [3,4]. As we shall see, this routing table can easily be implemented using optical processing.

Figure 1 provides an illustration of optical fixed-directory routing in a switching network. The network shown consists of five nodes, A through E. Seven stations, 1 through 7, are attached to the network at different nodes. Based on the destination address of the incoming data, a decision is made at each node as to which outgoing link the data should follow. For example, data arriving at node A with destinations 3 or 4 are routed through node B to node D, to which stations 3 and 4 are attached. Data arriving at node A with destinations 5 or 6 are routed through node C to node D, to which station 5 is attached, and to node E, to which station 6 is attached.

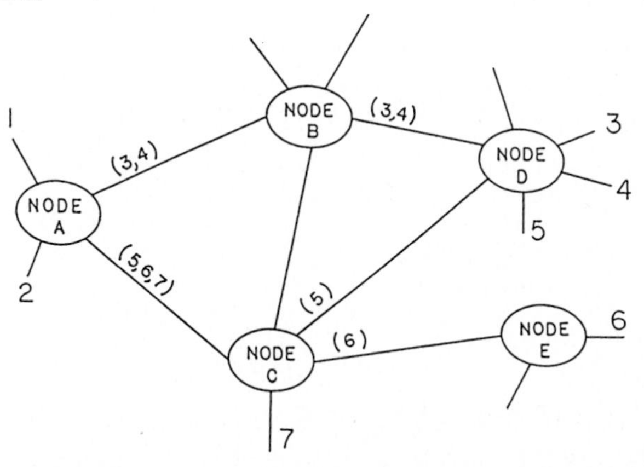

Fig. 1: Illustration of a switching network using optical fixed-directory routing.

Optically-processed fixed-directory routing is most suitable for a fixed network topology, where the connections at each node are set and left unchanged for periods of time, thus minimizing the required processing. Then the optical look-up table need only be altered when topological changes occur in the network due to node or link failure. For example, in the switching network shown in Fig. 1, if the link between nodes C and D fails, node C can be reconfigured to route data for destination 5 through node B.

To further minimize the complexity of the required optical processing, the destination address should be preserved along with the data during the routing process. In this way, the data can "self-route" itself through successive nodes in a switching network.

2.2. Design of an Optical Pipelined Routing Controller

A generalized photonic switching node consists of an optical switch and an optical routing controller, as shown in Fig. 2. An optical routing controller must perform the following functions: 1) recognize an address field, 2) determine the outgoing link (using an optical look-up table), 3) generate a control signal that will set the appropriate permutation of the photonic switch.

To route a packet in real-time, the controller must be able to accept a new input address in a time t that is less than or equal to the packet length, T_p:

$$t \leq T_p \tag{1}$$

At high bit rates, electronic processing would not be sufficiently fast to satisfy Eq. (1), resulting in a data flow bottleneck at the node input. With optical processing, t can be decreased substantially, increasing the allowable throughput of the node.

Once all the physical mechanisms for decreasing t have been exhausted, if Eq. (1) is still not satisfied, then parallel processing or pipelining can be used to further reduce t. By connecting a system of K processors in parallel, and sequentially allocating the input data to the individual processors, a K-fold increase in performance is obtained. However, it may be inconvenient to replicate the processing units K times, and to sequentially allocate the data to the K individual processors. Alternatively, a K-fold increase in performance is also obtained by forming a systolic pipeline, which partitions the processing into a sequence of K discrete processing stages, each of duration t_i. If Eq. (1) is satisfied for $t_i = t$ in each stage of the pipeline, then there is no data flow bottleneck.

As shown in Fig. 2, the routing controller can be divided into two pipelined stages: a linear

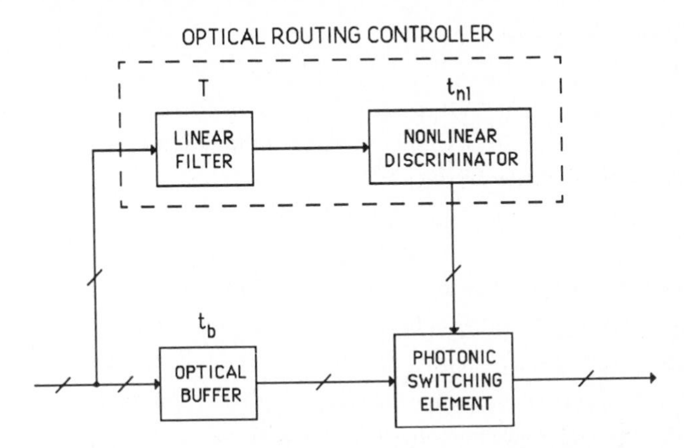

Fig. 2: Self-routing optical processor architecture.
The slash indicates multiple connections.

processing stage and a nonlinear processing stage. A correlator, consisting of fiber-optic delay lines, can act as an optical look-up table to recognize the destination address of each packet, as discussed in the next section. The maximum processing time of the correlator is T, where T is the length of the packet header, which satisfies Eq. (1) for the linear stage. The correlator output is then threshold-detected by a nonlinear optical logic element. To avoid a data flow bottleneck in the pipeline, Eq. (1) requires that

$$t_{nl} \leq T_p, \tag{2}$$

where t_{nl} is the full-width half-maximum of the impulse response of the nonlinear element. Switching requires synchronization between the control signal and the arrival of the information signal at the switch. An optical delay of length t_b is therefore required at the switch input, to match the time for the pipe to fill, that is,

$$t_b = \sum_{i=1}^{k} t_i \tag{3}$$

Summing the linear and nonlinear processing times, the required delay length is

$$t_b = T + t_{nl}. \tag{4}$$

If Eq. (2) and (4) are satisfied, then the pipeline generates a control signal in synchronism with the arrival of the data at the input to the switch. Though the pipeline introduces a propagation delay t_b in the switch, no data flow bottleneck occurs since switching is carried out in real-time.

3. OPTICAL ROUTING ALGORITHM

The routing of incoming data streams at a switching node requires recognizing an address. This process can be easily carried out optically by correlation. Optical address recognition using correlation was previously demonstrated using spread spectrum [5] and time-division encoding techniques [6]. These encoding schemes, and experimental demonstrations of optical fixed-directory self-routing for each of these techniques, are described below.

3.1. Optical Code-division Routing Algorithm

Spread spectrum code division (CD) has been investigated as a technique for coding a packet with its destination address.

To perform self-routing, each packet header of length T is encoded with a waveform s(t) that corresponds to a code sequence of M chips (chip duration is τ) representing the destination address of that packet (see Fig. 3).

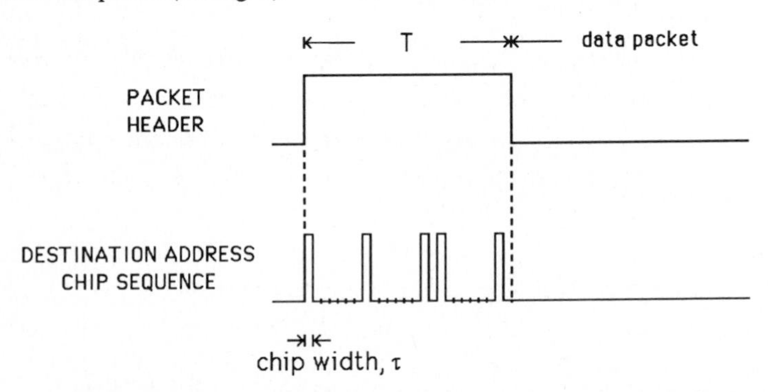

Fig. 3: Destination address encoding of a packet header with a spread spectrum chip sequence.
Number of chips, M=25; number of 1's, N=5.

At the switching node, an optical look-up table, based on an optical matched filter, correlates its own stored address f(t) with the received signal s(t). The controller output r(t) is

$$r(t) = \int_{-\infty}^{+\infty} s(z)f(t-z)dz \tag{5}$$

If the signal is to be switched, then s(t) = f(t), and Eq. (5) represents an autocorrelation function. If the signal is not to be switched, then s(t) ≠ f(t), and Eq. (5) represents a cross-correlation function. To maximize the discrimination between the signal to be switched and all other signals at the controller, it is necessary to maximize the peak of the autocorrelation function and minimize the standard deviation of the cross-correlation function [7]. This is accomplished by selecting a set of orthogonal code sequences. Increasing the processing speed, by using optical processing, allows a reduction in chip width and an increase in M. Increasing M yields an increase in the number of orthogonal sequences, and consequently, in the number of assignable addresses.

The set of code sequences that will yield adequate discrimination between the auto-and cross-correlation functions, depends on the nature of the correlation process used. A fundamental difference exists between optical correlation and conventional electronic correlation. Conventional electronic correlation can be based on electrical delay-lines which coherently combine tapped signals [8]. Though optical signals can also be processed coherently, this is not practicable at the present time, owing to the high frequency of the optical carrier. A more feasible optical correlation technique employs fiber-optic delay lines which incoherently combine tapped signals [9]. This results in a simple summation of optical power.

Thus conventional codes (e.g., Gold codes) which exhibit good orthogonality properties with conventional coherent correlation are not suitable in this case. It has been determined that the set of prime code sequences, which are derived from prime sequences of length N obtained from a Galois field GF(N), with N a prime number, exhibits good orthogonality properties with incoherent optical correlation [7]. The N distinct code sequences thus obtained have length $M=N^2$. The number of 1's in any code sequence, and therefore the peak of the autocorrelation function equals N. On the other hand, the cross-correlation peak equals the maximum number of coincidences of 1's in all shifted versions of any two code sequences. This maximum value does not exceed 2, independent of N. The prime code sequences have been chosen to represent destination addresses in an experimental demonstration of the optical control of a photonic switching element.

The optical routing controller consists of a matched filter (bank of fiber-optic delay lines connected to a fiber-optic summer) and a discriminator. The length of the optical delays match the positions of the pulses in the sequences representing the addresses to be switched.

Delays that correspond to the positions of the pulses in the sequences representing the addresses not to be switched are not connected to the fiber-optic summer. At most N(N-1)+1 delays are connected to the summer in the case where all N addresses are to be switched. A header, arriving at the optical routing controller and encoded with an address to be switched, will generate an autocorrelation peak at the summer output. The discriminator threshold-detects the autocorrelation peak and generates a gating pulse of duration T_p, the packet length, that sets the optical switching element in the switched state. On the other hand, a header encoded with an address not to be switched will not generate an autocorrelation peak. No gating pulse will be generated and the optical switching element remains in the unswitched state.

Due to the lack of optically controllable photonic switches, electro-optic switches can be used. In this case, an opto-electronic conversion is required at the optical controller output to set the electro-optic switch. It is assumed that the photodetector and preamplifier combination used for this purpose have bandwidth greater than $N^2/2T$ and introduce negligible propagation delay.

3.2. Optical Code-division Routing Experiment

The experimental setup for optical code-division routing is shown in Fig. 4. A 3-GHz $LiNbO_3$ integrated-optic waveguide modulator, of the Mach-Zehnder interferometric type, is used as an optical switching element.

Two alternating spread spectrum sequences representing alternating packet headers of length M=32 are produced by a word generator (7 padding zeros were added to the original code sequence to get M=32, a more convenient sequence length to use in the experimental setup). Sequence 1 (corresponding to destination 1) is contained in positions 0 to 31 and sequence 2 (corresponding to destination 2) is contained in positions 32 to 63. For simplicity of implementation, only packet headers are generated and transmitted; no data follows. The transmitted signal thus corresponds to alternating headers for destinations 1 and 2. The chip rate is 100 Mchip/s and the packet header rate is 3.125 Mbit/s. The two alternating spread spectrum chip sequences used in the experiment are shown in Fig. 5.

The word generator drives a 100-MHz, 830-nm semiconductor laser transmitter. The encoded headers at the laser output are directed to both the switching element and the optical routing controller.

The optical routing controller is an incoherent fiber-optic delay line matched filter with tap winding lengths matched to the positions of pulses in sequence 1. The output of the correlator is converted to an electrical signal using a 100-MHz silicon avalanche photodetector/preamplifier combination. Threshold detection of the autocorrelation peak is performed by triggering a pulse generator, which in turn generates a T=320-ns switching pulse. The output of the pulse generator is connected to the electrical gating input of the modulator.

A length of fiber, matched to the routing processing time of the controller, is inserted between the splitter and the optical input of the modulator. As previously mentioned, this delay insures that the packet does not arrive at the modulator before the switch is appropriately set.

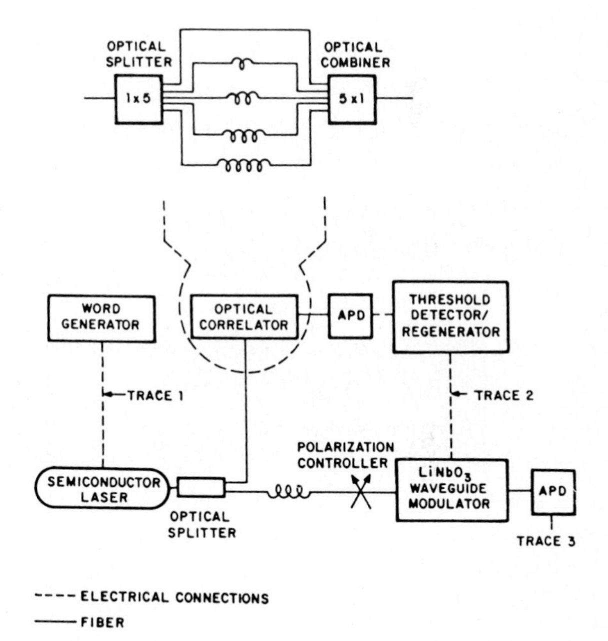

Fig. 4:

Experimental setup for the optical control of a photonic switching element with code-division address encoding.

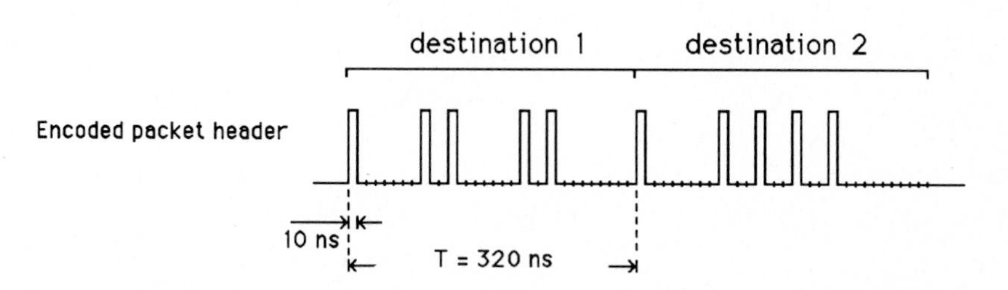

Fig. 5: Orthogonal spread spectrum packet-header sequences used in the experiment.
Destination 1: 10000000100100000001001000000000
Destination 2: 10000000010001000100010000000000
The last seven zeros in both sequences were added for experimental convenience.

Demonstration of routing through an optically controlled photonic switch is shown in Fig. 6. The alternating electronic spread spectrum sequences at the word generator output are shown in Fig. 6(a), trace 1. Sequence 1 generates an optical autocorrelation function at the output of the fiber-optic correlator (see Fig. 6(b)), which triggers a 320-ns electrical switching pulse at the pulse generator output (see Fig. 6(a), trace 2). This switching pulse opens the modulator so that sequence 1 passes through. Sequence 2 generates a cross-correlation function at the output of the fiber-optic correlator, which does not generate a switching pulse. The modulator does not open and sequence 2 does not pass through the modulator. The modulator output is shown in Fig. 6(a), trace 3.

Self-routing of packet headers through an 8x8 integrated-optic crossbar switch [10] was demonstrated using the same encoding technique [11]. In that experiment, 100 Mbit/s packet headers, encoded with 12.5 Gchip/s destination address sequences, were used to control the switch matrix.

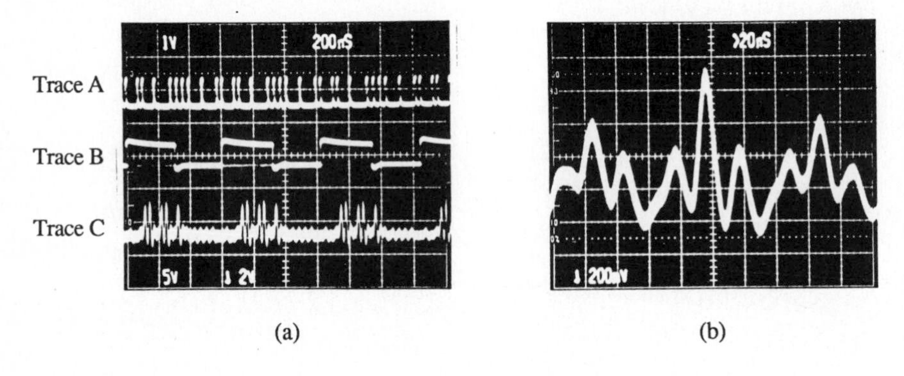

Fig. 6: Optical control demonstration of a photonic switching element using code-division address encoding
a) Trace A: alternating electronic packet-header sequences 1 and 2
Trace B: 320-ns gating pulses generated after recognition of packet-header sequence 1 by optical routing controller
Trace C: packet-header sequence 1 self-routed through photonic switching element
b) autocorrelation function for packet-header sequence 1, detected at the output of the correlator

3.3. Optical Pulse-interval Time-division Routing Algorithm

In the CD encoding technique described above, the destination address of each packet is represented by a chip sequence of length N^2, where N corresponds to the number of destinations. For a given packet header rate, as N becomes large, the generation of narrow pulses becomes increasingly difficult.

In this section, a pulse-interval time-division (TD) encoding technique for the optical control of a photonic switching element is reported. Here, the destination of each packet is encoded with a sequence of length N+1, representing one of N destination addresses. This technique requires approximately a factor of N less bandwidth that the CD encoding technique, and thus results in a more efficient utilization of the channel bandwidth. Though the pulse-interval TD encoding technique is more bandwidth efficient, the CD encoding scheme, with its capability to distinguish superimposed orthogonal sequences, can allow asynchronous concentration of several input signals to the same switch output.

A block diagram of optical pulse-interval TD routing of a photonic switch is shown in Fig. 7. As before, the photonic switching node consists of an optical switching element with two possible states, switched (cross) or unswitched (bar); an optical routing controller which decodes the destination address and sets the state of the optical switching element; and an optical buffer which matches the delay of the input packet to the processing delay of the optical routing controller.

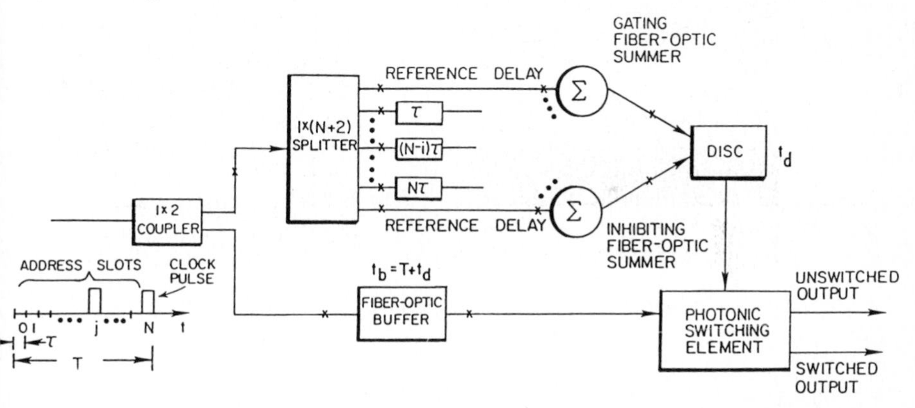

Fig. 7: Switching node architecture for the optical control of a photonic switching element using a time-division pulse-interval destination address encoding technique. Address i is switched if delay $(N-i)\tau$ is connected to the gating fiber-optic summer, whereas it is unswitched if that same delay is connected to the inhibiting fiber-optic summer

A data packet, containing a fixed number of bits, L+1, each of duration τ, is preceded by a packet header, of duration T; the length of the entire packet is given by $T_p=(N+L+2)\tau$. To perform self-routing, the packet header is encoded at the transmitter with the destination address of the data packet At the switch, part of the encoded optical signal is directed to the optical decoder, which reads the destination address of the packet header. If the data packet is to be routed to the switched output of the switching element, a gating pulse is generated that sets the switch in the cross state. If the data packet is to be routed to the unswitched output of the switching element, no gating pulse is generated, which sets the switch in the bar state.

Encoding is performed at the transmitter by dividing the packet header into a frame of N+1 bits of duration τ. The header bits are denoted i=0,...,N. The Nth bit is always occupied by a clock pulse. One of the remaining bits i=0,...,N-1 is occupied by a packet-address pulse, and that bit will be referred to as the "address". The packet format is shown in Fig. 8. Those addresses that correspond to the switched output form a subset, S, of all possible addresses i=0,...,N-1; that is, $S=\{i|\text{the } i^{th} \text{ address corresponds to the switched output}\}$. The complement of that subset, S^c, contains the addresses that correspond to the unswitched output.

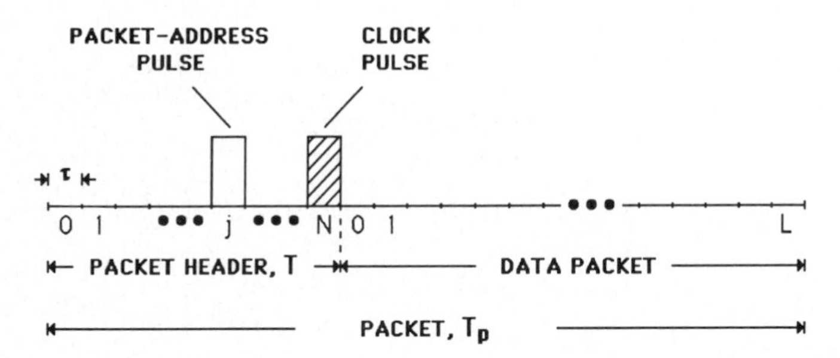

Fig. 8: The packet consists of a packet header, encoded using the pulse-interval technique, followed by the data packet.

The optical routing controller consists of a 1x(N+2) splitter, a set of N+2 fiber-optic delay lines, a "gating" fiber-optic summer, an "inhibiting" fiber-optic summer, and a discriminator. The set of fiber-optic delay lines is comprised of two equal reference delays and a set of N delays of length equal to $(N-i)\tau$ plus the reference delay, where i=0,...,N-1. The reference delays are connected to the gating and inhibiting summers. The remaining delays are connected to either the gating or the inhibiting summer, as follows: the delay $(N-i)\tau$ is connected to the gating fiber-optic summer if $i \in S$; the delay $(N-i)\tau$ is connected to the inhibiting fiber-optic summer if $i \in S^c$.

The situation where the delay $(N-i)\tau$ is connected to the gating summer ($i \in S$) is considered first. As illustrated in Fig. 9, this delay shifts the pulse from the ith bit to the Nth bit. The gating summer incoherently adds the shifted address pulse to the clock pulse in the Nth bit, resulting in a pulse of double amplitude. The discriminator threshold-detects the double-amplitude pulse and generates a gating pulse of duration T_p. The discriminator must be paralyzable so that a second double-amplitude pulse during the gating period does not cause retriggering. The gating pulse sets the switching element in the cross state, and the entire packet present at the input is routed to the switched output.

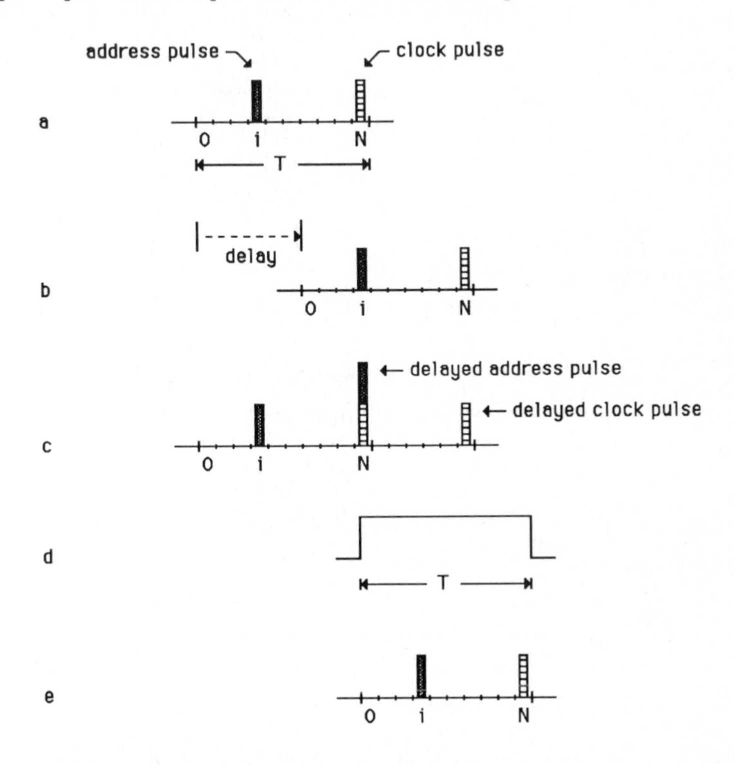

Fig. 9: Optical controller timing diagram. *Transmission and electronic delays are neglected.*

a. output of the reference delay.
b. output of delay (N-i)τ. The original header is delayed by the interval between the clock pulse in bit N and the address pulse in bit i.
c. output of the gating fiber-optic summer. A double-amplitude pulse results when the ith address is to be switched.
d. discriminator output. A gating pulse is generated when a double-amplitude peak is detected.
e. the encoded header is directed to the switched output of the photonic switching element.

The situation where the delay (N-i)τ is connected to the inhibiting summer ($i \in \mathbf{S^c}$) is considered next. As before, the delay shifts the pulse from the ith bit to the Nth bit. The inhibiting summer incoherently adds the shifted address pulse to the clock pulse in the Nth

bit, resulting in a pulse of double amplitude. The discriminator threshold-detects the double-amplitude pulse and is inhibited for a duration T_p. Again, the discriminator must be paralyzable, so that a second double-amplitude pulse during the inhibiting period, whether it be at the gating or the inhibiting summer output, does not change the discriminator output state. The absence of a gating pulse sets the switching element in the bar state, and the entire packet present at the input is routed to the unswitched output.

The inhibiting summer is required since the condition may occur where the address $i \in S^c$, and the interval between the address pulse and the clock pulse matches the interval between two connections to the gating summer. Without the inhibiting summer, this would result in the generation of a false gating pulse by the discriminator.

In the general case, more than one address is to be switched at any one time, so that several delays are connected to the gating summer.

Note also that the delay lines in the routing controller can also be arranged in serial rather than in the parallel configuration shown in Fig. 7 [12].

3.4. Optical Pulse-interval Time-division Routing Experiment

As in the experimental demonstration of optical control of a photonic switch using CD encoding, for simplicity of implementation, only packet headers were generated and transmitted; no data followed.

The experimental system is represented in Fig. 10. Single-mode fiber is used throughout, except where noted. A sequence generator produces a 100 Mbit/s pseudorandom packet header stream which drives an optical modulator A. The 3-GHz $LiNbO_3$ integrated-optic modulator is of the Mach-Zehnder interferometric type, and is preceded by a three-turn fiber-optic polarization controller, PC. The modulator gates a 100-MHz, 0.08% duty cycle, 1.3-μm optical pulse train, produced by a mode-locked Nd:YAG laser. The 10-ns time interval consists of 125 slots, each 80 ps in duration[13].

To encode the packet header, the modulator output is split into two paths. A delay D_3 in one path defines the end of the packet header by positioning the clock pulse in bit 124. The delay in the other path positions the pulse in one of the remaining 124 bits. In the present experiment, one of two address bits is selected by choosing the delay D_1 or D_2 with a 1x2 opto-mechanical switch, OMS. Delay D_1 positions the address pulse in bit 9, whereas delay D_2 positions the address pulse in bit 94.

At the input to the photonic switch, the encoded optical packet header is split into two paths:

one path is directed towards the photonic switching element; the other path feeds the optical routing controller. The photonic switching element, modulator B, is of the same type as modulator A, with one optical input and one optical output.

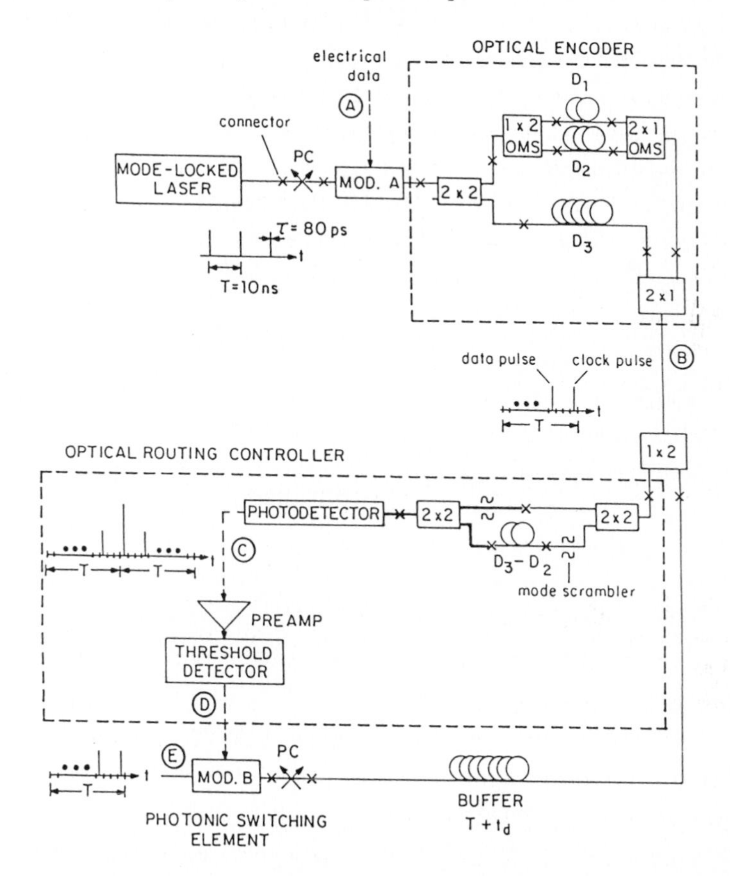

Fig. 10: Experimental setup for the optical control of a photonic switching element with time-division pulse-interval address encoding.
PC, polarization controller; OMS, opto-mechanical switch; MOD, electro-optic modulator.

In general, when several addresses correspond to the switched output, a set of connections must be made to the gating fiber-optic summer. However, in the particular case, demonstrated in this experiment, only one address corresponds to the switched output, only one connection needs to be made to the gating fiber-optic summer. The inhibiting fiber-optic summer is unnecessary in this case. The optical routing controller consists of a splitter, a set of parallel delays, a combiner, a photodetector, and a discriminator. The delay D_3-D_2 selects address 94 to be switched, whereas address 9 is not switched. The use of a multimode 2x2 combiner (thick lines in Fig. 10 represent multimode fiber), with a mode scrambler at each of its inputs, averages out interference effects between coherent pulses. The output of the combiner is incident on a 5.3-GHz Ge pn diode photodetector, followed by a 1-GHz preamplifier, and an electronic Schmitt trigger threshold-detection circuit. When a pulse is present in bit 94, it is added to the clock pulse in bit 124, resulting in a double-amplitude pulse at the photodetector output. The double-amplitude pulse exceeds the threshold of the Schmitt trigger, and a 10-ns electrical gating pulse is generated. This gating pulse opens the modulator, allowing the 10-ns header to pass through. When a pulse is present in bit 9 (or any bit other than bit 94), no gating pulse is generated, and the modulator remains closed. A fiber buffer of length $T + t_d$ is inserted before modulator B to synchronize the arrival of the header with the gating pulse at the modulator. Here, t_d includes the time delay introduced by the discriminator as well as the time delays due to the preamplifier and the photodetector.

The experimental results in the case of switched and unswitched packet header, as displayed on a 1-GHz bandwidth oscilloscope, are shown in Fig. 11(a) and Fig. 11(b), respectively. In Fig. 11(a), trace A, an electrical data sequence is shown that corresponds to the transmission of packets (point A in Fig. 10). Trace B shows the corresponding encoded optical headers (point B in Fig. 10). Each packet header has a clock pulse in bit 124 and an address pulse in bit 94; "0s" indicate that no packet are transmitted. The output of the address decoder is shown in trace C (point C in Fig. 10). The presence of the delay D_3-D_2 (30 bits x 80 ps/bit = 2.4 ns) at the optical routing controller retards the address pulse in bit 94 to the position of the reference pulse in bit 124, resulting in a double-amplitude peak in bit 124 after summation. As seen in trace D, each double-amplitude peak generates a T_p (= T) = 10-ns gating pulse at the discriminator output (point D in Fig. 10), which opens the modulator. The gated output of the modulator at point E in Fig. 10 is shown in trace E. The bit-error rates at the optical controller and switching element outputs were each measured to be less than 10^{-9}.
In Fig. 11(b), trace A, an electrical data sequence corresponding to the transmission of packets is shown. Trace B shows the series of encoded optical headers. Each packet header has a clock pulse in bit 124 and an address pulse in bit 9. The output of the address decoder is shown in trace C. Since the delay between bits 9 and 124 (115 bits x 80 ps/bit = 9.2 ns) is not equal to D_3-D_2, no double-amplitude pulse appears at the decoder output in trace C. Consequently, no gating pulse is generated (trace D), the modulator remains closed, and the encoded header does not appear at the modulator output (trace E).

Since, in the pulse-interval TD encoding technique, each packet carries its own time reference, no synchronization is required at the optical routing controller. The use of a self-clocked optical controller simplifies the network design. Indeed, on a given input link to the switch, packets can arrive asynchronously. (As in the case with any switch, conflicting routing information at multiple input ports would lead to a collision, unless a collision-avoidance protocol is used.) Furthermore, successive packets in a data stream need not have the same destination address.

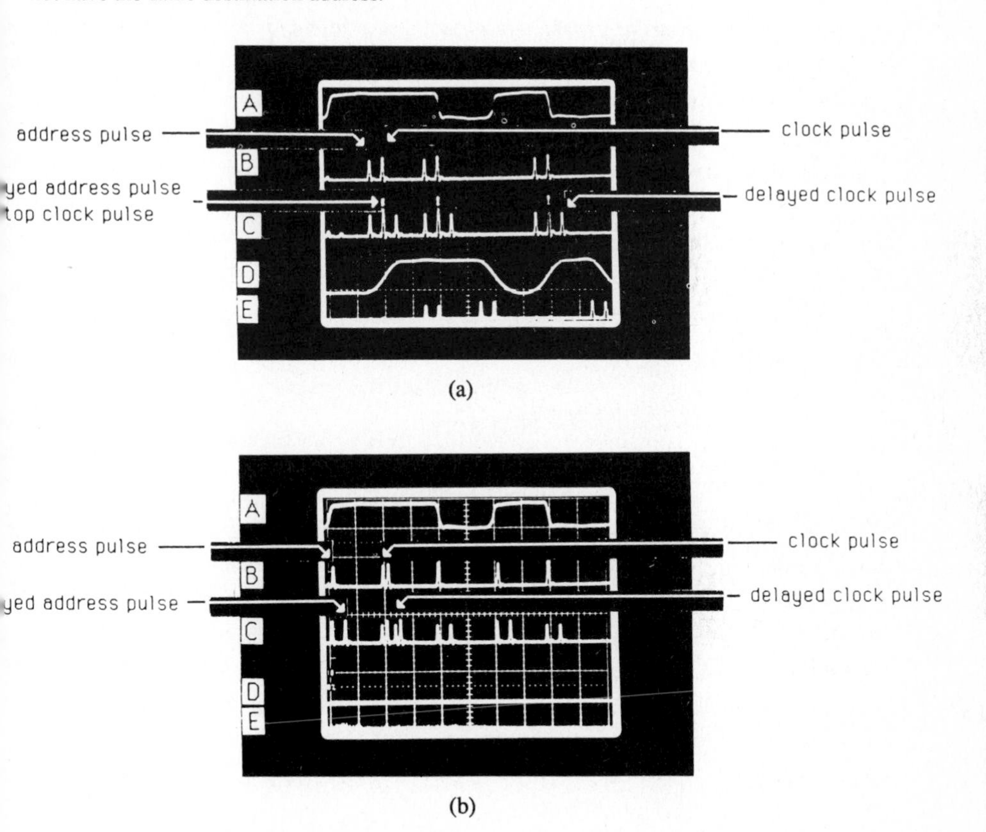

Fig. 11: Optical control demonstration of a photonic switching element using time-division pulse-interval address encoding.
a) switched data
b) unswitched data
Trace A: 100 Mbit/s electrical packet-header sequence
Trace B: encoded optical packet headers. 10-ns long packet header; 80-ps bit width; 125 bits per packet header.
Clock pulse in bit 124; address pulse in bit 94 (bit 9) in a) (b)).
Trace C: optical address decoder output tuned to bit 94
Trace D: discriminator output
Gating pulses in a) set photonic switching element in the cross state.
Trace E: photonic switching element switched output

4. CONCLUSION

For present-day transmission rates in fiber-optic packet- or circuit-switched networks, routing decisions are easily made with electronics. As the transmission rates increase, however, it will become more difficult to perform routing using electronic processing. This limitation can be overcome by using optical processing. Furthermore, with the anticipated development of opto-optic switches, switching networks can be all-optical if optical processing is also used to perform routing.

Several optical routing algorithms for real-time optical control of a photonic switch using time-domain optical processing have been reported. In each case, an optical routing controller, based on a look-up table architecture, allowed the self-routing of data through a switching node. The results of several experiments demonstrating optical control of a photonic switching element at 12.5 Gbaud were presented. In one experiment, 10-ns long packet headers, containing 12.5 Gchip/s code-division address sequences, were successfully optically self-routed through an 8x8 integrated-optic crossbar switch. In a second experiment, self-routing of similar packet headers was accomplished using 12.5 Gbit/s pulse-interval time-division address sequences.

In the future, optical processing for routing control may also be performed dynamically in two or three dimensions, using, for example, spatial light modulators and holographic optical elements. The development of digital optical logic elements may extend the realm of possible optical signal processing and computing functions far beyond that which can be achieved with passive structures. This could enable successively higher functional levels in the network, currently implemented in electronics, to be replaced by optical signal processing.

REFERENCES

[1] B. Saleh and M. Teich, Fundamentals of Photonics, Chapter 21, in press

[2] P. Cinato and A. de Bosio, "Optical technology applications to fast packet switching," OSA Proceedings on Photonic Switching (Salt Lake City, UT, March 1-3, 1989) (J.E. Midwinter and H.S. Hinton, Editors), Vol. 3, pp. 233-236

[3] M. Schwartz, Telecommunication Networks: Protocols, Modeling and Analysis, Addison-Wesley Publishing Co., 1987

[4] P.R. Bell and K. Jabbour, "Review of point-to-point network routing algorithms," IEEE Communications Magazine, Vol. 24, No. 1, pp. 34-38, 1986

[5] P.R. Prucnal, D.J. Blumenthal, and P.A. Perrier, "Photonic switch with optically self-routed bit switching," IEEE Communications Magazine, Vol. 25, No. 5, pp. 50-55, 1987

[6] P.A. Perrier and P.R. Prucnal, "Self-clocked optical control of a self-routed photonic switch," IEEE Journal of Lightwave Technology, Vol. 7, No. 6, pp. 983-989,1989

[7] P.R. Prucnal, M.A. Santoro, and T.R. Fan, "Spread spectrum fiber-optic local area network using optical processing," Journal of Lightwave Technology, Vol. LT-4, No. 5, pp. 547-554, 1986

[8] K.P. Jackson, S.A. Newton, B. Moslehi, M. Tur, C.C. Cutler, J.W. Goodman, and H.J. Shaw, "Optical fiber delay-line signal processing," IEEE Transactions on Microwave Theory and Techniques, Vol. MTT-33, No. 3, pp. 193-209, 1985

[9] M.A. Santoro and P.R. Prucnal, "Asynchronous fiber optic LAN using CDMA and optical correlation," Proceedings of the IEEE, Vol. 75, No. 9, pp. 1336-1338, 1987

[10] P. Granestrand, B. Stoltz, L. Thylen, K. Bergvall, W. Doldissen, H. Heinrich, and D. Hoffmann, "Strictly nonblocking 8x8 integrated optical switch matrix," Electronics Letters, Vol. 22, No. 15, pp. 816-818, 1986

[11] D.J. Blumenthal, P.R. Prucnal, L. Thylen, and P. Granestrand, "Performance of an 8x8 $LiNbO_3$ switch matrix as a gigahertz self-routing switching node," Electronics Letters, Vol. 23, No. 25, pp. 1359-1360, 1987

[12] P.R. Prucnal, M.A. Santoro, and S.K. Sehgal, "Ultra-fast all-optical synchronous multiple access fiber networks," Journal on Selected Areas in Communications, Vol. SAC-4, No. 9, pp. 1484-1493, December 1986

[13] P.R. Prucnal, D.J. Blumenthal, and M.A. Santoro, "12.5 Gbit/s fibre-optic network using all-optical processing," Electronics Letters, Vol. 23, No. 12, pp. 629-630, 1987

An Integrated-Services Digital-Access Fiber-Optic Broadband Local Area Network with Optical Processing *

Mario A. Santoro

AT&T Bell Laboratories
Holmdel, New Jersey 07733

This paper describes a medium access protocol proposed for an integrated services digital-access fiber-optic broadband LAN. The protocol uses a TDM frame format to allocate the various services. The time slots in the frame are divided between low-bandwith channels and high-bandwidth channels. Low-bandwidth channels occupy only a few of the time slots at the beginning of the frame. Voice and data are transmitted in the low-bandwidth channels using a synchronous CDMA scheme where the encoded sequences extend over many time frames but always occupy the same time slot. Since these services require a much lower rate than the high-bandwidth services, very long code sequences can be used. Thus, by selecting appropriate code sequences, many users can share the same channel. High rate services such as digitized video and HDTV are allocated the remaining time slots either on a dedicated or by demand basis. The proposed protocol uses optical processing to achieve the highest throughput for the network.

I. Introduction

Local area networks (LAN) were originally devised to interconnect various computer hosts and peripherals in order to use them more efficiently by sharing. However, the bursty nature of the traffic in computer networks renders the physical link idle for long periods of time. Therefore, other services can be accommodated on the same medium. The idea of sharing the physical link by both data and voice has been extensibly investigated and many proposals are being considered as standards for the network implementation [1]. The enormous bandwidth provided by the use of optical fibers as the physical link for LANs allows for the integration of other services such as digitized video, teleconferencing, fast data transfer and graphics, all of which require a high bandwidth. Furthermore, since the signal in the physical link is optical, it can be processed optically at much higher rates than if it were processed electronically.

In view of the above, this paper proposes an integrated services fiber optic broadband LAN with optical processing. The network can accommodate various type of traffic demands. It can be readily interfaced with similar networks to form a much larger (wide area) network. The physical layer of the proposed network is presented in section II. The media access protocol is described in section III. Section IV outlines the feasibility of the network. Section V examines ways to interface the proposed network to other networks. Conclusions are given in section VI.

---

* This work was performed while the author was with the Center for Telecommunications Research, Columbia University, as part of his thesis work.

II. Physical Layer

Two different synchronous accessing schemes for fiber optic networks were analyzed in [2,3] and demonstrated [4-6]: time division multiple access (TDMA) and synchronous code division multiple access (CDMA). A TDMA scheme is more suitable for services requiring a high transmission bandwidth (tens of Mb/s), like digitized (interactive or broadcast) video and fast data (file) transfer. CDMA, on the other hand, can accommodate a larger number of simultaneous users with low traffic demands and relatively low signal rates (tens of Kb/s), viz. voice and data communications. By implementing both TDMA and synchronous CDMA on the same physical link as proposed here, services with very different bandwidth requirements can be integrated in the network.

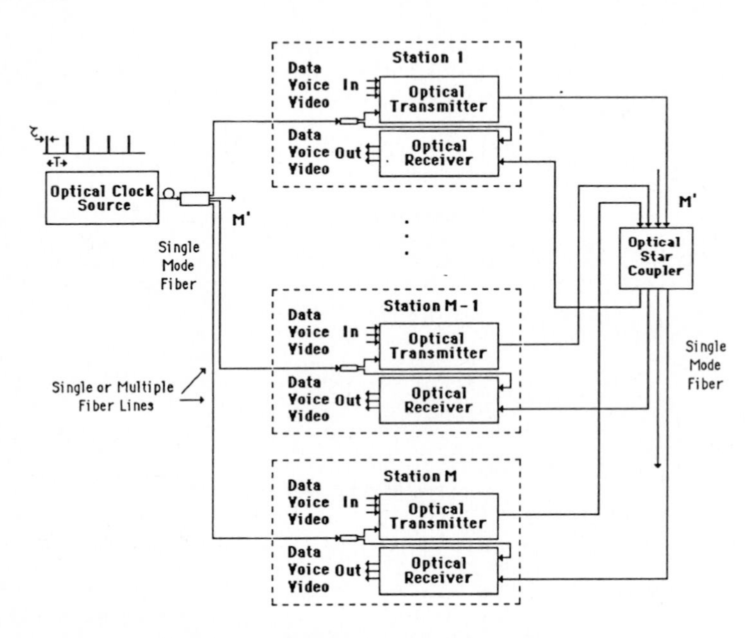

Figure 1: Network Configuration with M Stations using M' Lines

The proposed network topology is depicted in figure 1. A star configuration is chosen to allow the maximum number of users (M) for a given optical power budget. A central optical clock source is used to synchronize the network. Each station in the network receives the optical clock signal from the central optical clock source and distributes it among the various services the station has contracted. As explained in [2], the optical clock source generates very narrow optical pulses at a fixed rate, 1/T. The width of the pulses τ, and the repetition rate 1/T determine the maximum number of times slots S that can be allocated on a frame, i.e. $S = T/\tau$. The limitation in number of slots that can be practically allocated is discussed in section IV.

As indicated in figure 2, each time slot is assigned to either a high- or a low-bandwidth service. S_1 out of the S time slots are allocated for high-bandwidth services, where each one of these S_1 slots transmits a single high bandwidth channel. The rest of the slots, S_2, are allocated for low-bandwidth services. Low-bandwidth services are transmitted in these slots using a CDMA scheme. Therefore, each one of these slots can accommodate

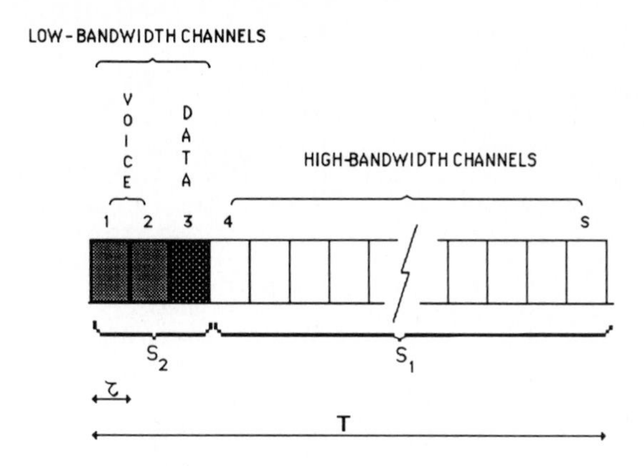

Figure 2a: Transmission Frame

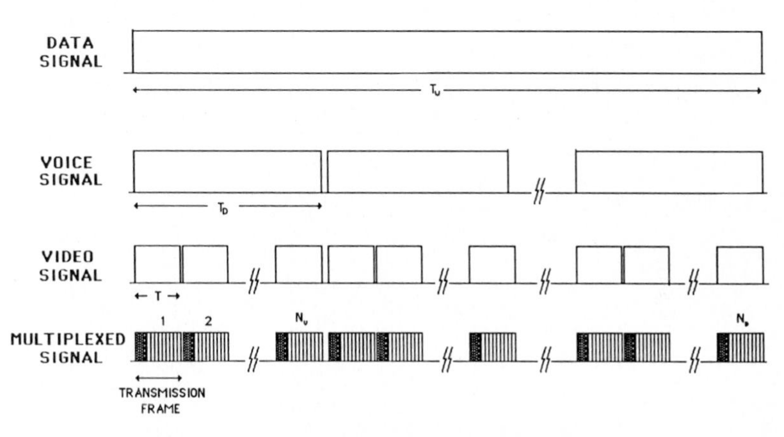

Figure 2b: Time Relation among Different Services and Transmission Frame

many low-bandwidth channels. The number of low-bandwidth channels that can simultaneously use a time slot depends on the number of chips per bit in the code sequence. In the proposed network, this number is determined by the ratio between the bit length of the low-bandwidth signal and the frame length T. Figure 2b shows the time relation among the different signals and the transmission frame. T_v and T_D are the bit length for voice and data signal respectively. Since digitized voice and data have different signal rates, the code sequences to use for each will have different numbers of chips per bit. These numbers are termed N_v for voice and N_D for data.

Figure 3 shows a typical transmitter contracting three different services. The signal from the optical clock is split between the video, voice and data channels. (The power splitting ratios among the three different channels is analyzed in the section on feasibility) For video channels, the digitized, electrical NRZ video signal is directly used to gate the optical clock using an electrooptic modulator. In effect, using this arrangement, the electrical NRZ signal is converted to an optical RZ signal with duty cycle τ/T. The optical signal is then appropriately delayed and sent to the network in the corresponding time slot (ϵ $\{S_1\}$). For voice and data, the electrical signal is first electrically encoded using either a ROM (read-only memory) or PLA (programmable logic array) to which both, intended receiver's address and the data signal are fed. The ROM or PLA generates the encoded signal with rate T_v/T for voice or T_D/T for data. This coded signal gates an electrooptic modulator. The variable optical delay then positions the signal in one of the S_2 slots.

The slot allocation can be either fixed or dynamically assigned depending on the network. Different media access protocols are discussed in the next section.

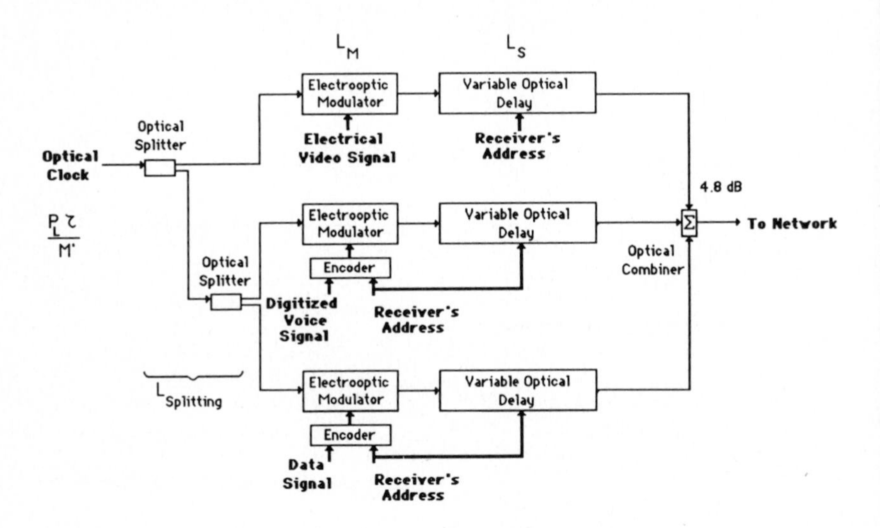

Figure 3: Transmitter

The receiver uses timely delayed optical clock pulses to detect the signal in the different time slots [2]. The diagram of a typical receiver is given in figure 4. To retrieve the signal in the high-bandwidth slots, the delayed clock pulse is directly added to the incoming optical signal. This results in the signal riding atop a clock pulse and a high-bandwidth channel must have a certain amplitude so that the composite signal is above the signal levels in the low-bandwidth (CDMA) slots. This problem is furtherly addressed in section IV.

The coded optical signals in the low-bandwidth slots are retrieved using delay-line fiber-optic correlators [2]. The output of the correlator is added to a delayed clock pulse to make the autocorrelation peak easily distinguishable from the signals in other slots. The composite signals are threshold detected, e.g., using optical bistable devices, and then converted back to electrical signals at the appropriate rate. Optical bistable devices can resolve energy differences of very short (picosecond) pulses. As before, the added clock pulse must be of sufficient high for the threshold detector to retrieve the intended channel.

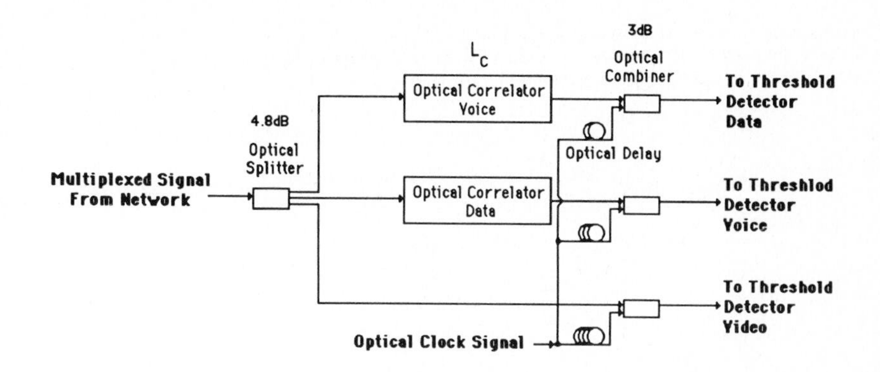

Figure 4: Receiver

III. Media access protocol

The stations in the network can contract various services. In a general network, each station will be provided with the basic voice and data channels for interactive traffic and will also be able to receive video signals. For a fixed assignment protocol, a station will be assigned a particular CDMA code sequence which will serve as its address. Any station wishing to transmit information (either voice or data) will select the address of the intended receiver and will use the CDMA code sequence to encode the information. The total number of simultaneous digitized voice or data channels that can be supported in the network is determined by both the number of chips per bit and the particular coding procedure used. For a prime sequence code used in a synchronous CDMA system the total number of addresses that can be assigned was found to be $(3N-2\sqrt{N})$ (see appendix A) where N is the number of chips per bit in the sequence.

As mentioned in the last section, the code sequence length (either N_V or N_D) is given by the ratio between the frame rate (1/T) and the low-bandwidth signal rate. High quality digitized video signals require 92 Mb/s transmission bandwidth [7]. According to ISDN standards [8], digitized voice requires 64 kb/s. Therefore, if a 92 Mbit/s frame rate is used, the chip rate for a voice channel N_V is 1437 chips/bit. This corresponds to prime code sequences with P = 37 and 68 padding zeros [4]. A data channel requires 16 kb/s. Hence the chip rate for data transmission N_D is 5750 chips/bit. The prime code sequence in this case will have P = 73 and 421 padding zeros. For the numbers mentioned above, and using equation (A5), 4033 possible voice addresses and 15841 data addresses can be assigned per time slot in the time frame. In order to prevent false threshold crossings due to possible overlapping between voice and data coded signals, different time slots are assigned for voice and data. Of the S_2 slots reserved for low rate signals the number of slots for voice channels and data channels will be chosen according to the particular needs of the network. Typically, the number of voice and data channels would be equal in number.

Broadcasted video signals can be received by any station through a channel selector, as shown in figure 5. One station in the network will be in charge of video signal broadcasting. This will be accomplished by time multiplexing the different video channels within the S_1 slots (high bandwidth channels). The receiver simply selects the clock delay that corresponds to the video channel chosen. In a network with 100 video channels, 3 slots for voice channels, and 1 slot for data channels, the optical pulse width required is 104 ps with a pulse repetition rate of 92 million pulses per second. These numbers are well suited for a medium size network (100 to 200 users). For a bit error rate (BER) of 10^{-9}, the maximum number of simultaneous users for a synchronous

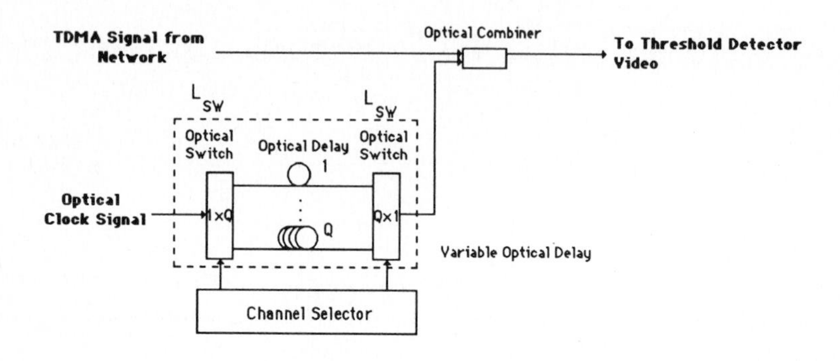

Figure 5: Video Channel Selector

CDMA system with prime sequence codes is 2P (see appendix B). Hence up to 74 voice channels per slot for a total of 222 voice channels and 146 data channels can be accommodated. The total number of assignable addresses is 12099 for voice and 15841 for data. The system performance will be adversely affected if a number of users greater than the number of prescribed channels tried to use the network simultaneously. Therefore a fixed assignment system will be limited in the number of subscribers.

The number of subscribers can be augmented by either increasing the number of slots in the time frame being allocated to low bandwidth services or by making the address assignment dynamic. In the latter case a control station will be in charge of the dynamic address assignment. The control station will poll the different subscribers in the network to see if any service is required. The station polled will either confirmed the request or forfeit its turn. If a service is required, the control station will identify the parties and assigned respective addresses according to channel availability. Polling will be done on a separate control CDMA channel. When not involved in any interactive communication, a station sets its receiver to a preassigned CDMA address (coding sequence) on the polling channel. When polled, the station will respond using the control station's address. A combination of fixed and dynamically assigned services would be more advantageous for large networks.

Interactive video or fast file transfer can be implemented using the high bandwidth channels. For fixed assignment, the station contracting either of these services will be allocated a time slot on which it can receive the high bandwidth signal. The transmitting station will select the slot corresponding to the intended receiver and will transmit the high bandwidth signal in that slot. The number of time slots required per time frame will depend on the number of stations contracting these services. Allocation of bandwidth could also be dynamic. In this case, a control station will allocate the slots according to the service demand. Communications between each station and the control station is done as before through a special control channel. Collisions arising from different stations trying to communicate to the same receiver in a fixed assignment network can be resolved in two different ways. For low rate signals, the reception of the acknowledgement will determine successful transmission, i.e. collisions are resolved at the higher levels of the OSI model. For high rate signals, the transmitting station will listen to the channel prior to transmission by setting up its receiver to that channel. If the channel is deemed idle, transmission will begin. Otherwise transmission is deferred for a random amount of time until the channel is freed.

IV. Network Feasibilities

The need to divide the optical energy among different services requires that the optical clock source provide enough energy to drive the whole network within acceptable performance limits. Based on figures 3 and 4, the power required for a given network performance is given by

$$P_{L_{dB}}\tau = P_{S_{dB}} + 20\log M' + L_M + L_{system} + L_{splitting} + L_S + L_C + 12.6 \quad (1)$$

where $P_{L_{dB}}\tau$ is the laser pulse energy in dB (P_L is the laser peak pulse power and τ is the pulse with), $P_{S_{dB}}$ is the receiver minimum detectable energy in dB, M' is the total number of fiber optic lines, L_M is the electrooptic modulator excess loss, $L_{splitting}$ is the loss introduced by the splitting of the optical clock signal at the transmitter, L_{system} is the total system excess loss due to fiber attenuation, splices/connectors, and various fiber optic components such as splitters and star couplers, L_C is the correlator excess loss for CDMA channels. L_S is the excess loss due to the variable optical delays, which is given by

$$L_S = 2L_{SW} + L_{splice} \quad (2)$$

for a variable delay implemented with switched delay lines (see fig. 5).

L_{SW} and L_{splice} are the excess losses for the switch and the splices respectively. 12.6dB account for the transmitter signal-combining (multiplexing) loss (4.8dB), the receiver signal-splitting (demultiplexing) loss (4.8dB), and the clock/signal-combining loss (3dB).

Notice that in equation (1), the power requirement is given as a function of the total number of fiber optic lines instead of the number of stations in the network to allow for some stations to have multiple input lines. This is required if a station wants to have more than one simultaneous session of the same category of service with other stations in the network. Hence,

$$M' = \sum_{i=1}^{M} (NL)_i \quad (3)$$

where $(NL)_i$ is the number of lines for the ith user.

The use of an electronic encoder in place of the optical encoder reduces the total power requirement compared to the synchronous CDMA system described in [2] by as much as 20 log N. An all-optical encoder could be implemented as shown in figure 6. A pulsed laser is used to generate a string of pulses of length T and rate equal to the repetition rate required by either voice signal or data signal. The coded sequence is generated using fiber optics delay lines [2]. The optical coded data is then used to gate an optical modulator either directly (photorefractive effect) or by converting it first to an electrical signal prior to using it as the gating signal (electrooptic effect). As is evident from the description, an optical encoder is not a cost-effective alternative. It requires an auxiliary pulsed laser with suitably high power output or some optical-to-electrical conversion.

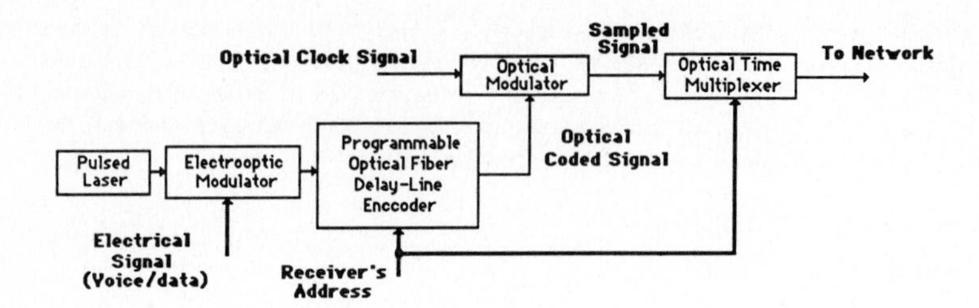

Figure 6: All-Optical Encoder for CDMA

Energy requirements for a typical network (as described in the last section) with 100 users can be calculated from equation (1). A single line per user is allocated, each line accommodating three services: video, voice and data. Assuming all the optical signals have the same peak power (energy), the splitting loss is 4.8dB. Losses due to connectors, splicers and fibers are neglected. The total excess loss in the system is taken to be 10dB allowing for some system margin. The electrooptic modulator excess loss is taken to be 5dB. Optical splitters/combiners and switches are assumed to have an excess loss of 2dB. A minimum detectable energy of 1fJ is assumed. The peak pulse energy required from the optical clock source is 14 nJ. This corresponds to a central optical clock source with 135 W peak power, 104 ps pulses. This values are within the range of commercially available mode-locked lasers.

Due to the nature of the optical processing scheme used, the optical power splitting among the various services in a transmitter (fig. 3) must be done in such a way that errors due to false threshold crossing are minimized. To calculate the proper signal splitting ratio the worst case of interference is considered. Let X be the peak power in a received TDMA signal and Y that of a received CDMA signal. The worst interference

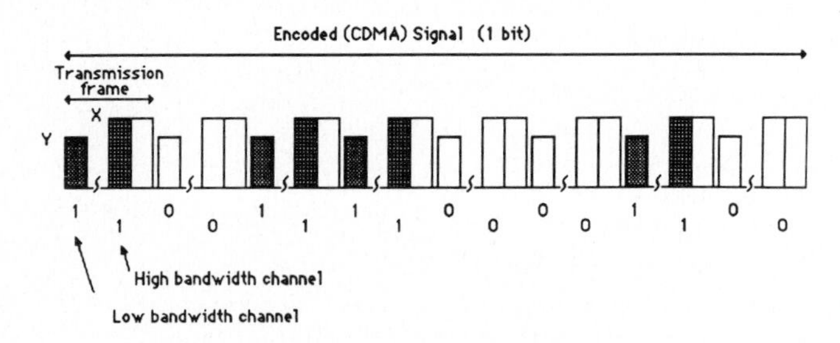

Figure 7a: False Threshold Crossing in CDMA Signal Detection

case for a CDMA signal corresponds to a bit sequence transmitted in a TDMA channel that exactly matches the coding sequence that is to be decoded, as shown in figure 7a. Therefore, by making the clock pedestal for CDMA signals at least as large as X, false threshold crossings are minimized.

False threshold crossings also occur for a TDMA signal when simultaneous CDMA signals in a given slot make the composite CDMA signal large enough (figure 7b). To minimize this error, the clock pedestal for TDMA signals is set at least equal to ρY, where ρ depends on the maximum number of simultaneous users per slot and on the code sequences used, but is not larger than P.

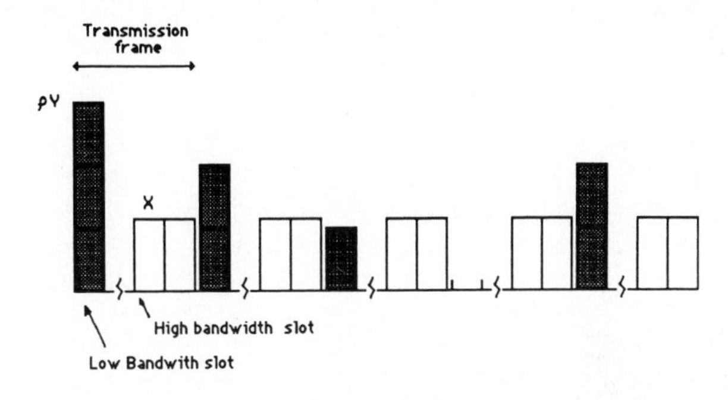

Figure 7b: False Threshold Crossing in TDMA Signal Detection

The ratio between signal and clock pedestal should be maintained well within the dynamic range of the threshold detector, i.e. neither should the combined signal approach saturation conditions in the detector nor should the difference between pedestal and combined signal be very large. This is accomplished by making the ratio between signal and pedestal equal in both TDMA and CDMA channels. To satisfy this condition, $X=\sqrt{\rho}Y$. The energy requirements for a given network performance will increase, since now the splitting loss will be given by

$$L_{splitting} = 10\log\,(2+\sqrt{\rho}) \qquad (4)$$

The power requirements for the clock pedestal are similar to those for the signals. But, instead of doubling the power output of the central laser, it would be advisable to use independent sources for the clock pedestal and the optical clock signal. Thus, possible signal fading due to coherence between the added signals is minimized. The sources could have two different wavelengths so that they can be distributed by the same fiber using wavelength division multiplexing. Provisions must be made to account for possible time delay differences due to dispersion in the fiber.

Finally, the total number of slots that can be allocated in a network will depend on the bandwidth of the optical fibers. Single mode fibers have bandwidth-distance products in excess of 100 GHz-Km. For networks with total fiber lengths shorter than 10 Km, optical pulses of width $\tau \geq$ 100ps will not suffer intersymbol interference due to pulse spreading in the fiber. Then the total number of slots in a time frame can be T/τ. For smaller pulses, the spreading will limit the closeness between the slots in the time frame. Hence the number of slots will be less than T/τ.

V. Network interfacing

Synchronous networks can be interfaced to other networks using internetwork gateways. A gateway must be able to identify the type of service been communicated, set it into the proper format and transmit it using the proper media access protocol.

The gateway will act as a station for either of the interfacing networks. The gateway station will be allocated special control channels which any station in either network can access. A station wishing to communicate with another station outside its local area network will request the service to the gateway station using a special CDMA channel. Channel allocation will be done by a polling scheme, as described in section III. Thus, the total power requirement is not largely increased (the number of lines allocated to the gateway station is kept to a minimum). The gateway station must bring the incoming optical signal power levels from one LAN to a sufficient power level before transmission in the other. The process can be done by converting the optical signal into an electrical signal. This electrical signal then drives an independent optical clock source synchronized to the clock of the network being accessed. Synchronization is done using the RF signal from the central clock in the accessed network. It can also be regenerated using a bistable laser [8]. The latter procedure provides an added advantage: the regenerated signal has only two levels (recall that CDMA slots carry multilevel signals). Thus, interference between TDMA and CDMA signals is minimized (power requirements are reduced accordingly) and since interference between different CDMA code sequences is reduced, more low-bandwidth channels are available. The former procedure allows interfacing with other non-optical networks and with networks with different media access protocols.

VI. Conclusion

An integrated services fiber optic local area network with optical processing was proposed. The network can accommodate a reasonable number of users (between 100 to 200). It can be interfaced to other networks so that the total total number of users could be larger. The network can readily be implemented with commercially available components.

REFERENCES

[1] IEEE Conferences in Local Computer Networks.

[2] P. R. Prucnal, M. A. Santoro, and S. K. Sehgal, "Ultra-fast All-optical Synchronous Multiple Access Fiber Networks," IEEE Special Issue of JSAC, pp. 1484-1497, December 1986.

[3] M. A. Santoro and P. R. Prucnal,"Asynchronous Fiber Optic Local Area Network Using CDMA and Optical Correlation." IEEE Proceedings, vol. 75, pp. 1336-1338, September 1987.

[4] P. R. Prucnal, M. A. Santoro and T. R. Fan, "Spread Spectrum Fiber-Optic Local Area Network using Optical Processing," IEEE/OSA JLT, vol. 4, May 1986, pp. 547-554.

[5] P. R. Prucnal, M. A. Santoro, S. K. Sehgal, and I. P. Kaminow, "TDMA Fiber-optic Network with Optical Processing," Electronics Letters, pp.1218-1219, November 6,1986.

[6] P. R. Prucnal, D. Blumenthal, and M. A. Santoro, "12.5 Gbit/s Fibre-optic Network Using All-Optical Processing." Electronics Letters, vol. 23, pp. 629-630, 4th June 1987.

[7] N.S. Jayant and P. Noll: Digital Coding of Waveforms. Principles and Applications to Speech and Video. Prentice-Hall Inc., Englewood-Cliff NJ, 1984.

[8] M. D. Decina and A. Rover, "ISDN: Architectures and Protocols," in Advanced Digital Communications and Signal Processing, Prentice-Hall Inc., Englewood-Cliff, NJ 1986.

[9] S. Suzuki, et al., "An Experiment on High-Speed Optical Time-Division Switching," IEEE/OSA JLT, Vol.4, July 1986, pp. 894-899.

Appendix A

Number of pseudo-orthogonal coding sequences for an expanded prime sequence code used in a synchronous system.

In principle, any cyclic version of a code sequence can be used in a synchronous system. A particular code sequence of length N can be cyclically shifted $N-1$ times, but not all shifted versions are pseudo-orthogonal. Since any two cyclic versions of a code sequence have the same autocorrelation function in form, the autocorrelation peaks must fall in different time positions to make these two cyclic sequences perfectly distinguishable. Therefore not all the shifted versions of a code sequence can be used.

A prime sequence code is constructed over a Galois Field $GF(P)=\{0, 1,..., j,..., P-1\}$, with P a prime number. The code sequences $C_x=(c_{x0}, c_{x1},..., c_{xi},..., c_{x(N-1)})$ are generating using the formula:

$$c_{xi} = \begin{cases} 1 & \text{for } i= x.j \ mod\ P + jP \ ; j=0, 1,..., P-1 \\ 0 & otherwise \end{cases} \qquad (A1)$$

with $x \epsilon GF(P)$. For example, for $P = 5$, $C_3 =$ (1000000010010000000100100) . The sequence length is $N=P^2$. The length could be extended by padding zeros at the end of the sequence without affecting the code characteristics.

Through computer analysis, it was found that the number of shifted versions of a code sequence which form a pseudo-orthogonal set depends on the number of zeros between the first and second ones in the original sequence. For the first prime code sequence ($x=0$) the number of zeros between the first and second ones is $P-1$. The number of pseudo-orthogonal shifted code sequences for the first code sequence is $2(P-1)+1$. This correspond to twice the number of zeros between the first and second ones, i.e. the sequence can be shifted forward or backwards $P-1$ times, plus the original sequence.

In general the ith sequence in a prime code has $P-1+i$ zeros between the first and second ones. Hence the number of pseudo-orthogonal shifted versions R_i per code sequence is

$$R_i=2(P-1+i)+1 \qquad (A2)$$

with $i=0,1,....,P-1$. The total number of pseudo-orthogonal code sequences that can be generated by cyclic shift of the sequences in a prime sequence code is

$$R = \sum_{i=0}^{P-1} R_i \qquad (A3)$$

Substituting $A2$ in $A3$ and operating, and by noticing that $N=P^2$, the total number of pseudo-orthogonal code sequences in an expanded prime sequence code is

$$R = 3N - 2\sqrt{N} \qquad (A4)$$

For an expanded prime sequence code where padding zeros are used, the total number of code sequences is given by

$$R = 3P^2 - 2P \qquad (A5)$$

The number of pseudo-orthogonal code sequences for a prime sequence code used in an asynchronous system is only $\sqrt{N}$. This represents an increase of $(3\sqrt{N}-2)$ times in number of subscribers for a synchronous CDMA network.

Appendix B

Number of simultaneous users for a given system performance in a synchronous fiber optic CDMA network using an expanded prime sequence code.

In [6], the performance (probability of error) of an asynchronous fiber optic network using prime sequence codes was plotted versus number of simultaneous users, for various code lengths. The signal-to-noise ratio used in the calculations depended on the variance of the interference noise which was found by computer simulation. For a synchronous network, and assuming the only source of noise is interference with other simultaneous users, a false crossing will occur when there was a "zero" transmitted and the interference level is equal the autocorrelation level for a "one". Any two code sequences from a prime sequence code have a crosscorrelation peak of one unit occurring time coincidently with the autocorrelation peak when the prime sequence code is used in a synchronous CDMA system. Since the autocorrelation peak for a transmitted one is P, up to $P-1$ interference users can be in the system simultaneously with the transmitting user and still transmission will have a required performance. Performance will not depend on the interference noise (provided that all users have the same power level) but on the receiver sensitivity and dynamic range.

For a prime code sequence the maximum number of code sequences is $P=\sqrt{N}$. For an expanded prime sequence code, the number of sequences is $3N-2\sqrt{N}$. Hence more users can be accommodated but when the number of simultaneous users is larger than P, system performance will depend on the particular code sequences in used. By computer simulation it was found that some code sequences in the expanded code has 0 crosscorrelation peak at the position of the respective autocorrelation peaks. By carefully selecting the code sequences, up to $2P$ simultaneous users can be accommodated. System performance will be adversely affected for some users if the number of simultaneous users is larger than $2P$.

PART 4

PHOTONIC SWITCHING

Review Of Photonic Switching Device Technology

A. Marrakchi, W. M. Hubbard, and S. F. Habiby
Bellcore
331 Newman Springs Road
Red Bank, NJ 07701-7040
USA

Abstract

Following a brief description of switching applications, the state of the art of some photonic device technologies is reviewed. Both refractive and diffractive switching elements fabricated in bulk or guided-wave form are considered.

1. Introduction

Now that the optical fiber is firmly entrenched in today's communication networks, it is only logical that efforts are sprouting around the world to keep the signals in an optical form throughout such networks. Transmission links with single-mode fibers have for all purposes unlimited bandwidth, although its current use is constrained by the necessity to convert photons to electrons (p-e conversion) in order to perform the switching function with existing digital electronic chips. These in turn have their own limitations. On the other hand, there are clear advantages in support of optical switching that derive from the physical nature of optics. These include speed, bandwidth, connectivity, and transparency to signal rates.

The same property that made optical transmission through fibers so successful is making switching difficult. Photons do not interact strongly with matter. Nevertheless, switching of a light beam can be performed by impacting one of its characteristics while it propagates through some material. Parameters that can be affected include the light amplitude, its polarization, its wavelength spectrum, the absorption it experiences and the refractive index it sees in the medium. Several optical phenomena have been already applied to switching. Among them are the electromagnetic effect whereby a magnetic field rotates the polarization of an optical beam (Faraday effect), the acoustooptic effect whereby an acoustic periodic excitation induces light diffraction, and the electrooptic effect whereby an applied electric field generates a phase modulation that can be used for switching (Pockels effect). As a result of the diversity of the parameters that can be modulated, photonic switching is generally sub-divided into time-, space-, and wavelength-division categories.

Even though there is no doubt that the evolution is towards an all-optical network, there are many factors other than technological that are slowing down this process. In this paper, we shall attempt to single out the materials and device technology factors that may influence the speed at which photonic switching will permeate today's communications. As previously mentioned, we are primarily concerned with optical switching devices that do not require a p-e/e-p conversion, although the switch control itself could be either optical or electrical. Following a brief review of switching applications and requirements in Section 2, mechanical switches will be described in Section 3, other refractive switching elements in Section 4, diffractive switches will be analyzed in Section 5, broadcast-and-select switching configurations will be described in Section 6, and some concluding remarks will be given in Section 7. The switch designs described in this paper are given only as illustrative examples, and were chosen either for their novelty or the issues they raise.

2. Typical Switching Applications

Each of the potential applications for photonic switching has its own set of requirements. The following list specifies the properties of switching elements which are of most concern:

- Bandwidth
- Switching Time
- Insertion Loss
- Crosstalk
- Power Consumption and Size
- Wavelength and Polarization Sensitivity
- Control and Temperature Sensitivity.

Other factors that need to be considered in the switch design include multiplexing of signals (either in the time or wavelength domain), and directionality of the channel, i.e., one- or two-way communication.

In the topology of switching networks, the most flexible design would connect any one of N inputs to any one of M outputs. In addition, the ideal network should be free of contention problems and allow any input to communicate with any output, at any time. Although switching networks can be built out of simple 2x2 switching elements, there is much to gain in using an NxN crosspoint switch which requires cascadable switching elements.

One application for photonic switching is that of protection for failed optical components. Simple mechanical switches are already commercially available for this application, and have been incorporated into communication systems operating under field-trial constraints. A protection switch does not need several switching elements in tandem, and hence the requirements on insertion loss are not as severe as for other types of switches. Since the signals at the input of the protection switch often have approximately the same power level, the crosstalk requirements may not be as severe either. Because the switch is ordinarily called upon to function only in the event of a failure in a working component, high switching speed is usually not a requirement.

Optical facilities are typically interconnected through manual cross-connect frames. With further development of some current state of the art optical devices, it is expected that these manual frames will be replaced by automatic cross-connect systems. These switches must function with a wide range of signal levels at the input ports. This places a tremendous burden on the crosstalk requirement, although recent advances in photonic amplifiers do somewhat relax this constraint. In certain applications it would be advantageous for a cross-connect to be able to switch fast, in others a few microseconds or even milliseconds would be acceptable. Nevertheless, the transparency available with photonic switches is a particularly attractive feature.

Certain types of applications require that a given path (or route) be established and dedicated to a particular channel for the duration of the communication. This is commonly referred to as "circuit switching". In this configuration the time required to establish the route depends on the application (telephony, teleconferencing, telecommuting, interconnects between processors, ...), and could be as large as a few hundreds of milliseconds. However, the required number of switching elements in some cases is large and should support thousands or hundreds of thousands of input and output ports. Signals will likely arrive at different power levels, thereby placing a strain on the crosstalk requirements. Many stages of switching will be needed in some architectures and this places difficult requirements on insertion loss, on insensitivity to manufacturing tolerances and on physical size.

In configurations such as the star or ring local area networks, subscribers are distributed around a transmission medium which carries more information than a given end-user needs. This information is generally composed of time-division multiplexed slots each having a beginning and an end tag, and a destination address. Such packets of information are interleaved, and require a switch that decodes the address and establishes a communication path for the duration of the time slot. In this type of switching, commonly referred to as "packet switching", the response time has to be quite fast in order to switch between channels without loss of information, or loss of bandwidth associated with delays. A subset of packet switching is "message switching". In this case, multiplexing is used to take advantage of the transmission link bandwidth, although the switching function does not have the same time constraints since parts of the message can be stored for later distribution. Message switching is commonly used in networks such as ARPANET.

Although it is clear that the opportunities for photonic switching are greatest in telecommunications, it is also clear that the constraints are most stringent there. Consequently, one should initially seek the best device/application match, which in some cases happens to be in optical interconnects.

3. Mechanical Switching Elements

The development of this type of switches, along with their excellent performance which derives from their simplicity, make them prime target for early incorporation in real-world systems. These switching elements have an electro-mechanical system that allows for voltage control of the switch state. They are usually divided into two categories, those that move the fibers themselves and those that move some other element (typically a prism) in order to steer the beam to the desired output fiber.

In the "moving-fiber" category, some recent experiments have been performed with fibers having a D-shaped cross-section (or D-fibers).[1] By making the cladding on the flat surface very thin, it is possible to obtain coupling of energy from one such fiber to another of similar design by placing the flat sides together over the appropriate coupling length.

Researchers at British Telecom Research Laboratories have calculated the properties of switches made from D-fibers and demonstrated a prototype of a switch point, schematically shown in Figure 1. The switching elements work by moving the fibers in a direction perpendicular to the flat surface. Theory and experiment agree that the switch is relatively insensitive to wavelength and state of polarization. The back-reflection caused by back-scattering is minimal (better than 60 dB). Crosstalk is about 40 dB with 10 μm separation in the "off" state and improves at a rate of 4 dB/μm. The coupling loss into the D-fiber is only 0.5 dB and the loss per meter of the fiber is 0.76 dB. One of the disadvantages is that even though the coupling length is only 2.3 mm, pratical considerations require about a 3 cm spacing between elements along the fiber. Thus a 16x16 crosspoint would be about 48 cm long. This is an inherent property of the design and there is little hope of significantly improving this result. In addition, the absolute value of the coupling loss in the "on" state is 30 dB but it is reasonable to expect that further work will result in significant improvement on this value.[2]

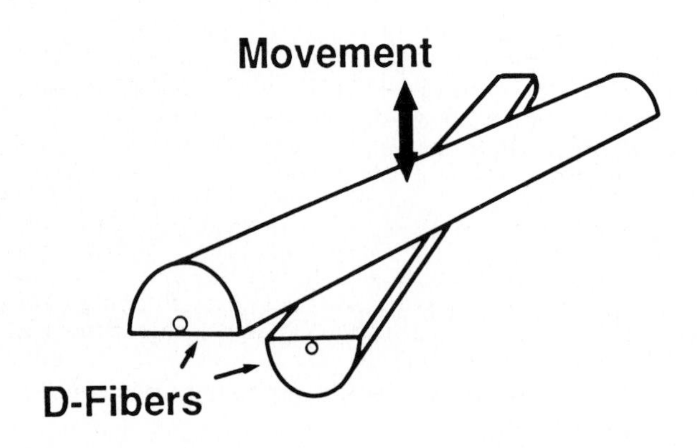

Fig 1: Moving D-fiber crosspoint [Ref. 1].

Although the performance of this switching element is quite promising, it appears however that the technological hurdles will be difficult to overcome. Typically, for most moving-fiber switches, the simplicity of the designs yields excellent results in terms of coupling loss, polarization and wavelength insensitivity, and bandwidth. But this is achieved at the expense of ease of control due to the complexity of the electro-mechanical devices, which in turn do not allow for fast set-up times.

In the "moving-beam" category, one type of commercially available single-mode fiber switch module is based on the refraction induced by moving a prism situated between input and output fibers.[3] Conceptually, this principle would also work with a mirror, and one would have a reflective rather than a refractive switch. Modules for 1x2, 2x2, and 2x4 switching are available. The switch state is altered in either latching or non-latching modes by applying a 5 V, 50 mA DC drive at a control terminal (allowing remote control). The switch module dimensions are typically on the order of 2x2.5x1 inches. The switching elements have a maximum insertion loss of 1.0 dB, which currently can be guaranteed by vendors, and are designed for operation at wavelengths between 1.2 and 1.6 μm. They are essentially transparent to signal rates, easily accommodating both analog and digital signals above several gigabits per second. Their switching speed is relatively slow (10-30 ms is a typical range), but practical for facility or protection switching applications. Switches have been field-tested that have better than 60 dB crosstalk, and a repeatability within ±0.02 dB.

4. Refractive Switching Elements

This Section includes photonic devices that rely on a uniform change of the refractive index or the optical absorption of a material to perform the switching function. Generally, most optical phenomena that are used involve an applied field to control the modulation. Even when magnetic fields are required, such as in the magnetooptic effect, these are usually induced with electrical currents. Some typical devices are described in the following, starting with bulk, then guided-wave switches.

4.1 Photonic Gate Arrays

Photonic gate arrays have on/off switching elements that are based either on polarization rotation or absorption change. They are very useful in switching configurations based on matrix-vector multiplication,[4] as shown in Figure 2. The input is a linear array of fibers. The incoming signal from each fiber is fanned out to all the gates in a column of the array. The output from these gates is collected for each row and fanned into one output in another linear array of fibers. By turning on or off the elements of the photonic gate array, it is possible to connect any incoming signal to any output fiber. It is obvious however that since this switching configuration is based on power division of the incoming light, this principle would be applicable only to a small number of fibers, unless amplification and level restoration is added to the gates. These arrays belong to a wider class of devices, the spatial light modulators, that can be either electrically or optically controlled.

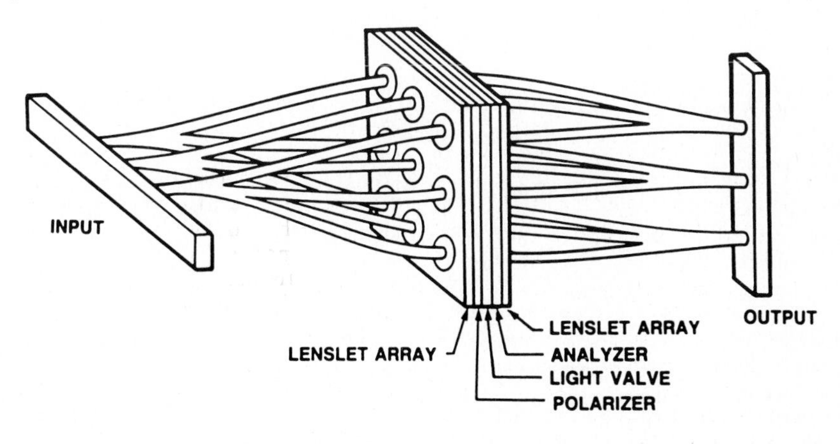

Fig 2: Photonic gate array switch.

Liquid Crystal Gates

Liquid crystals come in a variety of forms but the nematics are most notable for their application in displays. Recently, another class of materials has attracted much attention owing to their excellent performance. These are the ferroelectric liquid crystals that belong to the smectic C family. A typical cell is constructed by placing this material between two glass plates coated with transparent electrodes, usually indium tin oxide (ITO). If the cell is thin enough, on the order of 2 μm, only two states are allowed for the director of the molecules out of a cone of possible orientations.[5] These two "surface-stabilized states" have their optical axes at 45 deg. angle from each other and can be selected with voltages of opposite sign. Thus the device behaves as a programmable half-wave plate. For on-off amplitude modulation, the cell is placed between crossed polarizers. In one state, say for the positive polarity of the field, the incident polarization remains unchanged and is blocked by the analyzer (off-state), whereas in the other case, with the negative polarity, the polarization is rotated by 90 deg. and is transmitted through the analyzer (on-state). Consequently, these photonic gates are polarization sensitive and will suffer a minimum of 3 dB loss with randomly polarized light, although this drawback could be alleviated with special configurations such as the one that splits the incident beam in its two fundamental polarizations before switching them separately and recombining them at the output.

Among the advantages offered by the ferroelectric liquid crystals are the low drive voltage (typically ± 15 V for a 2 μm cell), the possibility of bistability in certain cases, the small response time (on the order of 10-100 μs), and the large contrast ratio (about 1500:1). Combined with the relative ease of fabrication of large arrays, these devices could find application in protection and facility switching, and in optical computing and interconnects.

Multiple Quantum Well Gates

Electroabsorption is defined as the change in the optical absorption induced by an electric field, usually supplied externally to the material. In this broad category, the Franz-Keldysh and the Stark effects can be seen as the precursors to the now common electroabsorption observed in multiple quantum wells near the band edge. The reasons behind the lack of interest in both the Franz-Keldysh and the Stark effects at the device level are related to their inefficiency in conventional semiconductors. On the other hand, electroabsorption has been observed at room temperature and with moderate applied fields in GaAs/AlGaAs MQWs.[6] In a typical configuration, the field is applied perpendicular to the layers and parallel to the optical wavevector. Increasing the applied field results in a shift of the band edge and excitonic peaks towards lower photon energies, with little broadening and some loss in the height of the peaks. Based on such an effect, one can easily build an amplitude modulator. Assuming that the device is operated just below the band gap, application of an electric field increases the optical absorption and hence reduces the transmitted intensity.

The interest in MQW modulators spurs from the fact that devices are made with a well-established solid-state technology, which can be extended to batch fabrication of large two-dimensional arrays of such devices. In addition, wavelength operation flexibility is possible with proper choice of the band gap in the III-V and II-VI semiconductors. However, so far, the performance has not been at the level where one might unequivocally choose this technology over other alternatives. The major drawback is that this is an absorptive effect, and hence, there will be energy dissipation requirements that might become stringent for large arrays. In addition, with typical applied voltages on the order of 10 V only a 2:1 modulation is achievable. Reflection-type devices have increased this value to about 10:1,[7] still far short from what would be required for minimal crosstalk. However, switching does happen at fast speeds (100-200 ps), and the devices are polarization independent. Hybrid bistability with low switching energies of about 18 fJ/μm^2 has also been achieved in these structures.[8] The switching time constant of this device is limited by the RC of the bias resistor and the capacitance of the structure, which results in a trade-off between switching time and switching power (about 1.5 ms for 670 nW and 400 ns for 3.7 mW).

Nonlinear Fabry-Perot Gates

An optically-addressed bistable gate is usually made of a nonlinear material placed between two highly-reflective mirrors, in a Fabry-Perot type of arrangement. The nonlinearity may be intrinsic or hybrid (i.e., externally controlled), and absorptive or refractive. Depending on these characteristics the bistable performance (speed and power requirements) can be made to match a particular application. These types of nonlinear bistabilities have been observed both in bulk crystals and in MQW structures. For an exhaustive study and review of bistability Ref. [9] is an excellent source of information.

For the fabrication of practical devices, it is desirable to have a material with an intrinsic dispersive bistability due to a fast electronic nonlinearity, although refractive index dispersion and absorption are most often intimately related through the Kramers-Kronig equations. Although the results with bulk GaAs were quite satisfactory, their best performance was achieved at low temperatures.[9] On the other hand, MQW structures do show bistability at room temperature. It is still too early to speculate on the outcome of this technology since the performance is constantly being improved with new designs such as the asymmetric Fabry-Perot cavity modulator which theoretically may achieve better than 20 dB contrast with less than 3 dB insertion loss at an operating voltage of about 10 V.[10]

4.2 Guided-Wave Switching

This technology is quite well advanced in its stage of development, to a point where some devices are already commercially available (from Crystal Technology, Amphenol, and BT&D for example). Some of the drawbacks are directly related to the type of substrate utilized for these integrated optics devices. Lithium niobate crystals have two characteristics that limit their usefulness in photonic switching. One is the fact that these materials are uniaxial, and hence will automatically have some polarization dependent behavior which will increase both the loss and the crosstalk. The second is the relatively small electrooptic coefficient, and hence long distances on the order of several millimeters are required for efficient switching, limiting this technology to matrices of about 8x8 inputs/outputs. Switching speeds have vastly improved by moving from the lumped to the traveling wave design. Nevertheless, for systems incorporation, polarization independence is critical, and hence recent work has concentrated on semiconductor substrates, and on a new design with promising results, the digital optical switch.[11]

Most electrically-controlled integrated optics switches on lithium niobate are of the interferometric type, and exhibit a sinusoidal response to the applied voltage. The digital optical switch is made of two input guides of unequal width, a junction where two modes are allowed, and two symmetric output guides. This asymmetric configuration is known to perform mode sorting. Adding an electrical control to break the symmetry of the output guides, allows switching between the two ports. In essence, the width of the guide is not only determined by the physical dimensions, but also by the refractive index in the transverse direction. Hence, by appropriately choosing the bias voltage, the excited mode at the input is transferred to either of the two output ports. Such a structure has been simulated using the beam propagation method, fabricated, and tested.[11] Some results are shown in Figure 3.

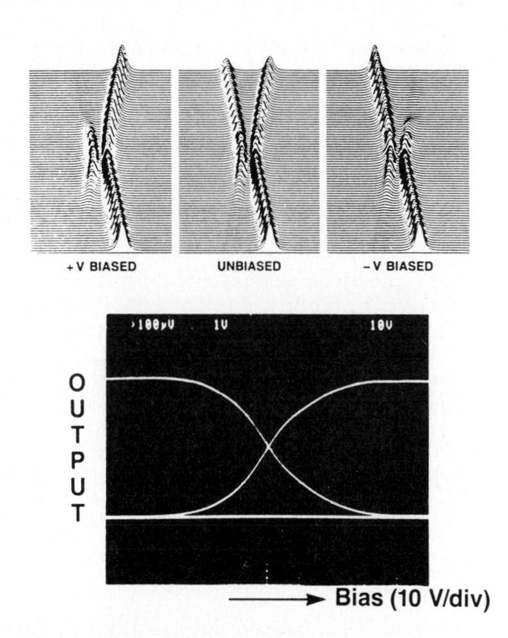

Fig. 3: Digital optical switching with asymmetric junctions [Ref. 11].

Another class of guided-wave switching devices utilizes light to control the state of a space switch. It is based on the optical nonlinearities of some materials, such as doped glass for example. Contrary to the switch described above which can be readily available and of immediate practical use, this class of all-optical switching elements is still in the early research phase. In one experiment, a dual-core fiber is used.[12] The advantages realized with these structures are two-fold. First, the long interaction lengths relax the requirements on switching power; and second, the low optical absorption of fibers in the near-infrared reduces the thermal effects that are detrimental to the switching performance. Figure 4A schematically illustrates the operation of such a device. As long as the input intensity is below a threshold, the signal comes out of output port 2 if the length of the interaction region is equal to a coupling length. This length is in turn determined by the optical path length which is a function of the refractive index. Increasing the incident intensity brings about changes in the refractive index, and hence the coupling length condition is spoiled and light starts leaking into port 1. At some intensity level, most of the signal exits from port 1. Figure 4B shows the relative output power from ports 1 and 2 as a function of incident peak power normalized to the critical power (solid curves are for cw and dashed curves for pulsed light). Improvements on the performance of this switch have been realized with the use of square optical pulses.[13] Because of the small nonlinearities (typically about 10^{-9} refractive index change for 1 W of optical power), these devices are not yet practical. However, this technology will certainly be extended to the novel organic materials that are being developed, and which show larger optical nonlinearities.

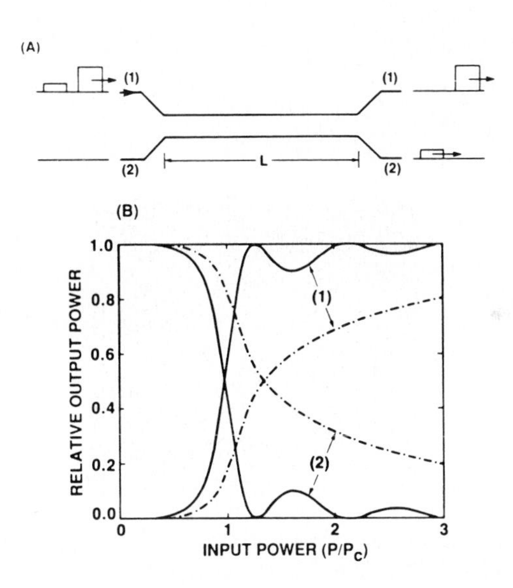

Fig. 4: Photonic switching in a dual-core fiber [Ref. 13].

5. Grating Diffraction Switching

A space-division switch is best implemented with an efficient, variable phase grating that can be easily and rapidly controlled. An optical beam incident on a grating can be diffracted to any one or many locations in a two-dimensional space by properly arranged gratings. To minimize the loss, Bragg (or bulk) diffraction is usually the logical choice, as opposed to

Raman-Nath (or thin) diffraction which produces many diffracted spots. In the following, two such technologies are described, one based on the acoustooptic effect and the other on the photorefractive effect.

5.1 Acoustooptic Diffraction

Changes in the density of a material caused by an acoustic wave induce variations in the refractive index, which in turn can be used to alter the characteristics of an optical beam propagating within such a material. This technology has already proven itself in some practical applications,[14–15] but again, stringent requirements have made it difficult to apply in photonic switching. Nevertheless, experiments are still being pursued to study the system aspect of a switching network based on acoustooptic diffraction. In one instance, a 4x4 multicasting switch has been realized.[16] Figure 5 shows schematically the layout of such a setup.

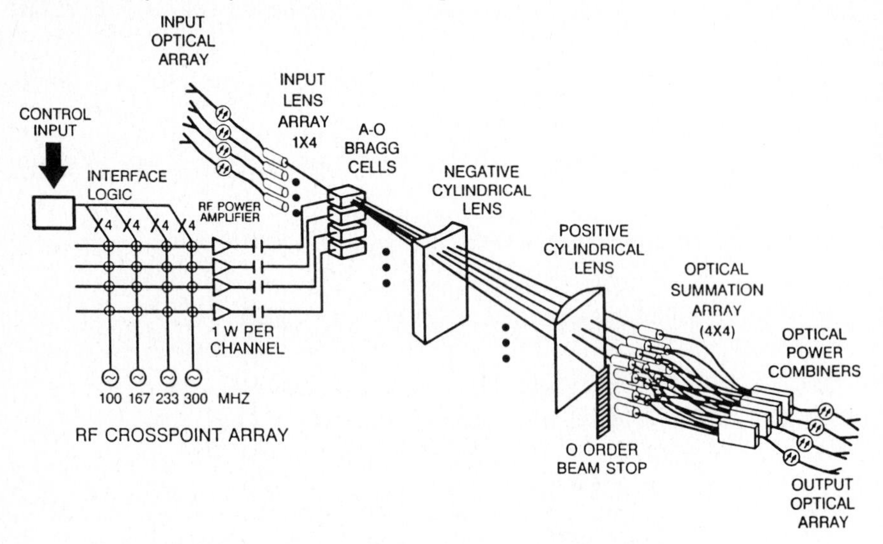

Fig. 5: A 4x4 acoustooptic space switch [Ref. 16].

Most of the performance characteristics of an acoustooptic switch are determined by the overlap of the optical beam and the refractive index perturbation. To maximize this overlap, GRIN-ROD lenses are used to collimate the beams at the output of the fibers in the input plane. They are again used at the input of the fibers in the output plane to collect most of the diffracted light. The experiment was performed at 0.83 μm, with multimode fibers. Acoustic gratings were written in a tellurium oxide crystal with RF signal generators at about 1 to 1.5 W of power. Switching could be performed at 50 ns speeds, with better than 24 dB crosstalk, a typical total insertion loss of 15 dB, and practically no polarization dependence.

5.2 Photorefractive Diffraction

The high sensitivity of photoconductive and electrooptic crystals, such as bismuth silicon oxide for example, in the visible spectrum (for the writing process), has allowed the simultaneous recording and reading of volume holograms to be achieved with time constants amenable to real-time operation. The holographic recording process in photorefractive materials involves photoexcitation, charge transport, and trapping mechanisms. When two coherent writing beams are allowed to interfere within the volume of such a crystal, free carriers are nonuniformly generated by absorption, and are redistributed by diffusion and/or drift under the influence of an externally applied electric field. Subsequent trapping of these charges generates a stored space-charge field, which in turn modulates the refractive index through the linear electrooptic (Pockels) effect and thus records a volume phase hologram. If both coherent writing beams are plane waves, the induced hologram will consist of a uniform grating. By changing the angle between the writing beams and their relative orientation, on can diffract an incident beam anywhere in a two-dimensional output plane. Such a result is shown in Figure 6, which is a time integration over several grating configurations.[17] Similar results have also been obtained by changing the wavelength of the writing beams rather than the geometrical parameters of the set-up.[18] Currently, research is being extended to the real-time multiplexing and individual control of several gratings simultaneously.[19] Although the success of this technology is also limited for switching applications, the prospect of being able to interconnect large arrays of devices opens up the field of neural networks in the context of optical computing.

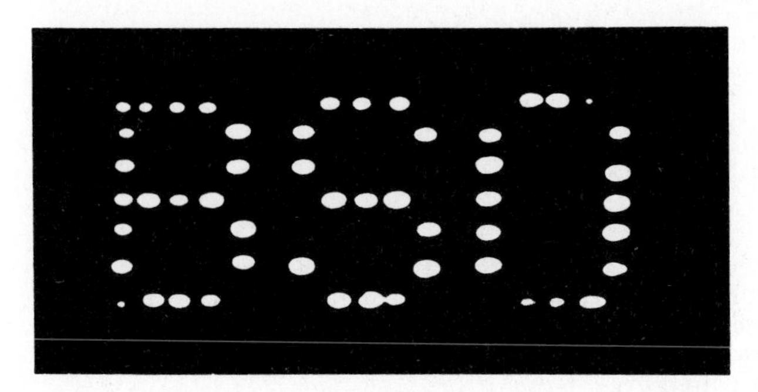

Fig. 6: Two-dimensional photorefractive switching [Ref. 17].

6. Broadcast-and-Select Switching

A rather different class of switching, which does not use crosspoints in the usual sense of the term, can be categorized as "broadcast-and-select". This category can be further divided into two types, wavelength-division and time-division. In the first type, the input ports assign a particular wavelength to the incoming signal and broadcast it to all of the output ports where the appropriate output port selects the wavelength destined for it.[20] This is somewhat analogous to radio-wave transmission and reception. In the other type (which is the time-domain analog of the first one), each input port broadcasts its data to all the output ports on a unique sub-time slot and the appropriate output port selects its data by picking from the proper slot.[21] Only the first type is further discussed in the following because of the similarities.

Wavelength-division switches

A schematic af such a design is shown in Figure 7. In this network, each channel is assigned a particular wavelength. All signals are then launched into a star coupler, whose output is distributed to all the receivers. At this end, the receiver tunes to the appropriate input channel by selecting the wavelength it wants to listen to. Realization of these networks depends critically on the ability to fabricate narrow-band lasers, and tunable receivers. The size of the switch depends on the power available from these lasers (or the availability of efficient photonic amplifiers) since there is an inherent 1/N loss in the N-star coupler.

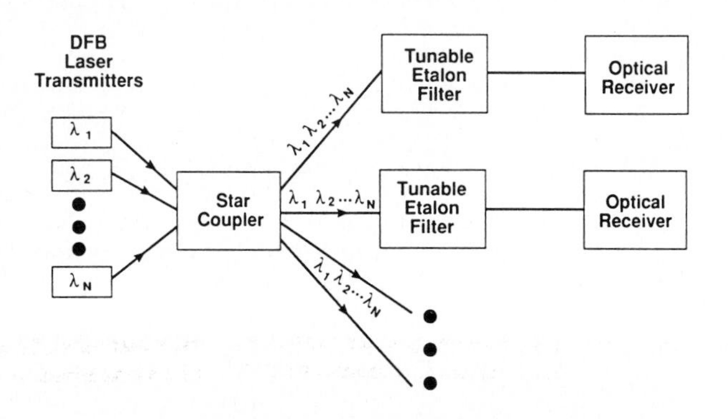

Fig. 7: Broadcast-and-select switching [Ref. 20].

In another implementation, rather than having separate wavelengths for each channel, the spectral width of the light pulse is used to encode both the address and the data.[22] In such a code-division multiple access system, self-routing can be achieved at very high speed, which makes it quite suitable for packet switching.

7. Conclusion

The photonic switching devices described in this paper represent only a fraction of the actual research done in this field. It would be easy to lay the blame on materials for the fact that photonic switching has not been yet incorporated in real systems. This would be only partly fair. At present, electronic circuits work at rates in excess of 10 Gb/s, and with normal evolutionary progress this figure is likely to double in the next few years. In addition, switch control which has already been addressed in electronic devices, has yet to be adequately considered for their photonic counterparts. Therefore, as far as speed, bandwidth, and control are concerned, it is reasonable to expect that electronic switches will continue to compete with photonic switches for a long time to come.

It is also believed that there is only a limited need for very fast photonic switches at present. Future services will probably broaden this need and add new constraints that will be more easily dealt with in the photonic rather than the electronic domain. Already, there exist certain applications, e.g., protection switching, in which photonic switches are beginning to find use. There is little doubt that this trend will continue to grow.

Acknowledgements

We thank C. A. Brackett, W. E. Stephens , A. M. Weiner, and Y. Silberberg for their critical review of this manuscript.

REFERENCES

[1] S. A. Cassidy and P. Yennadhiou, "Optimum Switching Architectures Using D-Fiber Optical Space Switches", IEEE JSAC, 1044 (1988).

[2] S. Cassidy and P. Yennadhiou, "D-Fiber Optical Space Switch Employing Electrostatic Fields", ECOC 88 Conference Digest, p. 276, 11-15 September, 1988, Brighton, UK.

[3] P. R. Strauss, "Optical Switches Herald a New Age of Enlightenment", Data Communications, p. 81, March 1988.

[4] A. R. Dias, R. F. Kalman, J. W. Goodman, and A. A. Sawchuk, "Fiber-Optic Crossbar Switch with Broadcast Capability", Opt. Eng. 27, 955 (1988).

[5] N. A. Clark and S. T. Lagerwall, "Submicrosecond Bistable Electro-Optic Switching in Liquid Crystals", Appl. Phys. Lett. 36, 899 (1980).

[6] D. S. Chemla, D. A. B. Miller, and P. W. Smith, "Nonlinear Optical Properties of GaAs/GaAlAs Multiple Quantum Well Material: Phenomena and Applications", Opt. Eng. 24, 556 (1985).

[7] G. D. Boyd, D. A. B. Miller, D. S. Chemla, S. L. McCall, A. C. Gossard, and J. H. English, "Multiple Quantum Well Reflection Modulator", Appl. Phys. Lett. 50, 1119 (1987).

[8] D. A. B. Miller, D. S. Chemla, T. C. Damen, A. C. Gossard, W. Wiegmann, T.. H. Wood, and C. A. Burrus, "Novel Hybrid Optical Bistable Switch: The Quantum Well Self-Electro-optic Effect Device", Appl. Phys. Lett. 45, 13 (1984).

[9] H. M. Gibbs, "Optical Bistability: Controlling Light with Light", Academic Press (1985).

[10] M. Whitehead, G. Parry, A. Rivers, and J. S. Roberts, "Multiple Quantum Well Asymmetric Fabry-Perot Etalons for High-Contrast, Low Insertion Loss Optical Modulation", Post-Deadline paper PD4-1, Photonic Switching, Technical Digest (Optical Society of America, Washington DC, 1989).

[11] Y. Silberberg, P. Perlmutter, and J. Baran, "Digital Optical Switch", Appl. Phys. Lett. 51, 1230 (1987).

[12] S. R. Friberg, Y. Silberberg, M. K. Oliver, M. J. Andrejco, M. A. Saifi, and P. W. Smith, "Ultrafast All-Optical Switching in a Dual-Core Fiber Nonlinear Coupler", Appl. Phys. Lett. 51, 1135 (1987).

[13] A. M. Weiner, Y. Silberberg, H. Fouckhardt, D. E. Leaird, M. A. Saifi, M. J. Andrejco, and P. W. Smith, "Use of Femtosecond Square Pulses to Avoid Pulse Break-Up in All-Optical Switching", Accepted in IEEE JQE special issue on Ultrafast Phenomena.

[14] M. Gottlieb, C. L. M. Ireland, and J. M. Ley, "Electro-Optic and Acousto-Optic Scanning and Deflection", Marcel Dekker Inc. (New York), 1983.

[15] N. J. Berg and J. N. Lee, "Acousto-Optic Signal Processing", Marcel Dekker Inc. (New York), 1983.

[16] P. C. Huang, W. E. Stephens, T. C. Banwell, and L. A. Reith, "Performance of a 4x4 Optical Crossbar Switch Utilising Acousto-Optic Deflector", Elect. Lett. 25, 252 (1989).

[17] G. Pauliat, J.-P. Herriau, A. Delboulbe, G. Roosen, and J.-P. Huignard, "Dynamic Beam Deflection Using Photorefractive Gratings in $Bi_{12}SiO_{20}$ Crystals", J. Opt. Soc. Am. B 3, 306 (1986).

[18] E. Voit, C. Zaldo, and P. Gunter, "Optically Induced Variable Light Deflection by Anisotropic Bragg Diffraction in Photorefractive $KNbO_3$", Opt. Lett. 11, 309 (1986).

[19] A. Marrakchi, "Continuous Coherent Erasure of Dynamic Holographic Interconnects in Photorefractive Crystals", Opt. Lett. 14, 326 (1989).

[20] C. A. Brackett, "Dense WDM Networks and Applications", Photonic Switching Conference Digest, p. 86, March 1-3, 1989, Salt Lake City, Utah.

[21] P. R. Prucnal and P. A. Perrier, "Demonstration of a High-Dimension Local Area Network Photonic Switch", ECOC 88 Conference Digest, p. 264, 11-15 September, 1988, Brighton, UK.

[22] A. M. Weiner, J. P. Heritage, and J. A. Salehi, "Encoding and Decoding of Femtosecond Pulses", Opt. Lett. 13, 300 (1988).

Photonic Switching Using Tunable Optical Terminals

Kai Y. Eng
AT&T Bell Laboratories
Crawfords Corner Road
Holmdel, N. J. 07733
U. S. A.

ABSTRACT

We discuss some of our focused research efforts aimed at exploring the technical feasibility of implementing an optical cross-connect switch suitable for use in a future Broadband National Network. Two specific techniques are described: Time Division Multiplex (TDM) and Frequency Division Multiplex (FDM), both are star-coupler-based systems. With technical bottlenecks in semiconductor mode-locked lasers and optical "AND" gates, the potential for the TDM method to realize a multi-Gb/s line-rate switch of large sizes (1000×1000) is limited. The FDM approach, on the other hand, is more promising. Its foremost technical issue is the design of a tunable receiver with a wide frequency range. This could be done either with a coherent receiver with a tunable laser as the LO, or with a tunable filter followed by a direct-detection receiver. Two recent laboratory demonstrations illustrating these two tunable receiver designs are described. In the coherent receiver experiment, a three-channel frequency-locked system with monolithic tunable lasers, each transmitting up to 700 Mb/s, was implemented, and the coherent receiver could cover a 1,000 GHz tuning range with its monolithic tunable laser (MQW DBR) LO. Power margin in the system substantiated feasibility for a 125×125 switch. In the tunable filter experiment, a four-channel system was provided with each transmitting up to 1.4 Gb/s. A tunable two-stage *fiber* Fabry-Perot filter was used preceding a direct-detection receiver to retrieve any channel of choice. It was polarization independent with a tuning range covering 15,000 GHz.

I. Introduction

Considerable progress has been made in optical Frequency-Division-Multiplex (FDM) technology. More recently, a number of experiments aimed at local network applications have been successfully demonstrated with the use of a star coupler to combine various FDM channels from the inbound lines for broadcasting onto each of the outbound lines [1-3]. Such a star topology has a fundamental power advantage over the bus architecture [4]. However, as in other distributed switching systems, a star network has many practical challenges due to the geographic separation of the users, e.g. synchronization, control of the FDM comb, polarization maintenance, channel reservation and so on. Most of these difficulties can be minimized when we shrink the FDM network inside a "box" and apply the same concept to build an optical switch. Such an approach has a potential for realizing a large size (1000×1000) switch at multi-Gb/s line rates, which in turn could have important applications in a National Broadband Network [5]. Here, we discuss some of the rationalizations behind choosing the optical FDM technique in our on-going optical switch experiments and also summarize our most recent laboratory demonstrations.

A Digital Access and Cross-connect System (DACS) is a switch providing distribution and access of trunk capacities between major hubs in a long-haul network. It is an important component in today's national network and will remain so in the network evolution to Broadband ISDN. Existing DACS fabrics are implemented with electronics operating in the kb/s to Mb/s range. Tomorrow's broadband services, on the other hand, will require widespread use of multi-Gb/s fibers, which in turn demand much higher speed and dimension capabilities on the DACS. Although electronics is expected to continue improving to meet these demands, the situation presents a tremendous opportunity for photonics to play a competitive role. This makes sense particularly when we take into account that the DACS is a static circuit switch with slow reconfiguration (Section 2). As such, if photonics could not be made a success in this application, then its prospect for solving other switching problems [6] would be dismal. Therefore, we devote most of our remaining discussions to this first test case of an optical DACS.

II. The Optical DACS

We depict in Fig. 1 an $N \times N$ cross-connect fabric for the DACS application. The input fibers carry signals originated from geographically remote locations. Each signal consists of many circuits in a time multiplexed format, i.e., repetitive frames of C slots each, and data in the same slot in successive frames constitute a circuit of a fixed capacity. If each slot contains only one bit, then the signal is said to be bit interleaved as adjacent bits in the signal belong to different circuits. Likewise, a byte interleaved signal has a slot size of eight bits. In any case, the purpose of the cross-connect system is to pull together specific circuits from different inputs and group them onto a specific output. As such, it makes sense to first demultiplex each input line into parallel signal streams at the cross-connect rate, i.e., the circuit rate at which cross-connections are required, and then use a (static) switch to rearrange all the circuits for appropriate multiplexing at the output (Fig. 1). Once the circuits are set, they generally remain fixed for a long period of time in the DACS application. A careful distinction should be made between the line rate and the cross-connect rate: The line rate refers to the speed of the incoming fiber line while the cross-connect rate is the circuit rate for cross-connections. In our discussion, the line rate is always greater than or equal to the cross-connect rate. If they are equal, the demultiplexers and multiplexers are obviously not needed. Conversely, if we just focus on the center switch by itself (with inputs from the demultiplexer and outputs to the multiplexer), it is operationally equivalent to a new line-rate switch with the demultiplexer output rate as the new line rate. It is also possible to combine the demultiplex/multiplex functions as part of the center switch for unequal line and cross-connect rates.

There is no distance limitation imposed in the DACS network. Consequently, the arriving optical signals are assumed to have gone through repeaters, and regeneration is assumed right at the entrance into the DACS. As shown in Fig. 1, the incoming optical signals are received and regenerated as serial high-speed data streams in electronic forms upon entry into the center switch.

An alternate approach to the demultiplex-multiplex technique is to use a time-multiplex switch (TMS), i.e., a space-division switch operated in a time-multiplex manner whereby a different interconnect pattern is provided on a slot-by-slot basis. For a bit-interleaved 1.7 Gb/s signal, each slot is a bit, or 4.7 nsec. In order to switch slot by slot, the switching time has be only tiny fractions of the 4.7 nsec duration, which is extremely difficult to achieve in most space-division fabrics. As for the byte-interleaved signals, the required switching speed is relaxed by eight times, but it remains very stringent.

III. Optical Switch Architectures: TDM versus FDM

There are three generic optical switching architectures for multi-Gb/s applications: Space Division Multiplex (SDM), Time Division Multiplex (TDM) and Frequency Division Multiplex (FDM). The SDM approach refers to the use of discrete space-division switch elements such as lithium niobate couplers. It is by far the most studied method with well-known characteristics [7]. Here, we concentrate on a brief discussion of the other two techniques (see [7] for more details) based on the use of a star coupler [9,10], and the emphasis is on their basic architectures, optical components and switch size limitations.

TDM Switching

A conceptual block diagram of an $N \times N$ cross-connect fabric using the TDM technique is shown in Fig. 2. The incoming signals are aligned at the inputs synchronously on a bit-by-bit basis. In order to carry out the time-division multiplexing, the bit interval is divided into (at least) N mini-slots, and each input bit is time compressed and inserted into one of these mini-slots so that the N input signals can in fact be time multiplexed together. This time-multiplex summation of the inputs is then distributed equally to N separate outputs. At each output, a receiver tuned to the same mini-slot over successive bit intervals would recover a particular channel of choice. A hardware arrangement to realize these operations is shown in Fig. 3 where a mode-locked laser in each input, say i, is used to generate periodic narrow (mini-slot wide) pulses at the data rate and coincident with the i-th mini-slot. This could be done either by separate mode-locked lasers with individual phase shift controls to provide proper time delays and also frequency locked to one another, or by using one high-power mode-locked laser with separate output delays. In either case, the narrow pulses are passed through an optical modulator gated by the corresponding input data, and then summed in the star coupler. At each output of the star coupler, an optical "AND" gate is required to pick out the data contained in a specific mini-slot, repetitive every bit period. This optical "AND" gate need not be a true AND gate because its (electronic) output could yield energy in the entire bit period to denote the presence of a data bit, instead of only a mini-slot duration. Finally, a conventional high-speed electronic receiver can be used to detect and recover the data.

It should be obvious that the system works exactly the same if all each optical "AND" gate, say j, is always tuned to its own j-th mini-slot, and the input narrow pulses are time delayed into various mini-slots based on destinations. In other words, the transmitters are tunable and the receivers fixed. The key advantage of the former technique of using fixed transmitters with tunable receivers is that broadcasting can be accomplished easily. We will continue our discussion with that as our baseline design for both the TDM and FDM approaches.

A study on the mode-locked laser and optical "AND" gate technologies [8] indicates that both of them are bottlenecks in achieving large switch dimensions. State-of-the-art semiconductor mode-locked lasers could yield pulses as narrow as 0.5 psec. However, in order to maintain synchronization, a pulse width of ≈ 1 psec seems to be a practical limit. As for the optical "AND" gate, a practical limit of a few psec's would already require major advances. In summary, substantial breakthroughs are needed to obtain DACS sizes in excess of a 100×100.

FDM Switching

A typical arrangement of using frequency tunable lasers in implementing a cross-connect fabric is shown in Fig. 4. In this diagram, the incoming signals are used to modulate separate lasers of different but fixed wavelengths. The optical signals are then combined in a star coupler. Switching is accomplished by having tunable receivers at the outputs. Each tunable receiver can be designed with either a coherent receiver with a tunable laser the local oscillator (LO), or a tunable filter plus a direct detection receiver. In both cases, there is a potential to realize DACS sizes significantly larger than that using the TDM scheme.

In the FDM arrangement of Fig. 4, it is necessary to maintain a comb of non-overlapping input laser frequencies. This is a well-known problem in the study of FDM star-based networks [1]. Here, by virtue of the close proximity of all the components, straightforward solutions are possible. For instance, a simple FP filter with free spectral range (FSR) equal to the desired channel spacing could be placed in one of the star coupler outputs as the reference to stabilize all the incoming laser signals. As the FP filter drifts, the entire comb would drift as a whole, and there is no need to establish any absolute reference.

The critical technologies in the FDM technique are the tunable lasers, coherent receivers and tunable optical filters. All these devices have advanced dramatically in recent years. We have been conducting on-going system experiments with such state-of-the-art optical components. In the next section, we summarize our recent results in both the coherent receiver and tunable filter approaches.

IV. Optical FDM Switch Experiments

Coherent Receiver Approach

The experimental set-up is depicted in Fig. 5 [11] where three transmitting laser signals (two closely spaced and one far apart) are summed in a 4×4 star coupler. One of the star coupler outputs is connected to a coherent tunable receiver which can receive any of the three transmitting signals. Another output is connected to a fiber Fabry-Perot (FP) filter (FSR = 13.8 GHz) providing a reference frequency comb for stabilizing the input laser frequencies [1,12]. The transmitting lasers are each dithered with a slightly different frequency for the purpose of channel identification. The frequency locking circuits can resolve dithering frequencies as close as 10 Hz apart, and the frequency locking system can support at least 1,000 channels.

The three single-mode transmitting lasers are all two-contact MQW DBR types, each tunable over $\approx$1,000 GHz in the 1.5 μm region [13]. The lasers were biased to produce a 0 dBm output into the isolators. One isolator was used for each laser (isolation $>$ 30 dB, 0.5 dB insertion loss). A 2×2 titanium diffused lithium niobate waveguide switch element was used as an external modulator providing on/off modulation. While the modulators could operate up to 3 Gb/s, our experimental data rate was limited to 700 Mb/s because of restricted receiver bandwidth as well as the broad LO laser linewidth. In our laboratory demonstration, the two widely spaced lasers (F_1 and F_3) were locked to channels separated by $\approx$600 GHz, and the in-between one (F_2) could be positioned just one channel away (13.8 GHz) from F_1 or in any channel between F_1 and F_3. Light power from the modulators was coupled into fibers at -16 dBm leading to the polarization adjusters and the star coupler. The star coupler was wavelength independent and had an excess loss of 0.5 dB.

A block diagram of the coherent receiver is shown in Fig. 6. This is a novel design using a tuned, balanced technique without equalizing circuits. The signal and LO are combined in a fiber coupler and directed onto two photodetectors in a balanced configuration. An inductive T-matching network is used to bring the detected (IF) signal through a tuned bandwidth of 1 GHz centered at 4 GHz to an FET. A frequency locking circuit is also included to adjust the LO frequency to maintain the detected IF signal centered at the tuned passband (4 GHz). The LO used was another two-contact MQW DBR tunable laser supplying about -4.6 dBm into the fiber coupler. At 600 Mb/s, we measured a receiver sensitivity (signal power at the fiber coupler input) of -42 dBm for a BER of 10^{-10}, and at 700 Mb/s, the receiver sensitivity dropped to -38 dBm for the same BER probably due to the excessive linewidths of the signal and LO lasers ($\approx$70 MHz each) causing spillover outside the tuned receiver bandwidth. Further increase of bit rates to 800 Mb/s seemed possible but more work was required before reliable measurements could be reported. Prior to this system experiment, the receiver was tested to 1 Gb/s with an external cavity laser as LO for a measured sensitivity of -42.5 dBm at a 10^{-9} BER.

Based on our experimental experience of using tunable receivers with sensitivity of -38 dBm at 700 Mb/s, and a practical capability of launching -12 dBm into the star coupler input, a power margin of 26 dB is available to account for the star coupler splitting and excess losses. This implies that sufficient power margin exists for implementing a 125×125 switch (at 700 Mb/s). The tuning range of the LO can easily support 125 channels with a channel spacing of 8 GHz.

Tunable Filter Plus Direct-Detection Receiver Approach

The experimental set-up is depicted in Fig. 7 [14] where four transmitting laser signals (two closely spaced and two far apart) are summed in a 4×4 star coupler. One of the star coupler outputs is connected to a direct-detection tunable receiver which can receive any of the three transmitting signals. Another output is connected to a fixed FFP filter (FSR = 13.8 GHz) providing a reference frequency comb for stabilizing the input laser frequencies, with the same stabilization method as in the coherent receiver experiment described above.

The four single-mode transmitting lasers are all two-contact MQW DBR types, as in the previous experiment. They were biased to produce a 0 dBm output into the isolators. One isolator was used for each laser (isolation $>$ 30 dB, 0.5 dB insertion loss). A 2×2 titanium diffused lithium niobate waveguide switch element was used as an external modulator providing on/off modulation at 1.4 Gb/s. In our laboratory demonstration, the two widely spaced lasers (F_1 and F_4) were locked to channels separated by $\approx$430 GHz; F_3 was midway between F_1 and F_4, and F_2 could be positioned just one channel away (13.8 GHz) from F_1 or in any channel between F_1 and F_3. Light power from the modulators was coupled into fibers at -9 dBm leading to the star coupler. The star coupler was wavelength independent and had an excess loss of 0.5 dB.

The direct-detection tunable receiver consists of a tunable two-stage (i.e., two in tandem) *fiber* FP filter [12] followed by a direct-detection receiver. The two-stage approach for increasing the overall filter finesse was first used by Kaminow et al [15], and our design is a variation suitable for the DACS application. Each of the FFP filters is *polarization independent* and is computer controlled for automatic tuning [16]. Their characteristics are superimposed in Fig. 8. The narrow filter is used to select the desired channel while the wide filter is needed to reject undesired signals admitted through other periodic peaks of the narrow filter. The tuning of the narrow filter was more critical and was done using the unique channel identification (dithering frequency), while the wide FFP filter positioning was less sensitive. The narrow filter was also placed preceding the wide filter in our experiment. The carrier-to-interference ratio in our setup was not optimized because the FFP filter samples did not have the optimal finesses. The overall finesse of the combined filters was measured to be $\approx$3,750, and the tuning range was $\approx$15,000 GHz. These were sufficient to enable our system experiment to proceed for concept evaluation. At 1.4 Gb/s, we measured a receiver sensitivity (signal power at the fiber coupler input) of -30 dBm for a BER of 10^{-9}, and at 1 Gb/s, the receiver sensitivity improved to -32 dBm for a 10^{-10} BER.

V. Discussion

Our on-going system experiments with the coherent receiver and tunable filter techniques have yielded valuable experience in understanding these important technologies for their potential, issues and challenges. The coherent receiver with a tunable laser LO offers better sensitivity and is integrable with a key potential for fast tuning. However, its present tuning range is somewhat limited. The *fiber* FP filter, although not integrable, is a compact component potentially suitable for future commercial applications. Its tuning range in our two-stage design is large enough to suggest the possibility of a large (1000$\times$1000) multi-Gb/s line-rate optical switch. But the power budget in such a system is of concern. With 30 dB of star coupler splitting loss and a receiver sensitivity of only $\approx$-30 dBm, the minimum launched laser power into each input has to be several mWs, which still leaves no margin for the FFP filter loss. One option is to use an optical amplifier to compensate for the FFP filter loss. This should be viable as amplifier technology is maturing rapidly. Therefore, the challenges ahead include: reducing the filter loss, increasing the launched laser power and the use of amplifiers.

VI. Conclusions

We have highlighted the DACS (Digital Access and Cross-connect System) application as an important first test case for photonic switching in future broadband networks. Two star-coupler-based techniques to implement an $N\times N$ cross-connect fabric have been discussed: Time Division Multiplex (TDM) and Frequency Division Multiplex (FDM). The critical technologies needed for a TDM switch are semiconductor mode-locked lasers and optical "AND" gates, with the latter as the more severe bottleneck. Substantial breakthroughs are required in these components before the TDM method could compete with the FDM approach, which is by far the most promising means to realize large DACS sizes (N). In the FDM technique, the foremost technical issue is how to design a tunable receiver with a wide frequency range. This could be done either with a coherent receiver a tunable laser as the LO, or with a tunable filter followed by a direct-detection receiver. State-of-the-art monolithic tunable lasers are limited to a range $\approx$1,000 GHz, but their performance is advancing rapidly. The alternative of using an optical tunable filter, such as a Fabry-Perot (FP) filter, offers a much wider tuning range. However, practical performance requirements (high finesse, compact size, low loss etc.) pose serious challenges, but again progress is being made continually.

We are pursuing research in both types of tunable receiver in our on-going system experiments. In a laboratory demonstration of the coherent FDM switch feasibility, three transmitting lasers were frequency locked onto peaks of a fiber FP filter providing the reference comb. These lasers were the *monolithic* MQW DBR type capable of tuning over a 1,000 GHz range and thus could be positioned in any of the designated channels within their tuning range. A coherent receiver using another *monolithic* MQW DBR tunable laser as the LO was used to recover any of the three transmitted signals. We measured receiver sensitivity of -42 dBm at 600 Mb/s and -38 dBm at 700 Mb/s, both at a BER of 10^{-10}. These measurements suggested that a 125×125 line-rate optical DACS could be feasible with this approach. In another experiment, a tunable two-stage *fiber* Fabry-Perot (FP) filter was used preceding a direct-detection receiver to achieve signal selection. In this demonstration, four transmitting lasers were frequency locked onto peaks of a fiber FP filter providing the reference comb. These lasers were the same monolithic MQW DBR type capable of tuning over a 1,000 GHz range. The two-stage fiber FP filter was computer controlled for tuning over a 15,000 GHz range (finesse $\approx$3,750). We measured a receiver sensitivity of -30 dBm at 1.4 Gb/s and a BER of 10^{-9}.

Acknowledgements

The author is grateful to many friends and colleagues for their help and contributions to various optical experiments - Mario Santoro, Adel Saleh, Tom Koch, Julian Stone, Bill Snell and David Saltzberg.

REFERENCES

[1] B. S. Glance et al., "Densely Spaced FDM Coherent Star Network With Optical Signals Confined to Equally Spaced Frequencies," *J. Lightwave Technol.*, November 1988.

[2] R. A. Linke, "Coherent Frequency Division Multiplexed Networks," *OFC '89.*

[3] L. G. Kazovsky and R. E. Wagner, "Multichannel Coherent Lightwave Technology," *ICC '88.*

[4] P. S. Henry, "Very-High-Capacity Lightwave Networks," *ICC '88.*

[5] K. Y. Eng, R. D. Gitlin and M. J. Karol, "A Framework For A National Broadband (ATM/B-ISDN) Network," submitted to IEEE ICC/Supercomm '90, Atlanta, GA, April 16-19, 1990.

[6] K. Y. Eng, "A Photonic Knockout Switch for High-Speed Packet Networks," *IEEE JSAC,* August 1988.

[7] R. C. Alferness, "Titanium-diffused $LiNbO_3$ Waveguide Devices," *Guided-Wave Optoelectronics,* Springer Verlag, New York, 1988, Chapter 4.

[8] K. Y. Eng and M. Santoro, "Multi-Gb/s Optical Cross-connect Switch Architectures: TDM versus FDM Techniques," to be presented in IEEE Globecom '89, Dallas, Texas, November 27-30, 1989.

[9] C. Dragone, "Efficient $N\times N$ Star Coupler Based On Fourier Optics," *Electronics Lett.,* 21st July 1988.

[10] A. A. M. Saleh and H. Kogelnik, "Reflective, Single-mode Fiber-optic Passive Star Couplers," *J. Lightwave Technology,* March 1988.

[11] K. Y. Eng, M. Santoro, T. L. Koch, W. W. Snell and J. Stone, "An FDM Coherent Optical Switch Experiment With Monolithic Tunable Lasers Covering a 1,000 GHz Range," paper PD10, *Photonic Switching Topical Meeting,* Salt Lake City, Utah, March 1-3, 1989.

[12] J. Stone and L. W. Stulz, "Pigtailed High-Finesse Tunable Fibre Fabry-Perot Interferometers With Large, Medium and Small Free Spectral Ranges," *Electronics Letters,* 16 July 1987.

[13] T. L. Koch et al., "High Performance Tunable 1.5 μm InGaAs/InGaAsP Multiple-Quantum-Well Distributed-Bragg-Reflector Lasers," *Appl. Phys. Lett.,* September 1988.

[14] K. Y. Eng, M. Santoro, J. Stone and T. L. Koch, "An FDM Optical Switch Experiment With A Tunable Two-Stage Fiber Fabry-Perot Filter Covering A 15,000 GHz Range," submitted for publication.

[15] I. P. Kaminow et al., "A Tunable Vernier Fiber Fabry-Perot Filter for FDM Demultiplexing and Detection," *IEEE Photonics Tech. Lett.,* January 1989.

[16] M. Santoro, D. Saltzberg and J. Stone, paper in preparation.

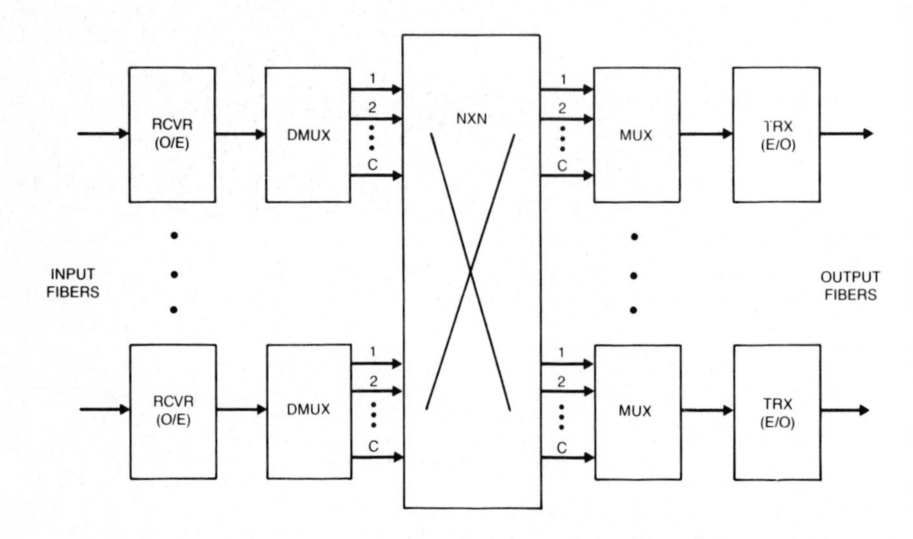

FIGURE 1 AN NXN CROSS-CONNECT FABRIC

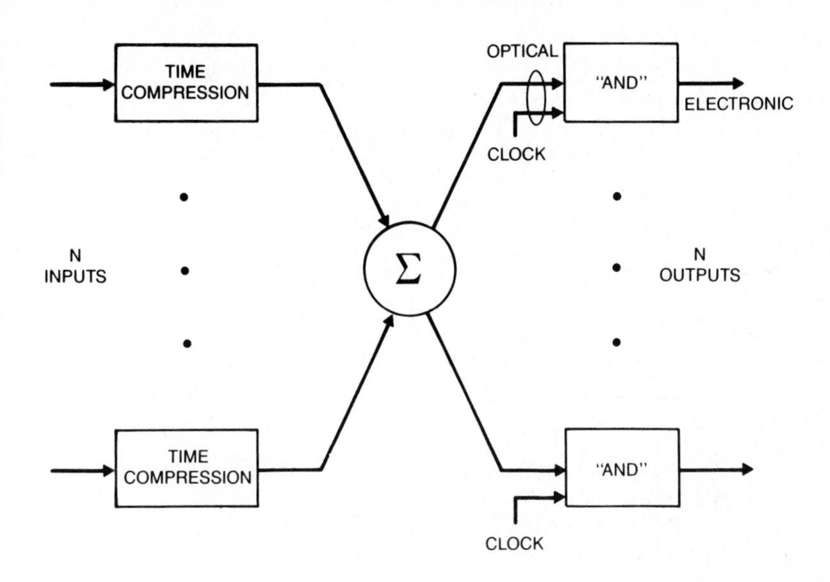

FIGURE 2 AN NXN CROSS-CONNECT FABRIC USING THE TDM TECHNIQUE

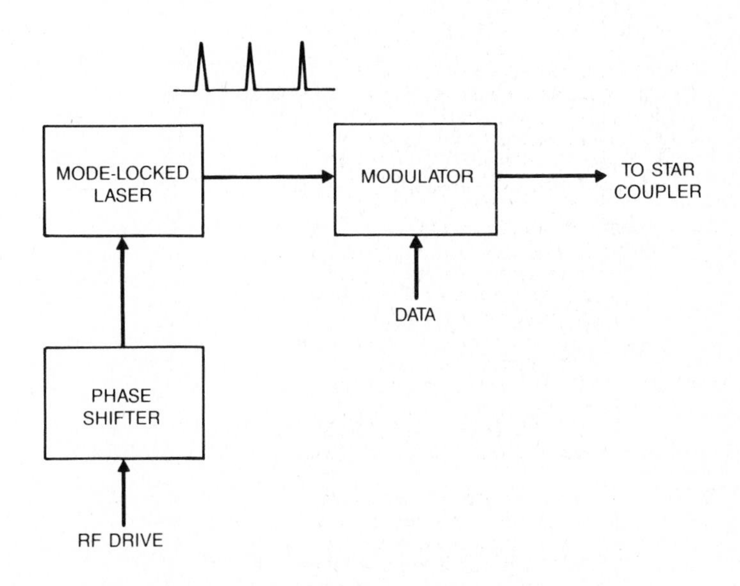

FIGURE 3 GENERATING TIME-COMPRESSED DATA PULSES FOR EACH INPUT TO THE STAR COUPLER

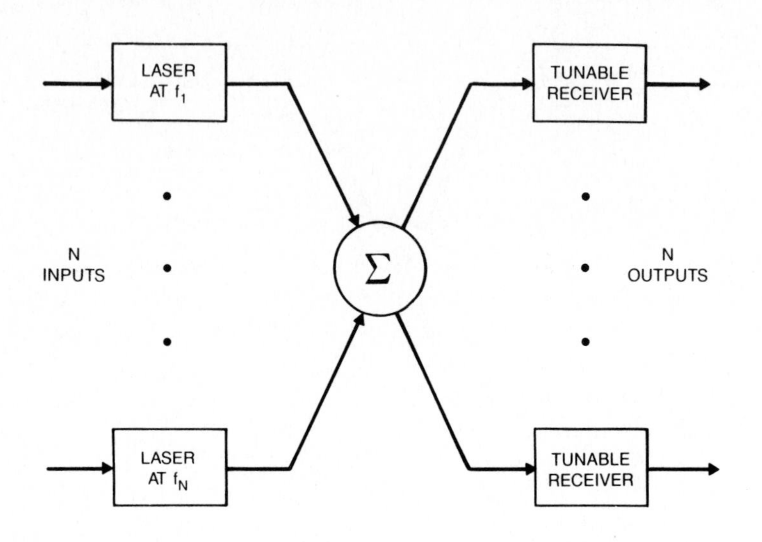

FIGURE 4 AN NXN CROSS-CONNECT FABRIC USING THE FDM TECHNIQUE

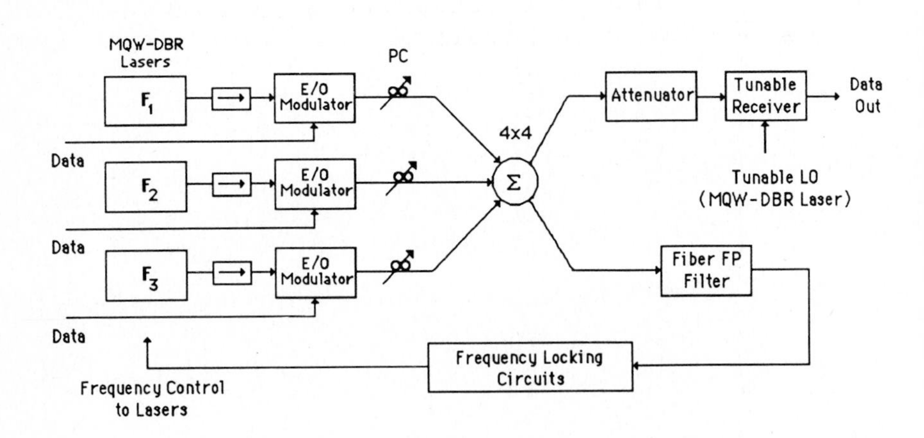

FIGURE 5 THE COHERENT FDM SWITCH EXPERIMENT

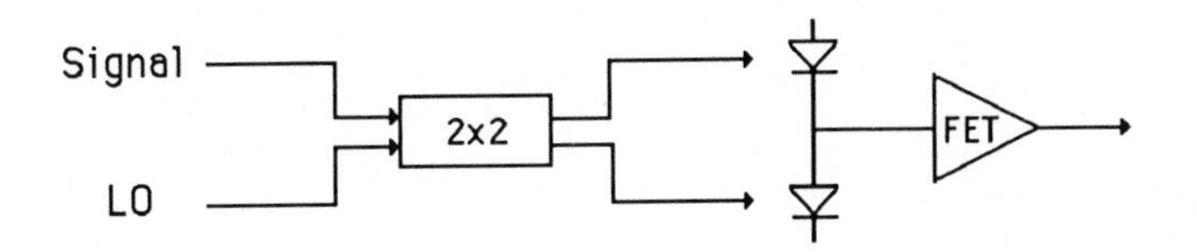

FIGURE 6 A TUNED, BALANCED PIN-FET RECEIVER

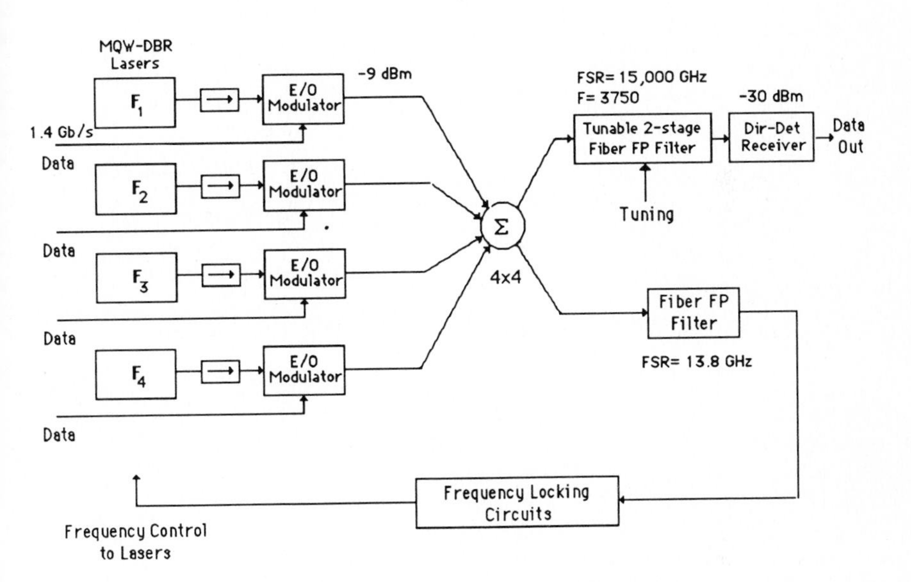

FIGURE 7 THE DIRECT-DETECTION OF FDM SWITCH EXPERIMENT WITH A TUNABLE FIBER FABRY-PEROT FILTER

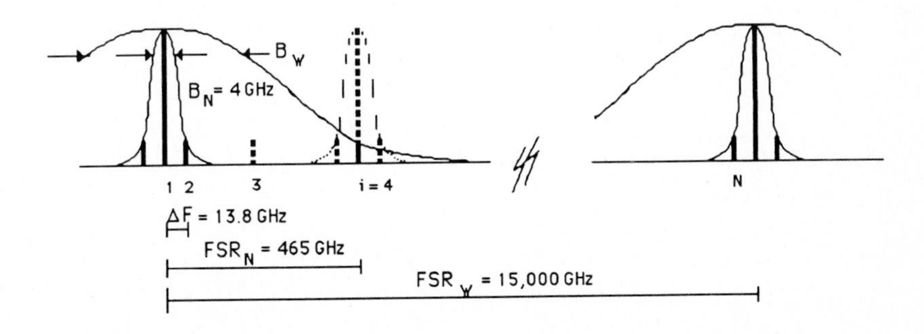

FIGURE 8 TUNABLE FILTER CHARACTERISTIC

PHOTONIC SWITCHING: A ROLE FOR SOPHISTICATED OPTICAL INTERCONNECTS?

John E Midwinter

University College London
Torrington Place
LONDON WC1E 7JE
UK

1. BACKGROUND

Much discussion in "optical" circles centres around the prospect of fibres to the home, the use of broadband services over them and the belief that this will generate a need for optical switching because of the high bandwidth used on the channels. Perhaps this is so, but noting that 625 line PAL is readily transmitted at a rate of about 70 Mbit/s and High Definition TV is expected to be carried on bearers of order 200MBit/s, and further noting that silicon circuitry can handle such line rates without undue difficulty, it is not obvious to this author that such an opportunity for "photonic switching" is real. Other factors seem to reinforce this view.

We should note that switching, unlike transmission, necessarily involves two distinct operations rather than one, in addition to the many ancilliary operation attendant on both. In switching, we must transport the data, just as in transmission, but before we can transport, it must associate with it a destination and establish the appropriate route. In the case of today's telephony switching, we are concerned with establishing a route of low bandwidth in a relatively long time (seconds) and then holding it for a longer time (minutes or sometime hours). By contrast, in a packet switch, routes may be held for very much shorter periods of time, (1 packet length) but are still of modest bandwidth. In wideband LANs, (FDDI for example), route establishment and data transmission are inseparable and are done in periods of nanoseconds, at the line rate. However, the number of potential destinations is limited so that the address header can be simple and the data rates are still within the range of today's chips.

Extrapolating from these services to the future, we note that the addition of broadband distributive services of the CATV network, although requiring large amounts of transmission, poses very little by way of a control problem. Typical programmes will be watched for long periods of time so that fast circuit setup and reset are not required and the material on offer is available from a finite library, not from the full customer base. Video-phone on the other hand is likely to involve much lower data rates, perhaps as low as 1Mbit/s, and will only generate a major traffic flow if there are large numbers of users. Again, it is circuit switched, requiring slow setup and reset and thus does not seem to pose a major problem for electronics.

The one area where it appears to this author that a problem of a different order is being posed is in the large central switching node of a broadband ATM network. From the preamble above, we have seen that a growing variety of services are attracting interest, ranging from the short burst packet-type communication, through many levels of circuit switched services. In addition, with the rapid growth in the use of powerful Workstations in business offices and factories, the stage is being set for a dramatic growth in computer-computer communication, leading to very "bursty" packet like messages but now of high bandwidth. The ATM type network seems ideally placed to handle to enormous traffic variety this picture envisages by packaging everything into a packet like format for transmission. As a result, we can envisage a day approaching when our fibre highways will be delivering data at 4Gbit/s entirely in packet format, requiring switches which can not only handle rates of 4Gbit/s per port but can deal with the control problem inherent in such traffic flow, namely the need to establish pathways "on the fly", meaning "within a few bit intervals. Moreover, since the destination data is likely to be contained within the packet header emerging from the fibre highway, it is desirable to find ways of exploiting it "in situ", since at these data rates, separating it for control purposes is likely to generate difficult problems of timing.

2. SELF-ROUTING & ATM NETWORKS

A class of self routing switches already exist that offer a good basis for solving this problem. One well known one is the Starlite switch, first described by workers at ATT-Bell Labs.[1] and implemented in silicon to switch packet type data at 1.5 Mbit/s. More recently, faster versions have been described operating at 150Mbit/s per input port. It is interesting therefore to examine the implications of extending this technology to very much higher data rates. Notice that in such a switch, it is implicit that the switch matrix is "intelligent" and that each "exchange bypass" or crossover switch can monitor the incoming data, recognise an address header when present and use the data it contains to set its own state to cross or bar. It is thus axiomatic that the switch can access the data flow at the line rate with logical processing circuitry and, unless buffer memory is used, act upon that information in real time. Looking at today's fast logic circuits, one discovers that the speed at which they can operate is controlled, not so much by the speed with which each discrete device can switch but more by the problems of maintaining synchronisation among many devices spread over substantial distances and the problems associated with moving data between chips. Since the history of optical fibre now shows with great clarity that optics is a supremely good communications medium, it seems sensible to examine whether some intimate fusion of optics and electronics might offer advantage, using optics primarily for communications and the electronics for logic.

3. ALL OPTICAL VERSUS ELECTRONIC LOGIC

Before doing this, let us examine the other option that is often proposed, namely optical logic coupled with optical connections or "wiring". A straight-forward comparison in terms of power-speed product shows that most optical logic devices are either slower or more power hungry than their electronic cousins. Many of the better devices prove, on examination, to be largely electronic in their operation albeit optically triggered, and these are generally inferior simply because they are bigger than the electronic nearest neighbour device. They are bigger because the wavelength of light is now comparable to the minimum feature size of ICs, in the region of 1 micrometer. However, to bring many beams of light into a large closely packed array of devices implies either the use of extremely sophisticated imaging optical systems or the use of devices large compared to a wavelength, say 10-20 micron across. Other studies have pointed out that the operating window for the all-optical bistable devices is very small except for those with intrinsic power gain, which again tend to be largely electronic in their operation.

These observations force us back to consider electronic logic again and to question more deeply where its real limitations lie. Simple calculations show that for nearest neighbour gate communication, it is very hard to see how metallisation can ever be improved upon using light. However, if we look at the problems of transmitting data across a large wafer, then we find that the power needed to charge the track capacitance and the interaction of resistance and capacitance giving rise to delay diffusion both start to make electronic communication more difficult. From the point of view of power consumption, an optical link might start to pay in if the capacitance of the transmit and receive devices was less than that of the metal track spanning the same gap. In terms of bandwidth, the optical link will always be superior although it may not be necessary and is thus only to be considered if the metal track is limiting in terms of bandwidth or crosstalk. Taking typical numbers, we find that this is unlikely to occur at distances of less than about 1mm and more probably will only apply for distances considerably in excess of this.

4. OPTICALLY CONNECTED ELECTRONIC ISLANDS

This observation thus suggests that a VLSI circuit should be partitioned into electronic islands of lateral dimensions equal to or less than the distance over which synchronisation and data transfer could be achieved and that optical interconnects might beneficially provide inter-island communication. Taking the numbers for today's circuits clocking at rates in the 10-100Mhz range, we find that the "island" dimensions considerably exceed the size of today's "chips", confirming that they work without optical interconnects. As the clock rate increases, we find that the optimum distance shrinks to the dimensions of a chip or less, implying that "optical pin-out" would be desirable. Extrapolating further to wafer-scale circuit complexities, we find that this approach would point to a partitioning of the wafer into synchronous islands with optical interconnects between them, assuming that the technology were possible. This would be an ideal structure to exploit imaging interconnections between the islands [2]. Note that similar conclusions have been drawn by Messerschmidt [3] but with the proposition that the synchronous islands should electrically connected, a design technique called meso-synchronous since all islands are clocked by a single clock albeit with different phases.

5. IMAGING OPTICAL FREE SPACE INTERCONNECTIONS

Much interest has centred upon the properties of light travelling in free-space with an imaging optical system to perform interconnection between sources and detectors or optical logic elements contained in a 2 dimensional planar array. A "simple" lens can image a large number of points to another position, providing in effect a parallel interconnection highway. Moreover, a suitable combination of two lenses provides imaging from one plane to another with the characteristic that the time delay for each pathway is identical (to femtosecond accuracy), in electrical terms a zero time-skew interconnect! Imaging optical interconnects can also implement simple wiring patterns other than parallel highways, such as butterflies and perfect shuffles [4].

However, whilst it is certainly possible in principle to image very large numbers of elements with very high resolution to their associated detectors, the practical reality is that very sophisticated lens systems are needed, together with extremely high precision and stability mounting systems that hardly seem compatible with normal electronic processing systems. The nature of the alignment problem becomes clear when it is realised that devices having the same dimensions as single mode fibre cores will be required to obtain power efficiency operation, but rather than aligning single elements, some proposals envisage the use of many thousands per lens. We doubt that this is realistic!

Our own thinking has been forced to focus on small arrays, perhaps of relatively tightly packed 8x8 elements, each with its associated optical system and providing I/O for a given area or island of circuitry. To solve the optical alignment and long term stability problems, it seems axiomatic that such elements must be solidly bonded to the circuit whose interconnects they provide and that this should be done using a technique broadly compatible with today's packaging technology. This seems to imply the use of solid optical overlays containing the necessary optical imaging elements in the form of phase holograms that would then be aligned and bonded into place by techniques and with accuracies similar to those used for mask alignment in microelectronics.

6. OPTOELECTRONIC SELF ROUTING SWITCH

Looking at the logical structure of a self routing switching matrix, (Fig.1) we find that it is composed typically of a two dimensional planar array of modules, each having two input and two output ports and capable of configuring itself to the cross or bar state on the basis of the address header data associated with the packet(s) entering its input ports. Each row of the units is then connected to the next row by a fixed wiring pattern of the butterfly, perfect shuffle or other related type. The resulting processor is thus of pipeline format, implying that total transit time is not of great significance and that throughput will be controlled by the upper limit clock rate that maintains synchronisation within each island and across a row. Since free space optical interconnects offer the possibility of interconnections of the perfect shuffle type but with very small time skew, it is natural to consider using optics for the inter-row wiring and electronic islands for each of the switching modules. Such an approach leads to rather small "islands" and to a requirement for one dimensional perfect shuffle interconnects which although possible in principle are difficult in practice to implement.

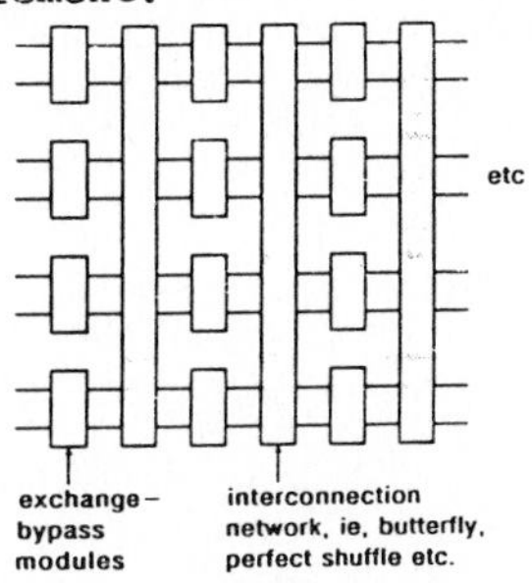

FIGURE 1. Schematic layout of a self routing switch.

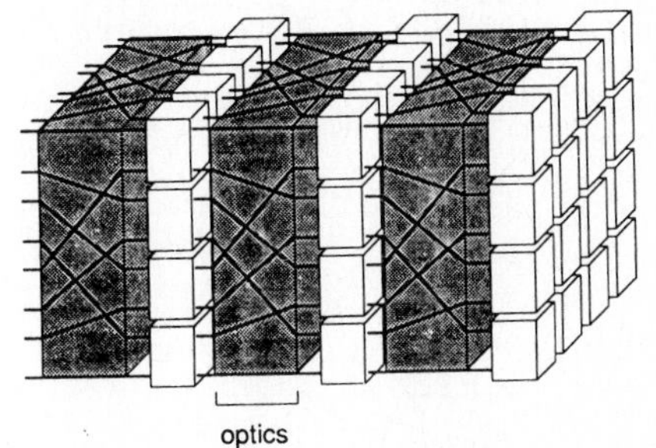

FIGURE.2 Two dimensional version of of Fig.1.

By repackaging the matrix into a form utilising 2 dimensional inputs with the rows of switching modules of the planar switch now placed in planes connected by two dimensional optical imaging interconnects, one is led naturally to a configuration in which the "electronic islands" are larger and the optical interconnect fully exploits the two dimensional nature of optical free space elements (See [5] and Fig.2).

If we follow this line of thinking to its logical conclusion and attempt to place in the optical interconnect a two dimensional perfect shuffle wiring pattern, then it transpires that the four segments of the input array must be interleaved as shown in Fig.3 a & b. Tying this back to the layout of a single electronic

8x8 2D Shuffle Start!

1	2	3	4	5	6	7	8
9	10	11	12	13	14	15	16
17	18	19	20	21	22	23	24
25	26	27	28	29	30	31	32
33	34	35	36	37	38	39	40
41	42	43	44	45	46	47	48
49	50	51	52	53	54	55	56
57	58	59	60	61	62	63	64

8x8 2D Shuffle Finish! (Showing top left quadrant positions)

1	5	2	6	3	7	4	8
33	37	34	38	35	39	36	40
9	13	10	14	11	15	12	16
41	45	42	46	43	47	44	48
17	21	18	22	19	23	20	24
49	53	50	54	51	55	52	56
25	29	26	30	27	31	28	32
57	61	58	62	59	63	60	64

FIGURE 3. a. 8x8 matrix before and b. after a 2D perfect shuffle.

island to embrace a complete layer of the switch, we recognise that a layout of the form shown in Fig.4 leads to the shortest electronic paths, freedom from electronic crossovers and an optical interconnection pattern that is immediately partitioned correctly for perfect shuffle implementation. Whether in practice this is the best way of proceeding or whether it will be better to wire the perfect shuffle connection pattern at the detector array depends upon the relative capacitances involved. The perfect shuffle interconnect presupposes 2:1 magnification, implying 4 times the minimum detector capacitance. On the other hand, wiring a perfect shuffle presupposes longer higher capacitance leads to the receiver input stage. Which will be more acceptable then depends upon many details of the particular technology used. In either case, the longest path length involved in the plane to plane interconnect of the switch after it has been packaged on a single wafer is carried by optics, with metal used only for the short links from detector to receiver and output driver to modulator.

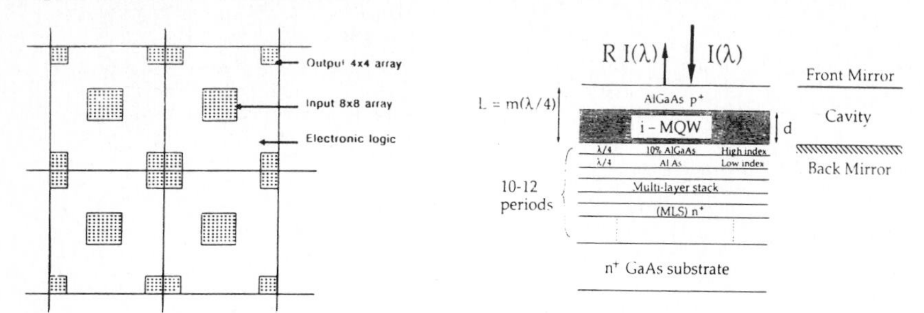

FIGURE 4 Proposed layout for an optoelectronic implementation of a 2D matrix as shown in Fig. 2.

FIGURE 5. Schematic structure of a high contrast MQW modulator.

7. POWER EFFICIENT OPTOELECTRONIC INTERFACES

All of these proposals rely critically for their implementation upon high quantum efficiency optoelectronic interfaces having very low capacitance and capable of being grown with high yield on silicon substrates. They also imply an ability to fabricate power efficient free space optical imaging interconnects capable of imaging with high spatial resolution from one area of a chip to another and doing so in such a way that they maintain alignment throughout a realistic working life.

For the interface elements, one device stands out as close to ideal. This is the high contrast MQW Modulator or Detector shown in Fig.5. By exploiting the structure of an asymmetric Fabry-Perot etalon, it is possible to obtain very high contrast and low insertion loss at modest drive voltage and reasonable wavelength and temperature stability when used as a modulator with very high quantum efficiency when used as a detector. These devices have been fabricated on GaAs [6] and Silicon [7] substrates, can be made in small closely packed arrays and offer very low capacitance (typically 10^{-16} F/sq.micron) which for a 20 micron diameter device projects to 0.03 pF. Driven at 50 ohms, that implies an upper frequency in excess of 600 GHz or conversely, an effective impedance of in excess of 10k ohm at 1Ghz. Note that metallisation on a 1 micron MOS circuit will also have an approximate capacitance of 10^{-16} F/sq.micron so that a good guideline for simple optical link planning will be that if the area of the modulator & detector is less than the area of metal line replaced, then the drive power should or could be less.

8. SUMMARY

We have discussed how optics can be used to overcome some of the bottlenecks in high speed digital processing systems, in particular to tackle the "pin-out" problem that currently limits the data throughput in a processor that is required to do minimal processing on a vast amount of data, such as a synchronous switching matrix. We have shown that most of the key elements already exist to allow an optimum mix of optical interconnects with electronic logic for processing, although the technology to assemble them is still very difficult. There appear to be excellent reasons for trying to solve these problems so that the design rules can be established that will allow the optimum mixture to be accurately defined and implemented. Once done, it appears that logical processing systems will once again take another leap forward in throughput.

9. REFERENCES

[1]. A Huang & S Knauer, "Starlite, a wideband digital switch", Proc.IEEE Global Telecomm.Conf., Atlanta, Georgia, USA, pp.121-125, Pub.IEEE (New York).

[2]. J E Midwinter, "Digital Optics, smart interconnect or optical logic", Physics in Technology, Vol.19, Pt.1, pp.101-108, Pt.2, pp.153-157, 1988.

[3]. D G Messerschmitt, "Communications in VLSI", 6th IEEE International Workshop on Microelectronics and Photonics in Communications", New Seabury, Cape Cod, MA, USA, June 7-9, 1989.

[4]. A W Lohmann, "What classical optics can do for the digital optical computer", App.Opt. Vol.25. pp.1543-1549, 1986.

[5]. M Taylor & J E Midwinter, "Two dimensional optical shuffle networks", Pub.Photonic Switching, Volume 3 in OSA Proceedings Series Optical Society of America, Washington, USA, 1989.

[6]. M Whitehead, A Rivers, G Parry, J S Roberts & C Button, "Low voltage multiple quantum well reflection modulator with on/off ratio > 100:1", Electronics Letters. Vol.25, pp.984-985, 1989.

[7]. P Barnes, P Zouganelli, A Rivers, M Whitehead, G Parry, K Woodbridge & C Roberts, "GaAs/AlGaAs multiple quantum well modulator using a multilayer reflector stack grown on a silicon substrate", Electronics Letters Vol.25, pp.995-996, 1989

PHOTONIC SWITCHING CAPACITY EXPANSION SCHEMES

Mitsuhito SAKAGUCHI and Masahiko FUJIWARA

Opto-Electronics Research Laboratories
NEC Corporation, Japan

Line capacity expansion in a photonic switching system is one of the most important issues to be achieved, to assure photonic switching systems into practical applications. This paper first describes the photonic switching network design for line capacity expansion. Then the present status of photonic switching devices and system experiments are reviewed. System technologies for expanding the line capacity for photonic switching systems are also discussed.

1. INTRODUCTION

Photonic switching is expected to play a key role in optical one link "transparent" networks over optical fiber transmission highways. The optical one link networks have design flexibilities to meet requirements for future optical telecommunication networks providing multi-media services, especially for video information services.

Conceptual studies on photonic switching were carried out in the 1970s, and some device oriented experiments have been made in the 1980s. A photonic time-division switching network was built, using optical-fiber delay line memories and $LiNbO_3$ photonic switches, in 1983 [1]. Since then, much interest has been focused on photonic switching studies, and various kinds of experimental results have been reported [2].

In the 1990s, photonic switching research trends will be divided into two directions: applying photonic switching technologies to practical telecommunication networks, and photonic processing technologies introduction to photonic switching systems.

Photonic switching capacity expandability is the most important measure for assuring achieving practical applications.

This paper describes the design for capacity expansion and discusses device oriented expansion schemes, as well as system oriented expansion schemes.

2. DESIGN FOR CAPACITY EXPANSION

2.1. Classification of photonic switching networks.

In general, a photonic switching system consists of photonic switching networks and control systems. Since no photonic logic circuit for the complicated switching traffic processing has yet been fully developed, the current photonic switching system is a hybrid system combining photonic switching networks and electronic control systems. There are three possible types in photonic switching network: space-division (SD), wavelength-division (WD) and time-division (TD) switching networks [3]. Figure 1 shows the structures of these photonic switching networks. Each individual photonic switching network requires different elements, as follows.

1) Photonic SD switching network is composed of photonic switch matrices.
2) A photonic TD switching network consists of time multiplexers/demultiplexers, time multiplexed space switches (highway switches) and time slot interchangers

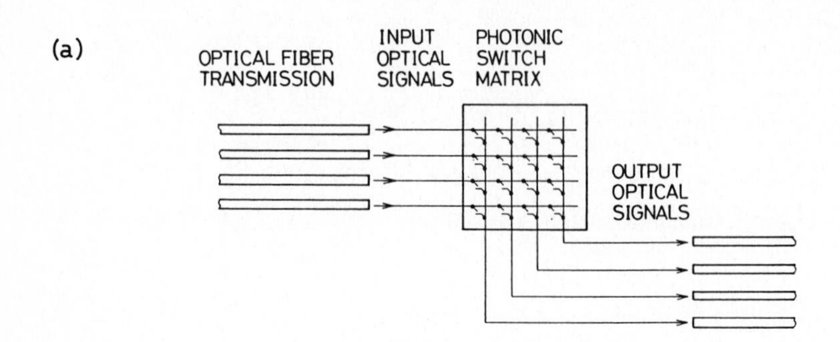

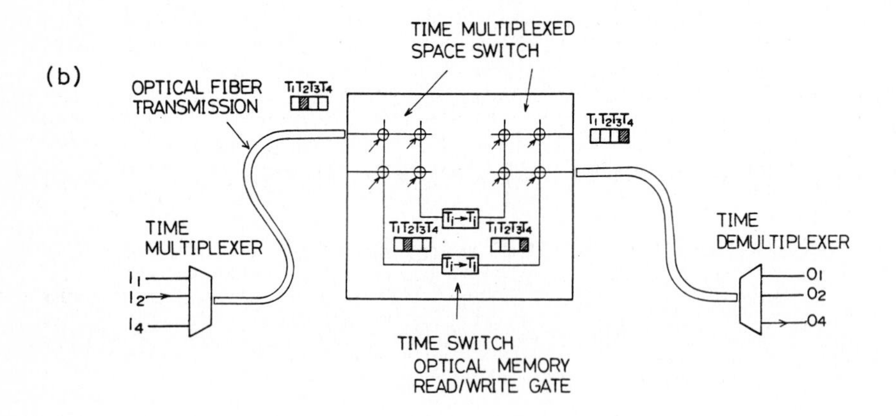

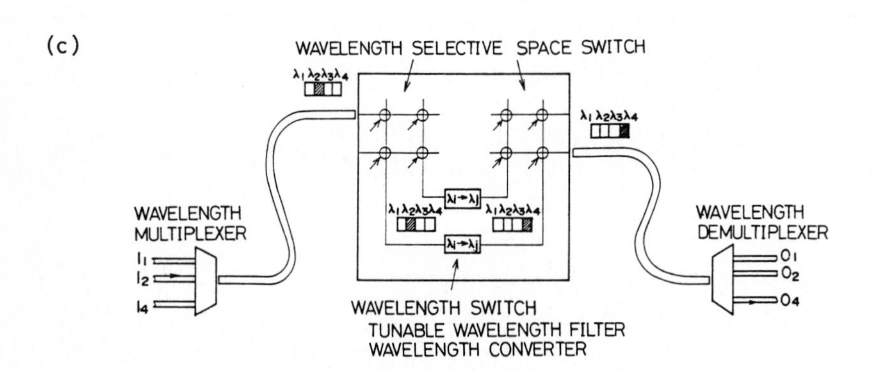

FIGURE 1
Photonic switching networks
(a)Space-division (SD) switching network
(b)Time-division (TD) switching network
(c)Wavelength-division (WD) switching network

(T switches). Time multiplexers/demultiplexers and time multiplexed space switches can be constructed with high speed photonic switch matrices. The key component of the T switch is the optical memory.

3) A photonic WD switching network is composed of wavelength multiplexers/demultiplexers, wavelength selective space switches and wavelength interchangers (λ switches [4]). A wavelength interchanger is constructed with tunable wavelength filters and wavelength converters.

The photonic SD switching network can be realized by simply using photonic switch matrices. In a photonic SD switching network, each photonic cross point guarantees transparent connection between two optical transmission media. Moreover, it does not require high speed switch control in case of circuit switched traffic application. Therefore, the photonic SD switching network is the most practical for realizing a broadband switching system in the early stage. However, the photonic SD switching network will encounter a serious problem, in that it requires a rapidly growing number of photonic switch matrices and a huge number of input/output connections, in order to expand the line capacity. This problem will be solved by using multiplexed switching networks, such as photonic TD and WD switching networks.

The photonic TD switching network has its inherent advantage in that it could provide good affinities with existing TDM optical fiber transmission systems. A high speed digital one link will be constructed, when the switching systems are combined with optical transmission systems. However, in a broadband photonic TD switching system, extremely high-speed operations in optical memories and time multiplexers/demultiplexers are required. Required rigid bit/frame synchronization is also a problem to be solved.

The photonic WD switching system has the following two advantages, in comparison with the photonic TD switching system: (1)the bit-rate independency for individual wavelength-division channels and (2)no need for high-speed operation in the switching control system for circuit switched traffic use. Therefore, the photonic WD switching network would be the most attractive for achieving a large scale broadband switching system [4][5].

2.2. Requirements for photonic switching devices

In general, the line capacity values for the photonic switching systems are limited by optical loss and crosstalk characteristics of the switching networks. Device limitations in time domain (operation speed) and wavelength domain (tuning range) would also restrict the line capacities of photonic TD and WD switching systems. For designing a large scale photonic switching network, these characteristics of the photonic switching network and devices should be examined. This section reports results obtained from an investigation on requirements for photonic switching devices.

(1)Photonic switch matrices

A photonic switch matrix is the key element for photonic SD switching networks. Switch matrices can be built up from the basic 2×2 crosspoints, according to different cascading or inter connect architectures [6]. The time or wavelength multiplexers/demultiplexers and read/write switches in a T switch also can be constructed with basic 1×2 or 2×1 elements, with wavelength selective or high speed switching characteristics. Therefore, architecture considerations for the photonic SD switching networks can also be applied to photonic TD and WD switching networks.

In choosing an architecture, several characteristics must be considered. The issue of connectivity is of prime consideration. The main factors related to the network scale expansion are optical crosstalk and optical loss. There are four ranks in connectivity: blocking, re-arrangeably non-blocking, wide-sense non-blocking and strictly non-blocking [6]. Although a strictly non-blocking switching architecture requires more cascaded stages of switches than re-arrangeably non-blocking architectures, such an

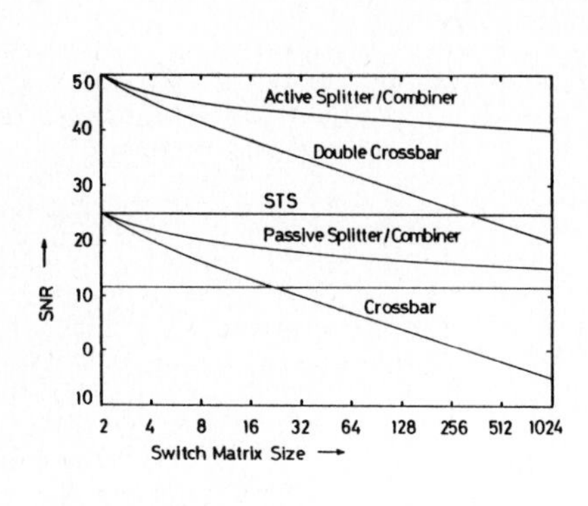

FIGURE 2
Variation of signal to noise ratio with switch network size.

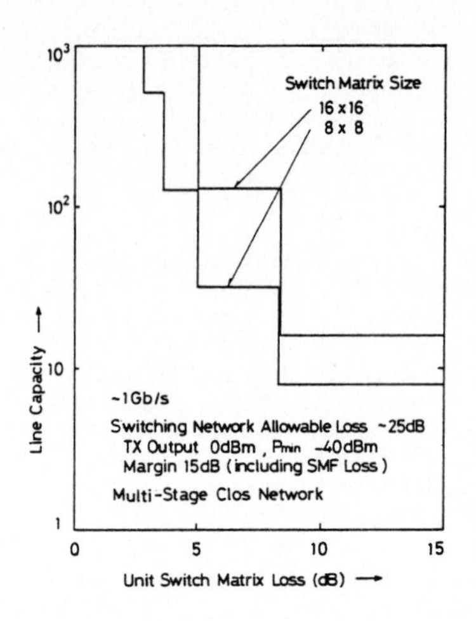

FIGURE 3
Line capacity vs. unit switch matrix loss.

architecture is obviously desirable in photonic SD switching networks. Several strictly non-blocking switching network architectures have been proposed as follows: 1)crossbar, 2)double crossbar[7] 3)splitter/combiner[8][9] 4)simplified tree structure (STS) [10][11]. Figure 2 shows the calculated results of the SNR degradation due to optical crosstalk, for strictly non-blocking switching network architectures. Also shown in the figure is an optical SNR required to achieve a bit error rate (BER) of 10^{-12} (11.5dB). The crosstalk value of each crosspoint was assumed to be -25dB. As shown, STS, splitter/combiner structure, and double crossbar structure have excellent crosstalk avoidance capability. An $1 \times N$ or $N \times 1$ tree structure, which are usually used in the multiplexers/demultiplexers and optical read/write switches, also has excellent crosstalk avoidance characteristic. When applying these architectures, there is no problem in regard to optical crosstalk up to switch size of 10^3. Switch matrices using these architectures can be used to construct a larger scale switching network, also by using a multi-stage network, such as a Clos network [12], without being subject to from significant degradation in SNR characteristic. Therefore, the inherent limitation to switch matrix dimensions is optical loss. The allowable loss for the switching network is determined by the loss budget allowed for the signal speed to be used in the network. Figure 3 shows the relation between line capacity and unit switch matrix loss. The multi-stage Clos network was assumed for the switching network structure. The following values were assumed in the calculation: signal bit rate=1Gb/s, allowable loss for the network=30dB (transmitter optical fiber output power=0dBm, receiver sensitivity=−40dBm, margin=15dB (including single mode optical fiber transmission line loss)). As shown in Fig. 3, unit switch matrix loss should be less than 5dB, in order to achieve a line capacity of about 10^3.

For achieving a large scale switching network, the use of optical amplifiers may be necessary to overcome the optical loss problem. A traveling-wave laser diode optical amplifier (TWA) [13] is very attractive for the application, because of its high gain, small dimensions and array structure capability. In case of introducing TWAs to the switching network, their spontaneous emission noise has to be taken into account. The structure of a photonic SD switching network with TWAs is depicted in Fig. 4 (a). Spontaneous noise effects in such a network were calculated using the switching network model shown in Fig. 4 (b), where the attenuated intensity-modulated signal due to each cascaded switch matrix loss L (including coupling loss with TWA) is

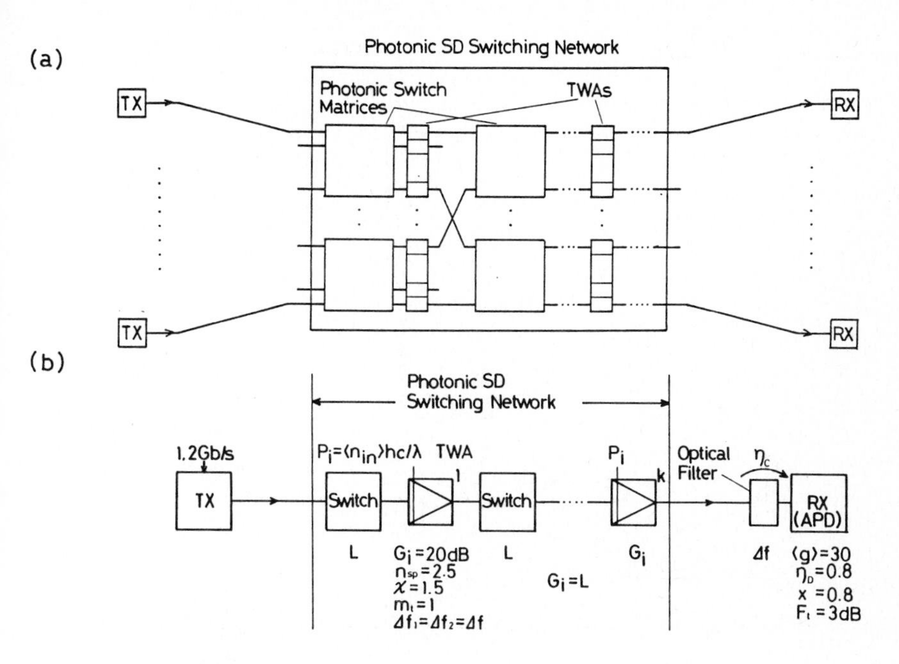

FIGURE 4
Photonic switching network with switch matrices and TWAs.
(a)Structure (b)Calculation model

recovered by individual TWA gain Gi (Gi=L=20dB). SNR calculation was made after Ref. [14], assuming values shown in Fig. 4 (b). The SNR just after the k-th TWA for a 1.2Gb/s signal is shown in Fig. 5, as a function of the number of cascaded TWAs, k. The parameters are optical input power to TWA (Pi) and bandwidth of optical filter (Δf) for spontaneous noise reduction. As shown, more than 20 stage cascaded connections in switch matrices and TWAs are possible with maintaining electrical signal SNR>23dB (corresponding to BER<10^{-12}), by keeping Pi above about −20dBm, even without an optical filter. For applying a photonic switching system to the LAN or MAN area (within several tens sq.-km area), there is no problem regarding keeping Pi above −20dBm, because optical fiber transmission line loss would be small. Another issue to be considered is the gain saturation due to spontaneous emission noise. Figure 6 shows the calculated results for mean noise power for the (k+1)-th TWA input. Assuming saturation input power for the TWA to be −15dBm, the numbers of cascaded connections in TWAs are restricted to less than 10, without an optical filter. However, the number is improved to about 20 by using a 20nm bandwidth optical filter. Assuming utilizing the Clos network, a 2048-line switching network can be constructed with only 9-stage connection in 8×8 switch matrices. The loss of a $LiNbO_3$ 8×8 switch matrix is about 10dB [23], even with current technologies. Therefore, it is possible to construct a photonic SD switching network with more than 10^3-line capacity.

In order to achieve a large line capacity switching network using multi-stage switching networks, it would be very important to develop the optical interconnection technologies. Polarization-independent characteristic is also very important for practical use, because it allows direct connections between the switching network and single mode optical fiber subscriber or interconnection lines without polarization controllers.

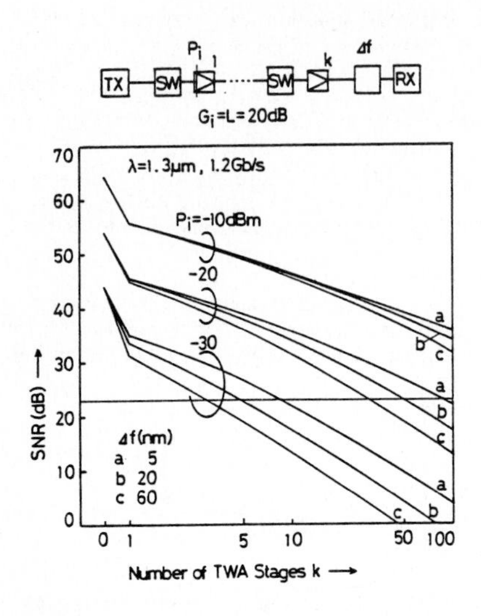

FIGURE 5
Signal to noise ratio just after the k-th TWA

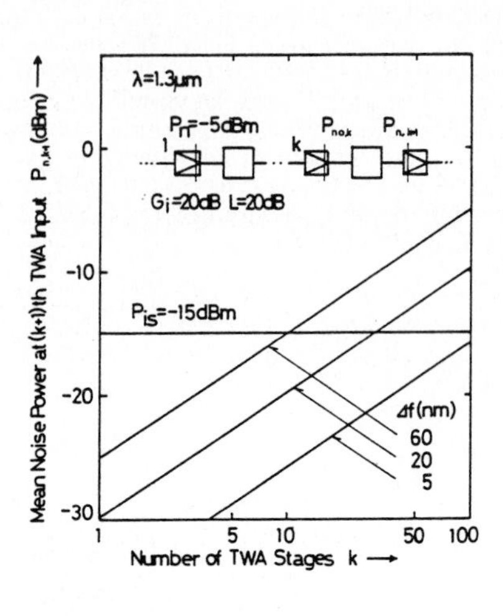

FIGURE 6
Mean noise power for the (k+1)-th TWA input

(2)Optical memories, Tunable wavelength filters

Important features for an optical memory are operation speed and required power for the operation (both optical and electrical). In a time slot interchanger, the product of TD multiplexity value and signal speed per channel cannot exceed the optical memory operation speed. It becomes more difficult to obtain large TD multiplexity with an increase in signal speed. For example, more than 15Gb/s operation speed is required for optical memories in order to obtain TD multiplexity of 100, for B-ISDN H_4 channels. Low power dissipation is also required to obtain a large scale integrated optical memory. Two kinds of optical memory have been investigated for application to photonic TD switching networks. One is the optical bi-stable device and the other is the optical delay line [15]. Optical bi-stable memories have the potential to reject noise accumulation. Therefore, SNR conditions would be relaxed in a time slot interchanger using a bi-stable memory. On the other hand, an optical delay line memory has inherent advantages, in that it has no limitation in operation speed and no power dissipation. However, it has no signal regeneration capability.

The important features related to the WD multiplexity expansion are wavelength tuning range and crosstalk characteristics of the tunable optical filter. The WD multiplexity value is determined by maximum wavelength tuning range of the filter and required WD channel separation for suppressing interchannel crosstalk. Therefore, optimum combinations of modulation schemes and tunable optical filters are important for achieving large WD multiplexity.

3. PRESENT STATUS OF PHOTONIC SWITCHING DEVICES

3.1. Switch matrices

At present, Lithium Niobate ($LiNbO_3$) switch matrices show the best characteristics in regard to matrix size and insertion loss. Several kinds of $LiNbO_3$ photonic switch matrix have been reported, using different architectures. 8×8 switch matrices using crossbar architecture [16][17] and re-arrangeably non-blocking architecture [18][19] were reported. 4×4 switch matrices with double crossbar architecture [7] for reducing optical crosstalk and improved crossbar architecture [20], which can reduce the number of switch stage, were also developed. Lower crosstalk and lower switching voltage, compared to crossbar structure, can be obtained in tree structure switch matrices. A 4×4 switch matrix with passive splitters/active selectors [21], suitable for broadcast mode applications, and a simplified tree structure (STS) 4×4 switches [11], were reported.

Polarization independent 4×4 and 8×8 switch matrices were recently reported, using tree structures [10][22][23]. Figure 7 shows the structure of a polarization-independent STS 8×8 switch matrix [23]. Switching voltage, insertion loss and optical crosstalk value are about 85V, less than 12dB and less than −18dB, respectively.

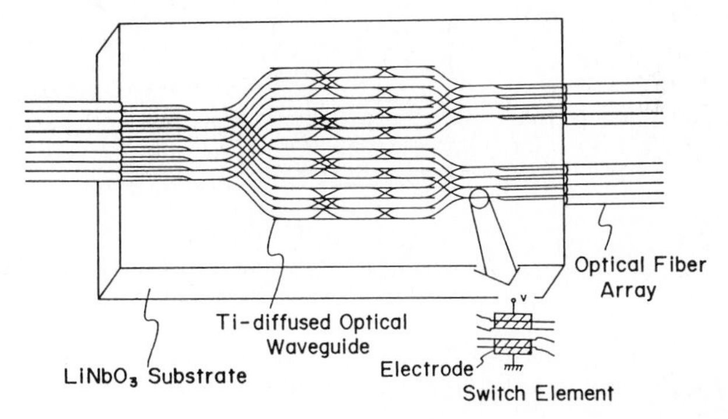

FIGURE 7
Polarization independent 8×8 switch matrix with simplified-tree- structure (STS)

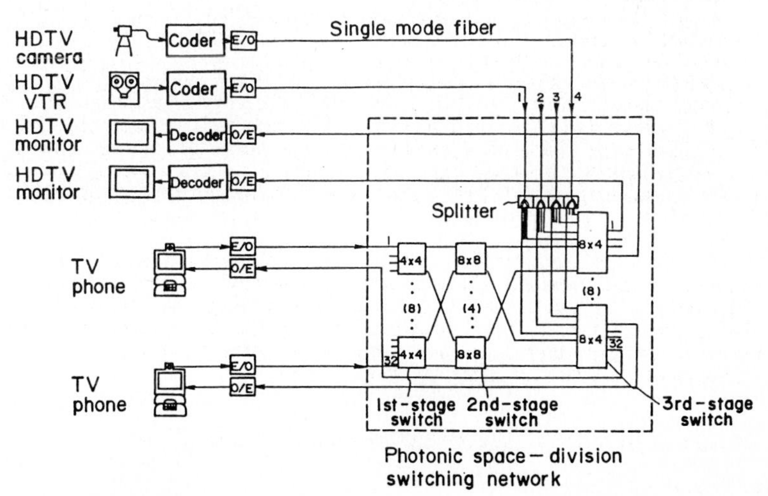

FIGURE 8
Blockdiagram of 32-line photonic SD switching system

A 32-line photonic SD experimental switching system [23] has been successfully demonstrated using the polarization independent $LiNbO_3$ 8×8 switch matrices. A blockdiagram of the system is shown in Fig. 8. This switching system can be used with conventional single mode optical fiber, because polarization independent $LiNbO_3$ 8×8 switch matrices are adopted in the switching network. This switching system provides TV phone communication services, as well as HDTV distribution services. For providing such integrated services, an 8×4 photonic functional circuit, shown in Fig. 9, has been developed. A passive splitter/active combiner structure 4×4 switch matrix, for point to multi-point connection, is integrated with the STS 4×4 switch matrix on a substrate. Because the STS switch matrices are adopted, crosstalk accumulation in the switching network is very small. Line capacity value is limited to 32, mainly due to switch matrix loss.

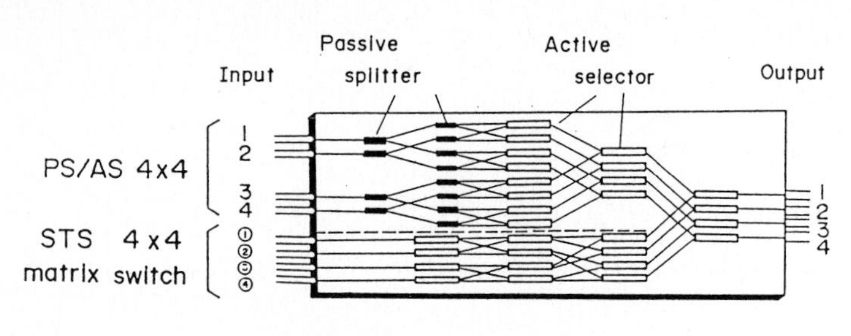

FIGURE 9
8×4 photonic functional circuit

Introduction of TWAs to the photonic SD switching network has been examined experimentally [24]-[26]. Figure 10 shows the set-up and results of the experiment made to demonstrate switch matrix loss compensation for with a TWA. The experiments were carried out at signal speed of 780Mb/s, which is the bit rate of the

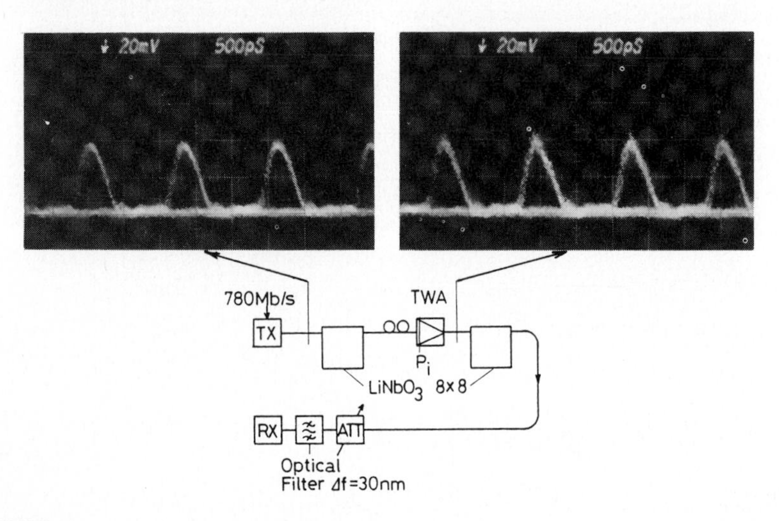

FIGURE 10
Experiments to demonstrate applicability of TWAs to photonic SDswitching networks.

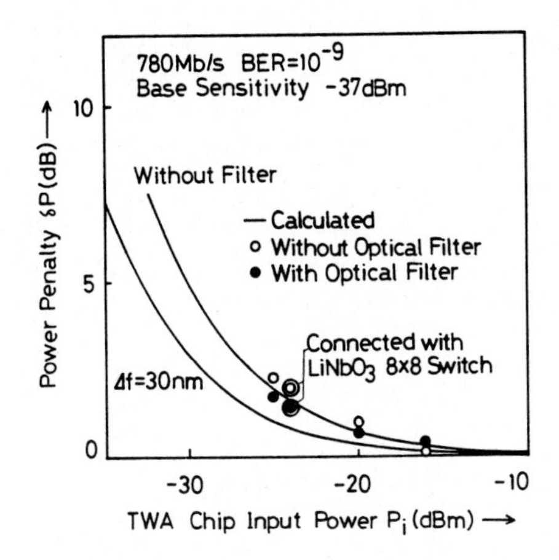

FIGURE 11
Power penalty as a function of TWA input power.

TCI (Time Compressed Integration) encoded HDTV. About 10dB switch matrix loss was completely compensated with a TWA. Figure 11 shows the measured the power penalty as a function of TWA input power Pi. The experimental results agree well with the calculated result based on the calculation model described in Sec. 2. This fact would confirm the theoretical prediction regarding line capacity expansion stated in Sec. 2.

Polarization independent switch matrices also have been fabricated, using semiconductor material. A 4×4 laser diode (LD) gate switch matrix was demonstrated, using SiO_2-TiO_2 planar waveguides on a Si substrate and a laser diode gate array [27]. The switch consists of four 1×4 optical splitters, a 16-element LD gate array and four 4×1 optical combiners. Switch dimensions are 10mm×25mm. Crosstalk of the switch matrix was estimated to be less than −13dB, from the measured extinction ratio of the LD gate. An 8mm long InGaAsP/InP 4×4 switch matrix, using total internal reflection due to carrier-induced refractive index change, was demonstrated [28]. Switching operation was obtained with about 120mA current injection. The insertion loss of the switch is rather high (about 20dB) with current technologies. However, semiconductor switch matrices have a potential for integration with LD optical amplifiers. Therefore, semiconductor switch matrices are expected to exhibit improved performance after the integration of the components and improvement of the fabrication process.

3.2. Optical memories

The world's first experiment in photonic TD switching was conducted using optical fiber delay line memories [1]. A 32Mb/s photonic TD digital switching was achieved with a 16dB optical loss margin. An optical fiber delay line memory has the following uniue features:
1) It's easy to store an extremely high speed signal and long bit stream.
2) It can store a WDM) signal simultaneously.

Taking advantage of these unique features, optical fiber delay line memories may be applied to high speed photonic TD or packet switching systems.

Bi-stable LDs (BSLDs) have been used in Japan in a 256Mb/s highway speed photonic TD switching experiment [29]. Figure 12 shows the device structure and the bi-stable

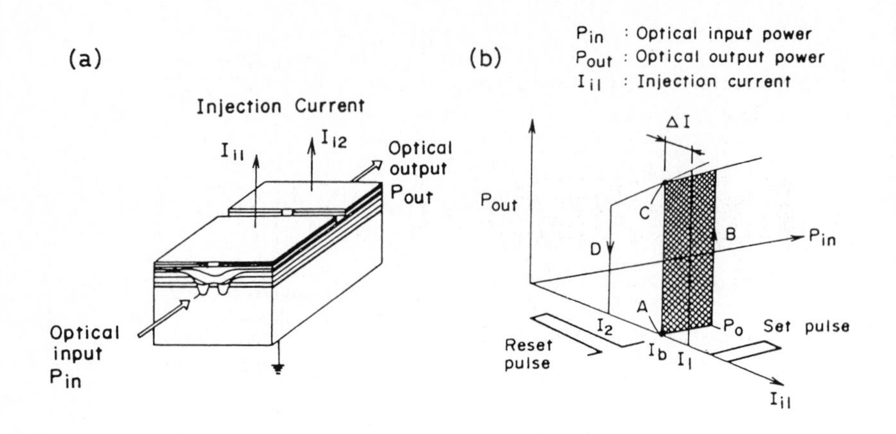

FIGURE 12
Bistable laser diode (BSLD) (a)Structure (b)Characteristics

characteristics. The turn-on triggering pulse width was 2ns at an optical power of 35μW. Turn-off was accomplished by decreasing the hold current by 5mA. A 2Gb/s operation in BSLD optical memories has been achieved [30]. However, the TD multiplexity value is still limited to less than 16 for B-ISDN H4 channels, even with the 2Gb/s operation speed. Therefore, the operation speed of the optical memory is required to be improved, in order to achieve a large scale broadband TD switching system.

To achieve a large scale integrated optical memory, assuring a low power consumption characteristic is important. 2-dimensional array optical memories with low power consumption have been successfully demonstrated using SEED [31] and VSTEP [32] devices.

3.3. Tunable wavelength filters

Although wide wavelength tuning ranges have been demonstrated using mechanically tuned filters [33][34], electrically tunable filters, such as DFB LD amplifiers [35]-[37] and acousto-optic filters [38][39], are desirable for large scale photonic WD switching system. An eight channel photonic WD switching system has been reported, using distributed feedback (DFB) LD filters [40]. The structure and characteristics of the phase shift controlled DFB-LD filter [37], used in the 8-channel photonic WD switching experiments, are shown in Fig. 13. A constant gain of about 25dB, throughout the 120GHz tuning range, was obtained. Optical crosstalk value is about −13dB. The WD multiplexity value for the wavelength switch using this wavelength filter is now limited mainly by optical crosstalk. Figure 14 shows the relations among required channel separation, Δλ, wavelength tuning range, W, and the WD multiplexity value, n. The out-band attenuation value X was used as a parameter. For the current phase shift controlled DFB LD filter, X and W values are 20dB and 10A, respectively. Therefore, n-value is limited to about fourteen. Improvements in tuning range and optical crosstalk of wavelength filters are the key factors for achieving large line capacity.

Acousto-optic filters have attractive features: broad continuous tuning range, narrow channel spacing (several-nm) and micro second order fast switching time. Three and sixteen channel photonic WD switching experiments were reported, using acousto-optic filters [38][39].

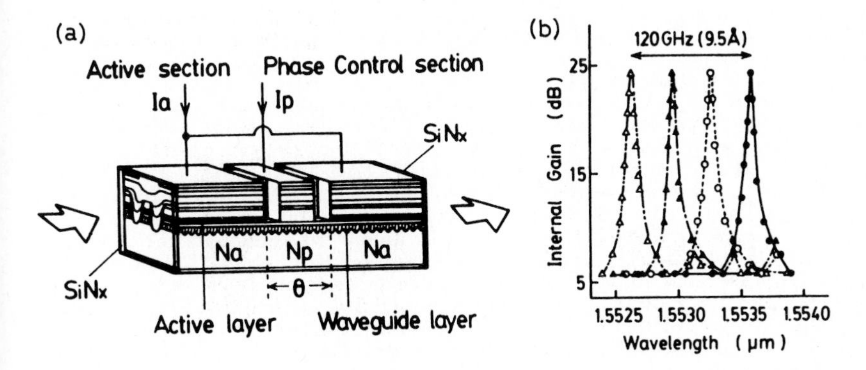

FIGURE 13
Phase shift controlled DFB LD filter.
(a)Structure (b)Characteristics

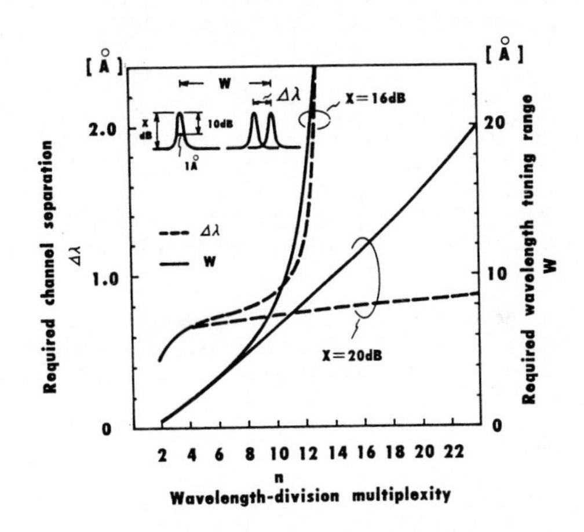

FIGURE 14
Relations among required channel separation, wavelength tuning range and WD multiplexity value.

In most of photonic WD switching experiments, wavelength conversion function was achieved using opto-electronic (O/E) and electro-optic (E/O) converters. To achieve a large scale switch, development of wavelength converters with opto-electronic integrated circuit (OEIC) format is desirable. Recently demonstrated wavelength conversion, using degenerate four wave mixing in an optical amplifier [41], is attractive to realize all optical wavelength converters.

4. SYSTEM ORIENTED EXPANSION

4.1. Hybrid switching systems

In electrical time division switching, large electronic digital switching systems make use of a time slot interchanger in combination with a time multiplexed space switch.

For example, a time-space-time (T-S-T) combination is used in telephone central office switches. Being analogous to the electronic switching systems, the practical photonic switching system will be a hybrid switching system, using two or three types of photonic switching systems. The recently proposed WD/TD hybrid switching system [42] is a good example of such hybrid photonic switching systems. Figure 15 shows one example of the photonic WD&TD hybrid switching system, using a wavelength & time (W&T) hybrid switch, which is the key component in the system. In the WD&TD multiplexer, nxm-channel input signals are WD (wavelength $\lambda_1 ... \lambda_n$) and TD (time slot $T_1 ... T_m$) multiplexed and sent to the W&T hybrid switch. In the W&T hybrid switch, the WD&TD multiplexed optical signal is split and parts thereof are led to individual tunable wavelength filters. Each tunable wavelength filter extracts a specific wavelength signal from the input signal, on a time slot by time slot basis. The output signal from the tunable filter is then converted to an electric signal and the electrical signal is applied to an electronic time switch (T switch), which interchanges time slots of the input signal. An electronic T switch output is used to modulate a tunable wavelength electro-optic converter, whose wavelength is also switched on a time slot by time slot basis. Finally, tunable electro-optic converter output signals are combined to form a W&T hybrid switch output signal. The WD&TD demultiplexer separates the input signal into nxm-channel output signals. In the W&T hybrid switch, the total multiplexity value is given by the product of WD and TD multiplexities. Therefore, a large multiplexity value can be achieved, even with current WD and TD multiplexity values. The WD multiplexity value of 8 has already been demonstrated, using phase shift controlled DFB LD filters [38]. TD multiplexity values of 4 or 8 for the broadband ISDN H_4 channel are easy to obtain with both electronic and photonic TD switching technologies. Therefore, total multiplexity values of 32 or 64 are expected to be achieved, even with the current technologies.

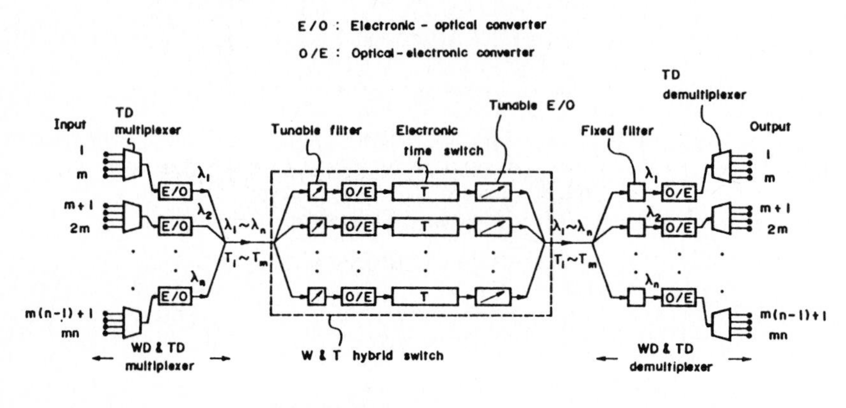

FIGURE 15
Wavelength & time hybrid switch

In the WD&TD hybrid switching system, fast wavelength tuning operation is required for the wavelength tunable devices. Figure 16 shows the results of fast wavelength tuning experiments, using a phase shift controlled DFB LD filter [43]. Four 100Mb/s electronic input signals were TD-multiplexed in a byte inter-leaved form with 6.7 ns guard time between adjacent channels. During the guardtime, the tunable filter accomplishes wavelength switching. In the experiments, the WD channel was switched sequentially within eight WD channels, from λ1 to λ8. Therefore, total multiplexity value in this experiment reaches 32. As shown, wavelength switching within guardtime was successfully demonstrated.

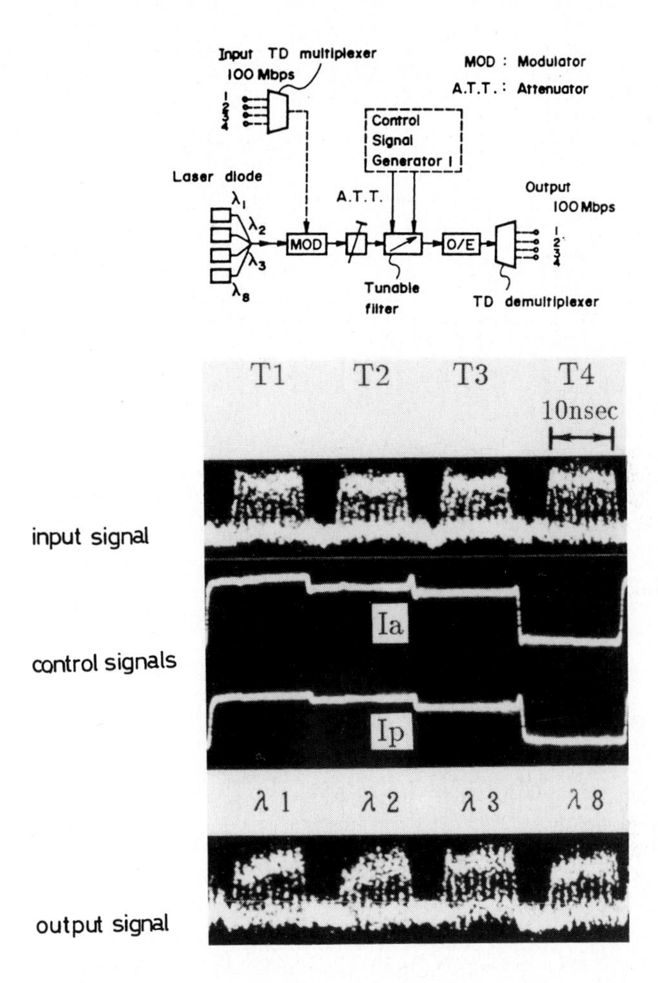

FIGURE 16
Fast wavelength switching experiment.

4.2. Introduction of coherent optical detection

Introduction of coherent optical detection technology to photonic switching systems will result in large scale photonic switching systems, because of receiver sensitivity improvement and fine frequency channel selectivity in coherent optical detection [44][45]. The receiver sensitivity improvement with coherent optical detection can increase the allowable loss value for the photonic switches and transmission lines. Therefore, a large scale network is possible with the coherent optical detection. Moreover, optical crosstalk can be rejected at a coherent optical receiver, if the optical frequencies of the signal and crosstalk slightly differ from each other. Optical crosstalk rejection can be achieved at the IF signal level, not in the optical signal level, utilizing IF filters with sharp cut off characteristic and high rejection-band attenuation.

Photonic SD and WD switching networks are applicable in coherent optical systems. Specifically, the photonic SD switching network is suitable for coherent optical systems, because it is able to preserve all the information the input optical signals have, including wavelength, phase and coherency. A photonic WD switching network is also suited to a coherent optical system, because practical application of dense FDM/WDM transmission would be possible in a coherent optical system.

Introduction of the coherent optical detection to the subscriber terminals in a photonic SD switching system is effective to achieve large line capacity (Fig. 17). Photonic SD switching experiments in a 100Mb/s optical FSK transmission system were carried out using $LiNbO_3$ photonic switch matrices. Receiver sensitivity improvement of 7.5 dB was observed in the transmission experiment through a photonic switch matrix. It was also shown that crosstalk components can be rejected at the receiver by introducing channel separation greater than 3GHz, even when the crosstalk power is ten times larger than the desired signal. From these experimental results, a photonic switching system with a line capacity exceeding 500-lines and whose transmission line length was over 20km, would be expected.

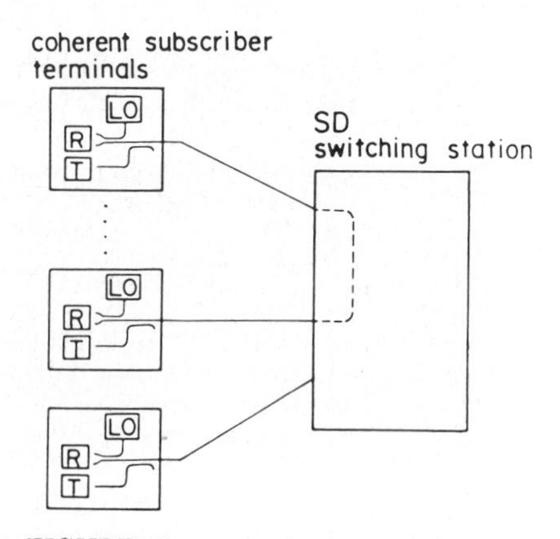

FIGURE 17
Coherent photonic SD switching network.

A coherent photonic wavelength-division (WD) switching system, utilizing a coherent optical detection is also proposed. In the proposed coherent λ switch, the tunable wavelength filter function is accomplished using coherent optical detection with a wavelength tunable local oscillator. Design consideration has shown that achieving a broadband MAN, with a line capacity exceeding 1000, is possible with the coherent photonic WD switching system. The switching function of the coherent switch was demonstrated in two channel wavelength synchronized WD switching experiments, using 8GHz-spaced, 280Mb/s optical FSK signals. Applicability of coherent optical detection to the photonic WD&TD hybrid switching system has been demonstrated, recently [47].

5. CONCLUSIONS

Present digital switching system expansion is mainly due to the advancement in Si monolithic integrated circuit technology. Photonic switching system practical use similarly depends upon the development of opto-electronic integrated circuit technology. Semiconductor opto-electronic device technology is following the path of Si monolithic integrated circuit technology quite closely, but lagging by about 20 years.

In the 1990s, there will be a chance to put photonic switching systems into practical use. Photonic switching systems should activate on exciting scenario, showing step by step hardware milestones and new applications, open to device development, and lead the advancement in opto-electronic integration for photonic switching.

REFERENCES

[1] Goto, H. et al., Tech. Dig. OFC'83, New Orleans, 1983, MJ6
[2] see for example Fujiwara, M. et al., Tech. Dig. IOOC'89, Kobe, July 18-21, 1989, 4, pp.6-7, 21A2-2
[3] Sakaguchi, M. and Goto, H., Tech. Dig. IOOC-ECOC'85, Venezia, Oct. 1-4, 1985, pp.81-88
[4] Suzuki, S. and Nagashima, K., Tech. Dig. Topical Meeting on Photonic Switching, Incline Village, March 18-20 1987, ThA2, pp.21-23
[5] Suzuki, S. et al., Tech. Dig. ICC'87, June 7-10, 1987, pp.1565-1569, 45.3
[6] Spanke, R.A., IEEE Commun. Magazine 25, pp.42-48, 1987
[7] Kondo, M. et al., Tech. Dig. IOOC-ECOC'85, Venezia, Oct. 1- 4, 1985, pp.361-364,
[8] Habara, K. and Kikuchi, K., Electron. Lett. **21**, pp.631-632, 1985
[9] Spanke, R.A., IEEE J. Quantum Electron. QE-22, pp.964-968, 1986
[10] Nishimoto, H. et al., Electron. Lett. **23**, pp.1167-1169, 1988
[11] Okayama, H. et al., IEEE J. Lightwave Technol. 7, pp.1023-1028, 1989
[12] Clos, C., Bell System Tech. J. **32**, pp.407-424, 1953
[13] O'Mahony, M.J., IEEE J. Lightwave Technol. **6**, pp.531-544, 1988
[14] Mukai, T. et al., IEEE J. Quantum Electron. **QE-18**, pp.1560-1568, 1982
[15] Goto, H. et al., Tech. Dig. Topical Meeting on Photonic Switching, Incline Village, March 18-20, 1987, pp.132-134, FD1
[16] Granestrand, P. et al., Electron. Lett. **22**, pp.817-818,1987
[17] Suzuki, S. et al., Tech. Dig. IOOC OFC'87, Reno, Jan. 19-22, 1987, WB4
[18] Duthie, P.J. and Wale, M.J., Electron Lett. **24**, pp.594-596, 1988
[19] Veselka, J.J. et al., Tech. Dig. OFC'89, Houston, Feb. 6-9 1989, p135, THB2
[20] Sawaki, I. et al., Tech. Dig. Topical Metting on Photonic Switching, Incline Village, March 18-20, 1987, PD6
[21] Bogert, G.A. et al., Tech. Dig. Topical Metting on Photonic Switching, Incline Village, March 18-20 1987, ThD3
[22] Granestrand, P. et al., Tech. Dig. IGWO, Santa Fe, March 28- 30 1988, PD3
[23] Suzuki, S. et al., Tech. Dig. Topical Meeting on Photonic Switching, Salt lake City, March 1-3 1989, FE1
[24] Thyl'n, L. et al., Tech. Dig. Topical Meeting on Photonic Switching, Incline Village, March 18-20, 1987, PDP8
[25] Tahara, T. et al., Tech. Dig. OEC'88, 3A2-2
[26] Fujiwara, M. et al., Tech. Dig. IOOC'89, Kobe, July 18-21 1989, 3, pp.182-183, 20D2-1
[27] Himeno, A. et al. Tech. Dig. GLOBECOM'88, Nov. 28-Dec. 1 1988,
[28] Inoue, H. et al., IEEE J. Selected Areas Commun. **6**, pp.1262- 1265, 1988
[29] Suzuki, S. et al., IEEE J. Lightwave Technol. **LT-4**, pp.894-899, 1986
[30] Tomita, A. et al., Tech. Dig. Topical Meeting on Photonic Switching, Incline Village, March 15-20 1987, FC2
[31] Miller, D.A.B. et al., Appl. Phys. Lett., **45**, pp.13-15, 1984
[32] Kasahara, K. et al., Appl. Phys. Lett. **52**, p.679, 1988
[33] Kobrinski, H., Electron. Lett. **23**, pp.825-826, 1987
[34] Frenkel, A. and Lin, C., IEEE J. Lightwave Technol. **7**, pp.615-624, 1989
[35] Magari, K. et al., IEEE J. Lightwave Technol. **24**, pp.2178-2190, 1988
[36] Numai, T. et al., Electron. Lett. **24**, pp.236-237, 1988
[37] Numai, T. et al., Appl. Phys. Lett. **54**, pp.1859-1860, 1889
[38] Shimazu, Y. et al., IEEE J. Lightwave Technol., **LT-5**, pp.1742-1747, 1987
[39] Cheung, K.W. et al, Electron. Lett. **25**, 375-376, 1989
[41] Nishio, M. et al., Tech. Dig. ECOC'88, Brighton, UK, Sept. 11-15 1988, part 2, pp.49-52 [41]Grosskpof, G. et al., Electron. Lett. **24**, pp.1106-1107, 1988
[42] Suzuki, S. et al., Tech. Dig. GLOBECOM'88, Nov. 28-Dec. 1 1988, pp.933-937,

29.2
[43] Nishio, M. et al.; Tech. Dig. Topical Meeting on Photonic Switching, Salt Lake City, March 1-3, 1989, pp.98-100, ThE5
[44] Fujiwara, M. et al., Tech. Dig. Topical Meeting on Photonic Switching, Incline Village, March 15-20, 1987, pp.27-29
[45] Fujiwara, M. et al., Trans. IEICE Japan E72, pp.55-62, 1989
[46] Fujiwara, M. et al., Tech. Dig. ECOC88, Brighton, UK, Sept. 11-15 1988, Part1, pp.139-142
[47] Shimosaka, N. et al., Tech. Dig. ECOC'89, Gothenburg, September 10-14 1989, WeA15-2

PHOTONIC SWITCHING SYSTEMS USING COHERENT OPTICAL TRANSMISSION TECHNOLOGIES

Masahiko FUJIWARA, Katsumi EMURA,
*Syuji SUZUKI, Michikazu KONDO and Shigeru MURATA

Opto-Electronics Research Laboratories
*C&C Systems Research Laboratories
NEC Corporation, Japan

Photonic switching and coherent optical transmission would be key technologies for realizing future all optical broadband networks. This paper reports results of studies on integrating photonic switching systems and coherent optical transmission technologies. Introducing coherent optical transmission technologies to photonic switching systems will result in large scale photonic switching networks because of receiver sensitivity improvement and fine wavelength channel selectivity of coherent optical detection. Several photonic switching systems utilizing coherent optical transmission technologies are proposed and fundamental switching functions are demonstrated.

1. INTRODUCTION

Broad-band networks providing various kinds of services, such as video telephony and video broadcasting, have received increasing attention in not only local-area networks (LANs) but also in metropolitan-area networks (MANs) and wide-area networks (WANs). In order to achieve such networks, both broad-band transmission and switching technologies are required. Optical fiber transmission can provide a long span transmission link with extremely high information carrying capacity. Recent progress in optical fiber transmission has already made broadband terrestrial and undersea trunk lines possible. Coherent optical transmission would further enhance transmission distance, information carrying capacity, and application area of the optical transmissions [1][2]. Optical technology is also important for broadband switching [3]. Photonic switching systems are expected to have advantages over conventional switching systems for use in exchanging broadband signals, because of their bit-rate independent crosstalk characteristics. Consequently, optical fiber transmission, especially coherent optical transmission, and photonic switching would be key technologies for achieving all-optical broadband networks. In such networks, switching and routing functions,as well as signal transmission, will be accomplished in the optical domain.

This paper presents results of a study on the integration of photonic switching and coherent optical transmission which would be important for achieving future all optical broad-band networks. Detailed investigation on features of integrating photonic switching and coherent optical transmission are discussed in Sec. 2. Rest of the article describes some examples of the photonic switching systems utilizing coherent optical detection technologies.

2. INTRODUCTION OF COHERENT OPTICAL DETECTION TECHNOLOGIES TO PHOTONIC SWITCHING [4][5]

Coherent optical transmission, using heterodyne and homodyne detection techniques, has two excellent features in comparison with conventional direct detection optical transmission. One is the potential for improved receiver sensitivity. This means that coherent optical transmission is foreseen for use in realizing very long span transmission. The other feature is the fine wavelength channel selectivity. This feature allows optical frequency-division-multiplexing (FDM) or wavelength-division-multiplexing (WDM) of a large number of optical channels with very narrow wavelength separations. It could be used to access the vast wavelength bandwidth available from single mode optical fibers(SMF). It also allows the use of optical amplifiers such as laser diode (LD) amplifiers [6], without being adversely affected by their quantum noise. For this reason, coherent optical transmission is also likely to be important in subscriber loop or LAN applications [7][8].

A photonic switching system is expected to have advantages over conventional switching systems, especially for use in exchanging broadband signals. There are three possible photonic switching networks: time-division (TD), space-division (SD) and wavelength-division (WD) switching networks [3]. Since various different demands are placed on the broadband services, networks should ideally be transparent both in terms of band-width capability and modulation format. Photonic SD and WD switching networks could easily comprise transparent networks for the circuit switched traffic, without requiring rigid data formatting and bit/frame synchronization. Furthermore, photonic SD and WD switching networks are applicable in coherent optical systems. Specifically, photonic SD switching network, using photonic switch matrices, such as the $LiNbO_3$ guided-wave switch matrices [9][10], is suitable for coherent optical systems, because it is able to preserve all the information the input optical signals have, including wavelength, phase and coherency. A photonic WD switching network is also suited to a coherent optical system, because practical application of dense FDM/WDM transmission would be possible in a coherent optical system.In this paper, application of coherent optical transmission technology to these photonic SD and WD switching networks will be considered.

In general, the line capacity value for photonic switching networks are limited by optical loss and crosstalk characteristics of the switching networks. Introduction of coherent optical detection technology will result in large scale switching networks, because of receiver sensitivity improvement and fine wavelength channel selectivity in coherent optical detection. The receiver sensitivity improvement with coherent optical detection can increase the allowable loss value for the photonic switches and transmission lines. Therefore, a large scale network is possible with the coherent optical detection. Moreover, optical crosstalk can be rejected at a coherent optical receiver if the optical frequencies of the signal and crosstalk slightly differ from each other. Optical crosstalk rejection can be achieved IF signal level, not in optical signal level, utilizing IF filters with steep cut off characteristic and high outband rejection. Therefore, high crosstalk rejection characteristics would be expected. These features,

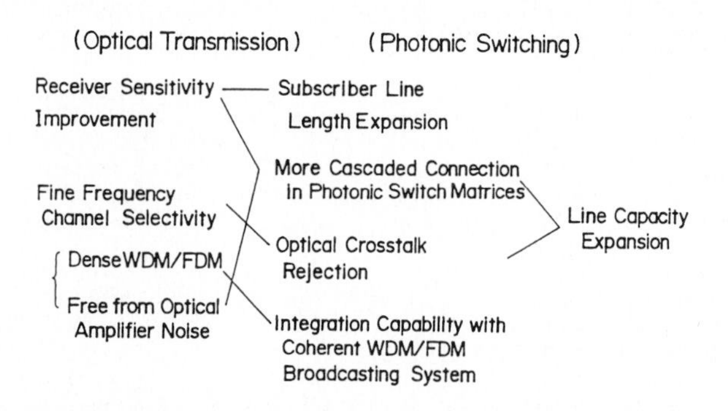

TABLE 1 Features of coherent photonic switching system.

together with the features inherent in coherent optical detection, are shown in Table 1. In the following sections, some examples of photonic switching systems utilizing coherent optical detection will be described.

3. COHERENT PHOTONIC SD SWITCHING SYSTEM

3.1. System description

Photonic SD switching system is the most practical photonic switching system, because it can be simply constructed with photonic switch matrices [11][12]. Using coherent optical transmission technologies, a large scale photonic SD switching system would be expected. Figure 1 shows the coherent photonic switching system with coherent optical detection. In the system (coherent photonic SD switching system), each subscriber terminal has coherent optical transmitter and receiver(s), and the switching function is accomplished at the SD photonic switching station, using photonic switch matrices.

Introduction of coherent optical detection technology to photonic SD switching networks will remove the constraints of distance and line capacity. The line capacity value is restricted by both the insertion loss and the optical crosstalk value of the photonic switch matrices [12]. An increased span loss margin with coherent detection allows more cascaded connections in photonic switch matrices. Figure 2 shows the calculated results of line capacity expansion with receiver sensitivity improvement. In the calculation, the allowable photonic switching network loss with direct detection was assumed to be 22dB [12]. Multi-stage Clos network structure [13] was assumed in the switching network. With experimentally attainable receiver sensitivity improvement (~10dB), large line capacity expansion can be expected. The increased span loss margin also can be allotted to transmission line loss. In that case, networks which are located at great distances from each other could be directly linked to each other. This would remove the constraint of the distances involved in realizing all optical broadband networks. Optical crosstalk, accumulated at every photonic crosspoint causes an additional power penalty. Therefore, it is necessary to limit the line capacity value, in order to guarantee holding the power penalty to within an allowable value. For example, the line capacity value was limited to within 32, assuming a -15dB crosstalk value, to guarantee incurring only a 2dB power penalty caused by optical crosstalk [12]. Using coherent optical detection, crosstalk can be

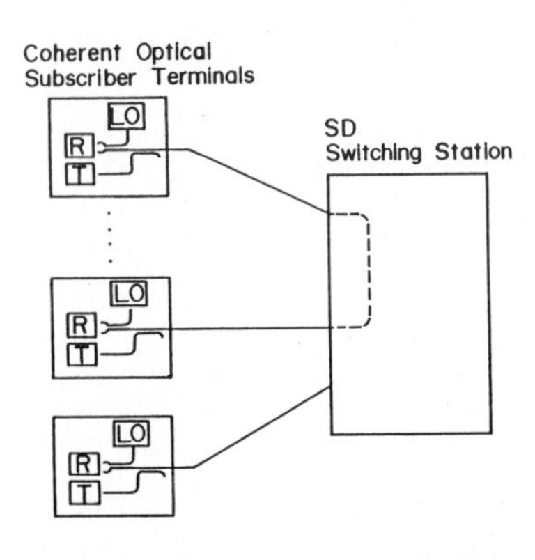

FIGURE 1
Photonic SD switching system with coherent optical detection technology

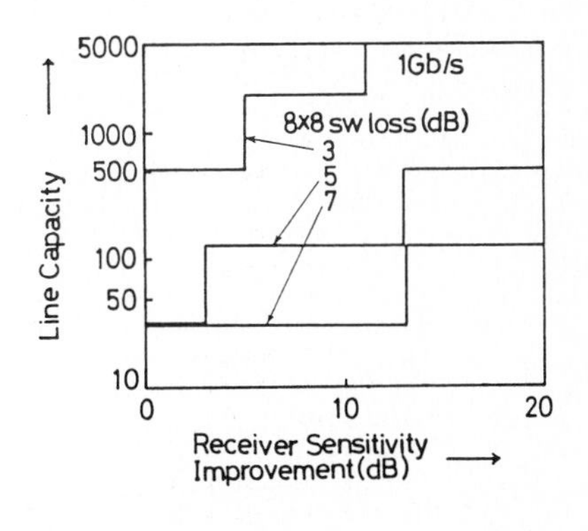

FIGURE 2
Line capacity expansion with coherent optical detction technology in a photonic SD switching network.

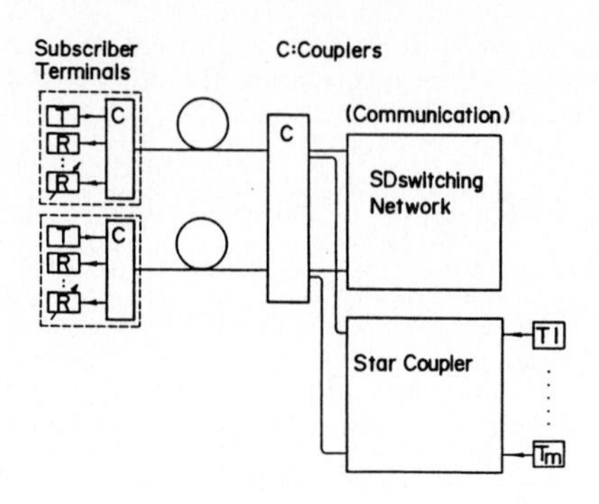

FIGURE 3
Coherent Integrated services system.

rejected at the receiver end by introducing slightly different emitted wavelengths to each subscriber. Therefore, the constraint necessitated by optical crosstalk, in regard to line capacity, can be removed when coherent optical detection is incorporated.

In addition, broadcasting services with coherent FDM/WDM technology could easily be integrated into photonic SD switching networks utilizing coherent optical detection. Figure 3 show a blockdiagram of such a coherent integrated services network. It will be divided into two subsystems. One subsystem is the photonic SD switching network described above, which supports the wideband communication services. The other containing the central transmissive star coupler handles the wideband broadcasting services, whose channels are multiplexed using coherent FDM/WDM technology. A broadcasting channel can be selected by tuning the local laser in the subscriber terminal. Broadcasting services, as well as communication services of very large number of channels, would be offered to all of the subscribers.

3.2. Experiments

In order to demonstrate the advantages described above, photonic SD switching experiments were carried out using a $LiNbO_3$ switch matrix [10] and an optical FSK heterodyne transmission system [14]. The experimental set-up is shown in Fig. 4. Two 1.55 μm wavelength phase tunable DFB-DC PBH LDs with flat FM response [15] were used as FSK transmitter light sources (LD_1, LD_2). At the receiver terminal, a dual filter detection system with a balanced receiver were employed. Each transmitted signal was a 100Mb/s optical FSK signal with frequency deviation of 800 MHz. Measured bit error rate characteristics through a $LiNbO_3$ photonic switch matrix and long SMFs (22km, 100km) are shown in Fig. 5. Receiver sensitivity as high as -54.5dBm(BER$=10^{-9}$), which is 7.5 dB higher than direct detection, was obtained for the receiver. Sensitivity degradation caused by insertion of a $LiNbO_3$ photonic switch matrix and SMFs could not be observed.

The inter-channel interference characteristics were measured with LD1 and LD2 being driven simultaneously. Figure 6(a) shows power penalty change with channel separation $\Delta f(=f2-f1)$, which was monitored in IF signals (Fig. 6(b)). The parameter in Fig. 6(a) is optical crosstalk value(CT) at the entrance of the receiver. In the case of 10dB CT, the crosstalk signal reached the receiver with ten times larger optical power than the desired signal. A crosstalk component with CT$>$10dB might be comprised of optical reflected power from the photonic switching network. When Δf was above 3GHz, no interference effect was observed even with CT at 10dB. This result clearly shows that constraints, necessitated by optical crosstalk, in the number of subscriber lines can be ignored in a photonic SD switching network with coherent detection.

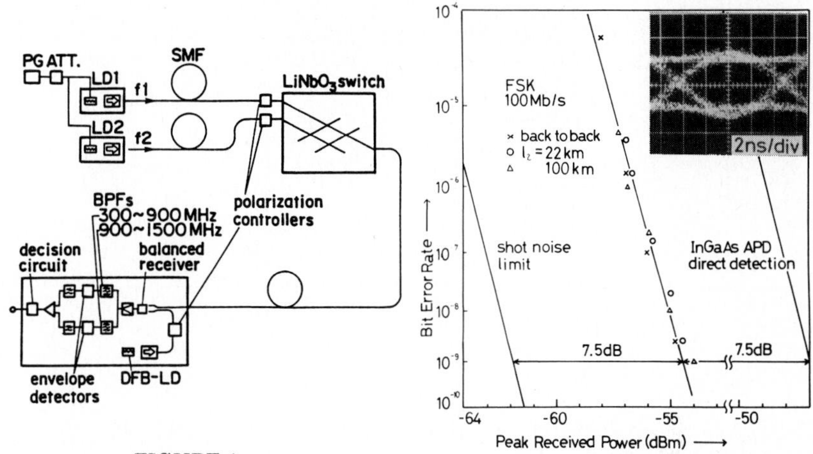

FIGURE 4
FSK transmission system photonic SD switching experiment set-up.

FIGURE 5
Bit error rate vs. peak received power characteristics and demodulated 100Mb/s waveform.

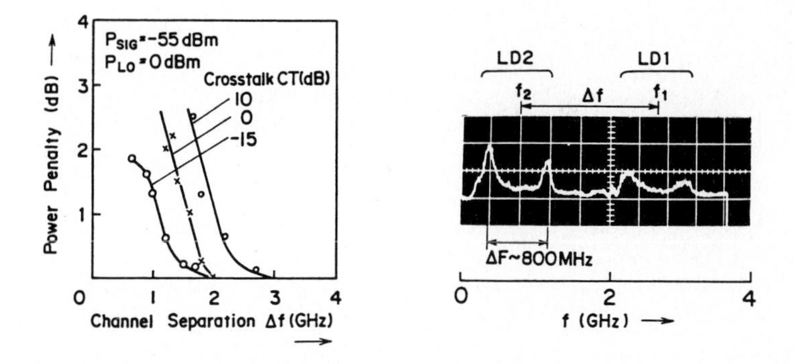

FIGURE 6
Interchannel interference characteristics (a) and IF spectrum (b).

A brief design consideration, based on the experimental results, has been made. From the results described above, the crosstalk problem can be eased in photonic SD switching systems utilizing coherent optical detection technologies. Therefore, only loss budget problem is to be considered. Loss margin exceeded 60dB in the experimental system at a bit-rate of 100Mb/s (receiver sensitivity -45.5dBm and optical fiber input power > 5dBm). Figure 7 shows the relation between transmission line length and number of subscriber lines. A 4 dB loss for the couplers for bi-directional transmission and 0.25dB/km SMF loss were assumed. A photonic SD switching system whose line capacity exceeding 500-lines and whose transmission line length was over 20 km, would be possible in this system, assuming 5dB 8x8 optical switch insertion loss.

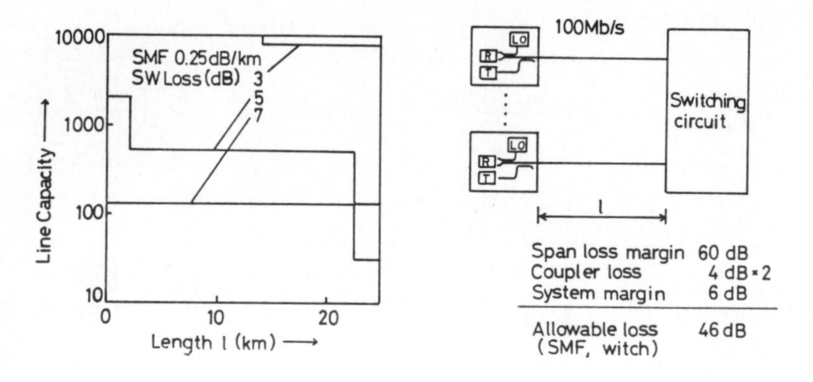

FIGURE 7
Line capacity vs. subscriber line length in coherent photonic SD switching system.

4. COHERENT PHOTONIC WD SWITCHING SYSTEM

To realize future broadband WDM technologies are attractive, both for transmission and switching. An optical broadband network architecture, using photonic WD switching systems and WDM optical transmission systems, was proposed [16]. Various photonic WD switching systems have already been proposed and demonstrated. Two kinds of wavelength switch, which accomplish wavelength interchange, and a wavelength selective space-switch have been demonstrated, using acousto-optic deflectors and DFB LDs as tunable wavelength filters [17]-[19]. Also proposed and demonstrated are the passive wavelength routing systems [20]-[22] and WDM passive-star systems [23]-[25]. This section proposes and demonstrates a novel photonic WD switching system utilizing the coherent optical detection technologies [26].

4.1. Coherent wavelength switch

The key element for the photonic WD switching system is the wavelength switch (λ switch) [16], as a time switch is a basic component in the time-division(TD) switching system. The λ switch is used to accomplish wavelength interchange. Figure 8 shows the structure for the proposed coherent photonic WD switching system, which consists of a wavelength multiplexer (MUX), wavelength demultiplexer(DMUX) and a coherent λ switch. In this system, the FSK modulation format is considered because FSK system is most practical from laser linewidth and modulation consideration. In Fig. 8, lightwave signal paths are indicated by solid lines and electrical signal paths by dashed lines.

The traffic generated by each user(s_1, s_2....s_n) is wavelength-division-multiplexed in the MUX, using the equally spaced comb of wavelength λ_1, $\lambda_2 \cdots \lambda_n$. The MUX may be remotely located close to the users and the MUX can terminate a number of users. The WDM signal from the wavelength multiplexer is transmitted to the coherent λ switch by a SMF transmission line. In the λ switch, an input WDM signal is split and parts thereof are led to individual coherent optical receiver with tunable wavelength local oscillator(LO). At every coherent optical receiver, the desired signal channel can be selected and demodulated by tuning the LO wavelength with control signals from the switch controller. Each demodulated electrical signal from the coherent optical receiver is then applied to a single wavelength LD to create an output optical FSK

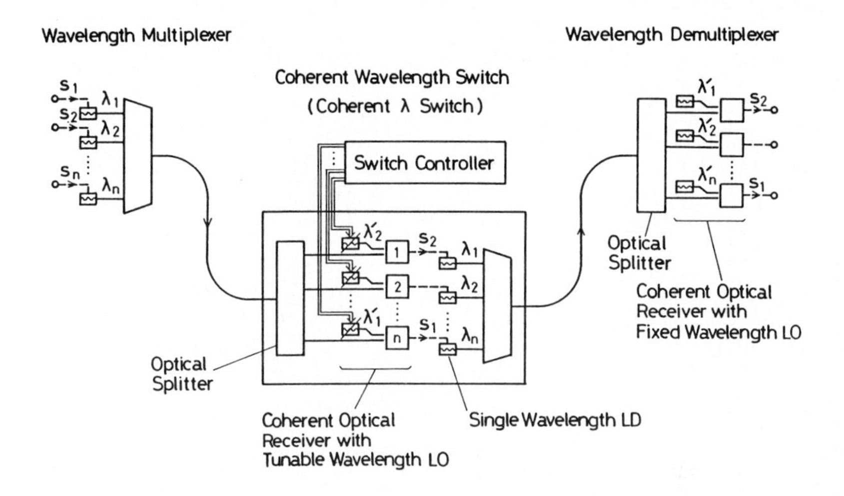

FIGURE 8
Coherent WD switching system.

signal. The wavelengths of LDs are pre-assigned also to be λ_1, $\lambda_2 \cdots \lambda_n$, as shown in Fig. 8. This process accomplishes wavelength interchange. By way of example, it can be seen in Fig. 8 that the coherent optical receiver 1 and n select λ_2 and λ_1 wavelength channels, respectively. As a results, wavelength interchange from λ_2 to λ_1 and λ_1 to λ_n are performed using the coherent optical receiver 1 and n. Any kinds of one to one wavelength interchange are possible by controlling channel selections in coherent optical receivers. Moreover, a multicast function can be achieved by controlling all coherent optical receivers to select the same wavelength channel. On the outbound side of the coherent λ switch, the wavelength interchanged optical signals are multiplexed to form an output WDM signal and sent to the wavelength demultiplexer. There, a switched signal for each user is obtained using the coherent optical receiver with fixed wavelength LO.

In the coherent λ switch, low crosstalk switching for dense WDM signals is possible because of fine wavelength channel selectivity of coherent optical detection. The receiver sensitivity improvement with coherent optical detection can increase the allowable loss value for the λ switches and transmission lines. Therefore, a large scale network is possible with the coherent photonic WD switching system. Moreover, a large-capacity switching system can be constructed with multi-stage connection of the λ switches, with fewer number of WD channels. Figure 9 shows the three-stage photonic WD switching network (λ^3 switching network), using wavelength multiplexers and demultiplexers (MUXs,DMUXs) in inter-stage networks[16]. The inter-stage network can provide each λ switch with potential connectivity to every next stage λ switch. Since, the switching network shown in Fig. 9 is equivalent to a multi-stage Clos network[16] using nxn switch matrices, a non-blocking switching network can be constructed with this network. Utilizing wavelength interchange function of the λ switches, this kind of network allows the re-use of wavelength channels in different switching stages. This leads to a large-capacity switching system with fewer wavelength channels. Consequently, all WDM input lines to the switching network can use same wavelength channels λ_1, $\lambda_2 \ldots \lambda_n$). Line capacities for single λ switch is expressed as n, which is the same for the WDM passive-star network [23]-[25]. Line capacities for λ^3 and λ^5 switching networks are expressed as n^2 and n^3, respectively [13][16]. Therefore, extremely large line capacity can be achieved using multi-stage connection with λ switches even with limited number of WD channels.

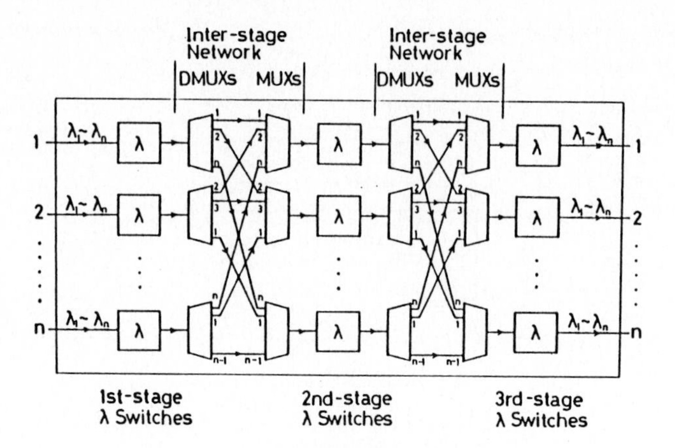

FIGURE 9
Multi stage WD switching network (λ^3 switching network).

4.2. Design consideration of broadband MANs

The proposed coherent WD switching system is very attractive for realizing a broadband Metropolitan-Area-Network (MAN). The structure of MAN, utilizing the coherent WD switching system as a transit switch(TS), is illustrated in Fig. 10. User signals are multiplexed into coherent WDM signals at local switch (LS) line interfaces and transmitted to the TS. The TS is constructed with multi-stage connection with coherent λ switches, as shown in Fig. 9. In the coherent WD TS, wavelength interchange is carried out. The feasibility of such a network is demonstrated through the design, taking into consideration the following three items: (1)Number of wavelength channels, (2)optical power level, and (3)wavelength synchronization.

(1)Number of WD channels
The number of WD channels n is determined by LO light source tuning range and required channel separation for suppressing inter-channel crosstalk. Using 1.55μm wavelength tunable DBR LD, over 44Å (550GHz) continuous wavelength tuning has already been demonstrated [27]. Theoretical calculation and experiments have shown

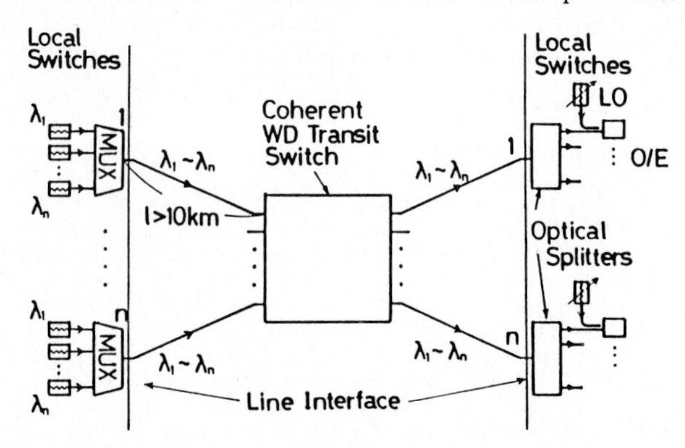

FIGURE 10
Broadband MAN utilizing coherent WD transit switch.

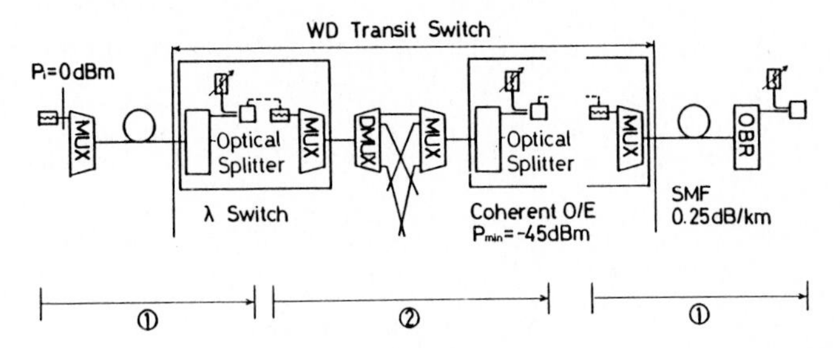

FIGURE 11
Two divided parts of the switching system for optical power level consideration.

that required channel separation is about 10GHz, considering optical FSK dual filter detection of the 1.2Gb/s signal [28]. Therefore, more than 32 ($=2^5$) WD channels are very easy to obtain. As a result, line capacity values reach 1024 and 32768, for λ^3 and λ^5 switching networks, respectively.

(2)Optical power level diagram
In this study, receiver sensitivity was assumed to be -45dBm. This value is possible for the 1.2Gb/s CPFSK system. Therefore, allowable loss for the λ switches and transmission lines was 39dB (fiber input power=0dBm, system margin=6dB). MUXs and DMUXs in the inter-stage network were assumed to consist of $m=\log_2 n$-stage cascaded connections with 2x1 (2x1) M-Z MUX/DMUX elements. The MUXs in LSs and outbound-side of coherent λ switches were assumed to composed of 2x1 M-Z elements or 2x2 optical couplers(excess loss=0.2dB). Optical power level was examined, dividing the switching system into the following two parts, as shown in Fig 11.

(a)LS line interface-coherent receiver in λ switch: Figure 12 shows the relation between Single Mode optical Fiber(SMF) length l and number of WD channels n. The parameter is MUX element or 2x2 coupler loss L_{MUX}(dB/stage) in the LS or coherent λ switch. Even when commercially available optical couplers are used as the MUX in LS or coherent λ switch ($L_{MUX}=3.2$dB), 15km SMF transmission is possible, using 32 WD channels. The 15km transmission line length is large enough for achieving broadband MANs. Longer transmission line length would be expected if low loss MUX elements are obtained, as shown in Fig. 12.

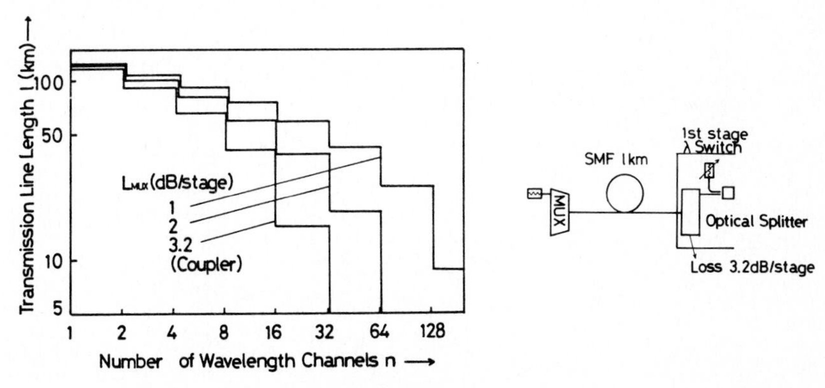

FIGURE 12
SMF transmission line length l vs. number of WD channels n.

(b)Inter-stage connection: The inter-stage connection for 32 WD channels can be achieved within the loss budget, if MUX/DMUX elements with L_{MUX} of less than 1.5dB/stage are available. With the current technologies, L_{MUX} value is about 3-4dB/stage [29]. However, optical loss in inter-stage network can be reduced by reducing the number of MUX/DMUX stages and increasing the number of interconnection lines. Figure 13 shows an example of such inter-stage networks. Optical amplifier with wide wavelength bandwidth, such as a travelling-wave LD optical amplifier, can be applied to multi-channel simultaneous amplification without any sensitivity degradation, if the amplifier input optical level and the WD channel separation are appropriatly set [30]. Therefore, introduction of such an optical amplifier can increase the allowable loss value for the inter-stage networks. With only a 10dB optical gain, inter-stage connections for 32 WD channels are possible, using current MUX/DMUX elements.

(3)Wavelength synchronization [16]
To achieve multi-stage WD switching networks as shown in Fig.9, individual wavelengths for input and output WDM signals of each λ switch should be exactly the same. Wavelength locking of WDM optical sources using the "reference pulse method" [31] can be applied to achieve wavelength synchronization [26]. This is demonstrated experimentally in the next section.

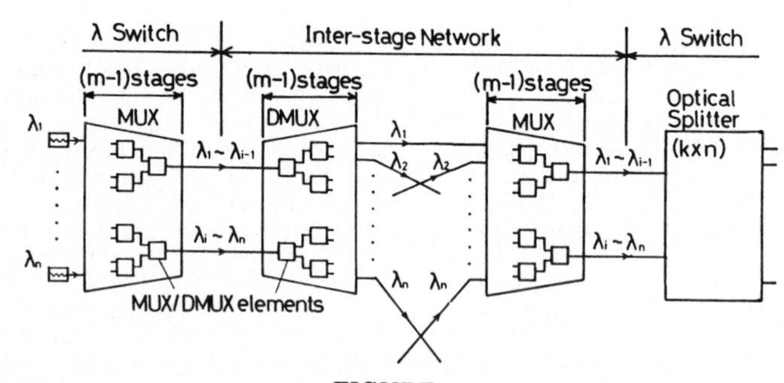

FIGURE 13
Inter-stage connection with reduction in the number of MUX/DMUX element stages m.

4.3. Experiments

A 2 channel wavelength-synchronized switching experiment was carried out to demonstrate the switching function of the coherent λ switch. Figure 14 shows the experimental switching system diagram. The experimental switching system consists of a wavelength multiplexer, a coherent λ switch and a wavelength controller for wavelength synchronization. The "reference pulse method" was applied for wavelength synchronization. Both the direct output from the sweep LD and the output via the Fabry-Perot resonator(F-P) were distributed to the wavelength multiplexer and the λ switch. LD wavelengths were controlled so that the generation time for the beat pulses between the transmitter LD lights and the sweep LD light coincide with that for the F-P output pulses [31], both in the wavelength multiplexer and the λ switch. As a result, each wavelength of the corresponding LDs in the wavelength multiplexer and the λ switch was stabilized to the same respective resonant frequency of the F-P resonator. Through this process, wavelength synchronization is achieved between the input and output WDM highways for the λ switch. In the experiments, the optical frequency discrepancies, between corresponding LDs in the wavelength multiplexer and the λ switch, have been controlled to less than 2.7% of the frequency spacing (8GHz) [15]. 1.55μm wavelength tunable DBR LDs were used as the transmitters, LOs and sweep LD. Transmitted and

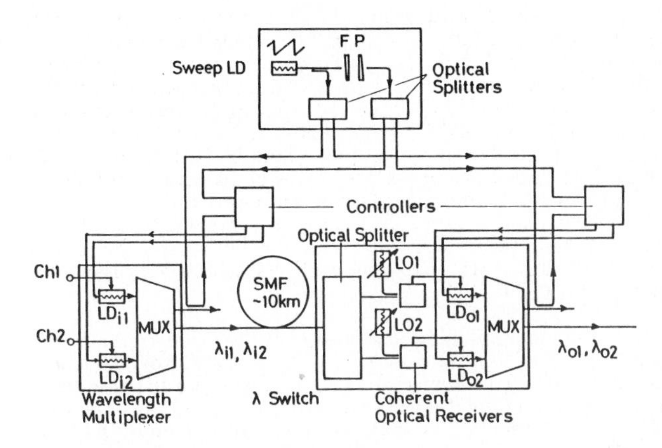

FIGURE 14
Experimental coherent WD switching system.

switched signals are 280Mb/s optical FSK signals with frequency deviation of 1GHz. Frequency separation for the WDM signals was set to be 8GHz, which is sufficient to avoid inter-channel crosstalk up to 400Mb/s [28]. Coherent optical receivers in the λ switch consist of balanced receivers and FSK single filter detection systems [32] with 400MHz-1.4GHz passbands. Beat spectral linewidths, between two DBR LDs, were around 30MHz, which is sufficiently narrow for FSK single filter detection. SMF couplers were used as MUXs and optical splitters.

Continuous wavelength tuning range of the DBR LD, used as a local oscillator(LO), was 20.6Å (257GHz). Therefore, 32 WD channels with 8GHz separation can be selected using this local oscillator. The receiver sensitivity value of about -44dBm (bit error rate $=10^{-10}$) was obtained for each receiver in the coherent λ switch. The obtained receiver sensitivity value was comparable to the value assumed in the design consideration.

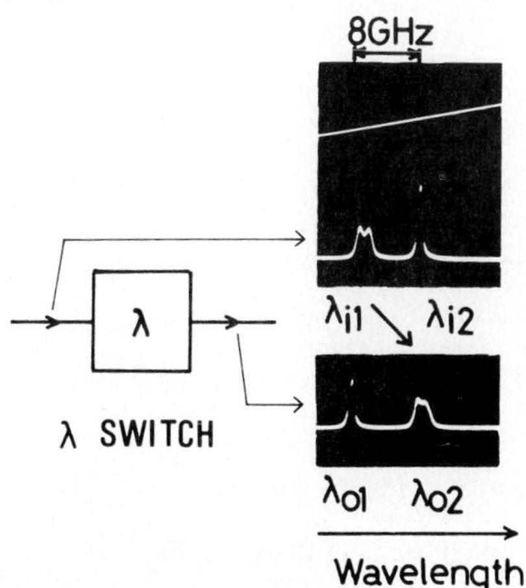

FIGURE 15
WDM signal spectra of input and output WDM signals for λ switch.

Optical spectra of the input and output WDM signals for the coherent λ switch, measured using a scanning Fabry-Perot interferometer, are depicted in Fig. 15. In this case, only the LD_{i1} was modulated, where LO_2 was tuned to select the channel 1 signal. Switching between synchronized WD channels is clearly observed. In the wavelength switching experiments, no bit error rate degradation due to optical crosstalk has been observed, in each coherent optical receiver, even under simultaneous two channel operation condition.From these experimental results, it can be argued that a 280Mb/s coherent WD switching system with 32 WD channels would be possible, using the experimental switching system.

5. APPLICATION TO THE PHOTONIC TIME-DIVISION AND PACKET SWITCHING SYSTEMS

The coherent λ switch technologies described in the previous section can be applied to photonic time-division (TD) or packet switching systems [33][34] by fast wavelength switching in LO wavelength. Especially, the introduction of coherent WD switching technologies to photonic TD switching system will result in a unique hybrid switching system. This section describes this coherent photonic wavelength-division and time-division (WD&TD) hybrid switching system [35]-[37].

5.1. Coherent wavelength and time hybrid switch

The practical photonic switching system will be a hybrid switching system, using two or three types of photonic switching systems. Recently proposed the WD/TD hybrid switching system [35] would be a good example of such hybrid switching systems. A photonic wavelength and time (W&T) hybrid switch, which interchanges WD&TD multiplexed signals, is the key component of the system. In the W&T hybrid switch, the total multiplexity value is given by the product of WD and TD multiplexities. Therefore, a large multiplexity value can be achieved, even with current WD and TD multiplexity values. To obtain large WD multiplexity further, wavelength filtering based on coherent optical detection is attractive as described in the previous section.
Figure 16 shows the structure of coherent W&T switch. The input WD&TD multiplexed optical signal is split by an optical splitter. In each coherent optical receiver, WD channel is selected from each split WD&TD multiplexed signal, by switching the local oscillator (LO) wavelength, time slot by time slot. Time slots of

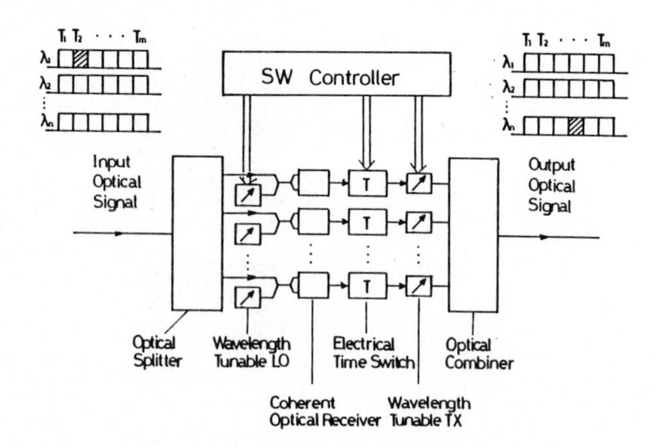

FIGURE 16
Coherent photonic W&T hybrid switch.

each selected TD multiplexed signal are exchanged in an electrical time switch. The exchanged TD multiplexed signals are used to modulate TX LDs, whose wavelengths are switched between adjacent time slots. Through this process, both time slot and wavelength interchanges are accomplished.

5.2. Requirement for wavelength switching

The key technology for coherent W&T switch is wavelength switching of LO and TX LDs. The following characteristics of LO and TX LDs should be considered for the W&T switch.

(1)Fast and wide-range wavelength switching
Continuous wavelength tuning range of the DBR LD is large enough for attaining WD multiplexity of 32. In addition, W&T switch requires high switching speed as well as wide tuning range in wavelength switching. A byte-interleaved or bit-interleaved TD multiplexed signal format would be used in the photonic WD&TD hybrid switching system. For the byte-interleaved signal format, the applicable TD multiplexity for a W&T switch is determined by switching guard time which is ultimately limited by wavelength switching time. TD multiplexity for bit-interleaved RZ signal format is directly restricted by the attainable wavelength switching time. The relations between TD multiplexity and wavelength switching time for 100Mb/s signals, both for bit- and byte-interleaved signal formats, are depicted in Fig. 17. For the byte-interleaved signal format, 10 percent guard time was assumed. Figure 17 indicates that larger TD multiplexity is obtained for byte-interleaved signal format. To achieve TD multiplexity of 4 for 100Mb/s signals, less than 2ns wavelength switching time is required, even using a byte-interleaved signal format.

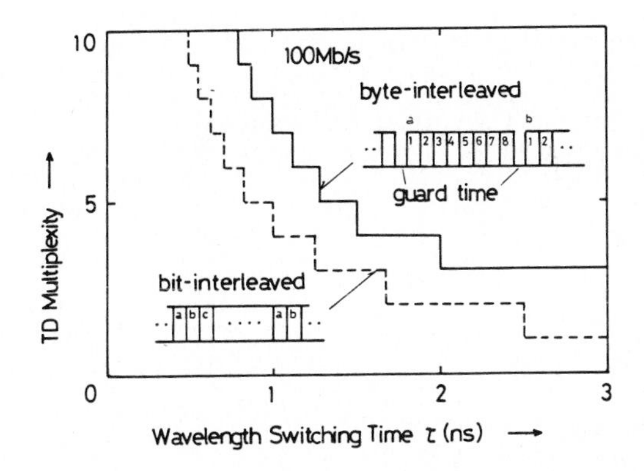

FIGURE 17
Relation between attainable TD multiplexity value and wavelength switching time.

(2)Rapid settling and stable holding of wavelength
Wavelength of the wavelength tunable LD is required to be stable during wavelength holding period from just after the wavelength switching. This requirement for stability would be severe and very important in coherent W&T switch in comparison with the W&T switch using tunable optical filters. In the following experiments, this problem will be discussed.

5.3. Experiments

Experiments have been carried out to demonstrate the feasibility of coherent W&T

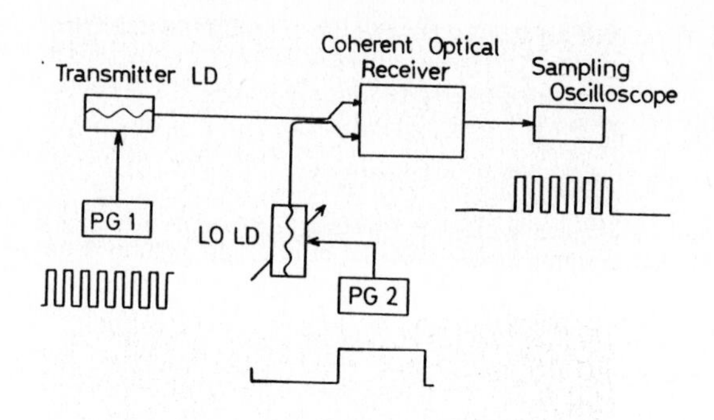

FIGURE 18
Experimental setup for fast WD channel selection.

switch. The experimental setup is shown in Fig. 18. The wavelength tunable DBR LDs [27], with FM response equalization up to about 1GHz, were used as a transmitter LD and a LO LD. The transmitter LD was FSK modulated by a pattern generator (PG1) with 400Mb/s fixed pattern (1010....). The LO LD wavelength was switched between 2 wavelengths according to the control signal from another pattern generator (PG2). One of the switched wavelength was set to select the transmitter LD's signal. The coherent optical receiver was the same as used in the experiments described in the previous section.

Figure 19 shows a time resolved spectra for the DBR LD output, driven with 400ps rise time switching signal. A 12.5Å wide-range wavelength switching was accomplished with switching time of as short as 1.8ns. Transient response of wavelength switching was very smooth. Switching times were almost independent on switching wavelength-ranges and wavelength swing directions. The obtained 1.8ns switching time is short enough to achieve 4 TD multiplexity for byte-interleaved 100Mb/s signals.

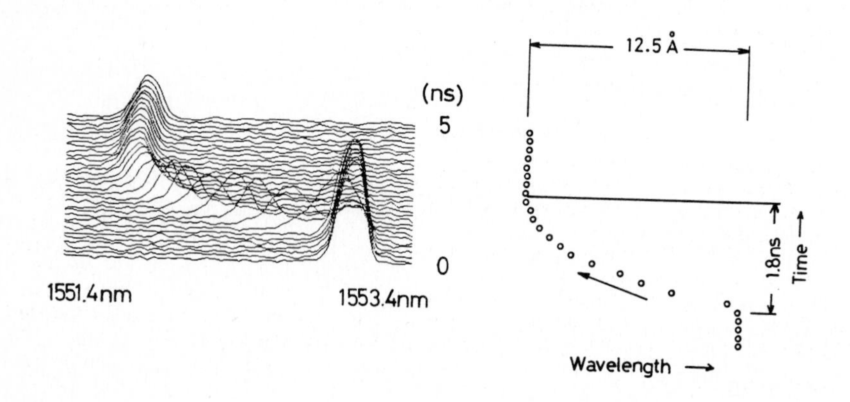

FIGURE 19
Dynamic wavelengthg switching response.

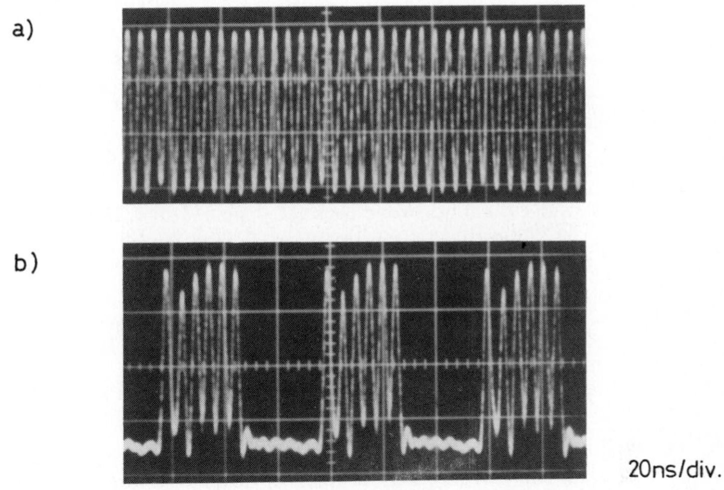

FIGURE 20
Demodulated signals with (b) and without (a) wavelength switching.

In the wavelength selection by coherent detection, the switching wavelength range was set at 6-7Å, which corresponds to the wavelength range of over 8 WD channels with 0.64Å (8GHz) separation. Demodulated signals using the high speed wavelength switching LD, with and without LO wavelength switching, are shown in Fig. 20. The figure shows that wavelength switching is accomplished completely during 1bit duration set as guard bit of 400Mb/s 1010 pattern. None of eleven bits (10101010101) during wavelength holding period are lost. This result shows that coherent detection, which follows fast and wide-range LO wavelength switching, was successfully conducted. Wavelength and linewidth were seemed to be kept constant during the holding period. These results confirm the feasibility of coherent photonic W&T switch.

6. CONCLUSION

The integration of photonic switching and coherent optical transmission technologies, which would be important fo realizing future all optical broadband networks, was theoretically considered. Introducing coherent optical transmission technologies to photonic switching networks will result in large scale switching networks due to receiver sensitivity improvement and fine wavelength channel selectivity of coherent optical detection. Some examples of the integrated swithcing networks were proposed and the fundamental switching functions were demonstrated. The future broadband networks will require a variety of technologies, including both electronic and photonic technologies, for suporting many kinds of wideband survices. It is therefore important to consider the integration of technologies, as described in this paper.

ACKNOWLEDGMENT

Thanks are tendered to S. Yamazaki, N. Henmi, N. Shimosaka and M. Nishio for their co-operations to this work. The authors would like to express their gratitude to M. Sakaguchi, T. Ishiguro, K. Minemura, N. Nishida, K. Watanabe K. Kobayashi and T. Suzuki for their continuous guidance and encouragement.

REFERENCES

[1] Wagner, R.E. et al., IEEE J. Lightwave Technology LT-5, pp.429-438, 1987
[2] Stanley, I.W. et al., IEEE J.Lightwave Technol. LT-5, pp.439-451, 1987
[3] Sakaguchi, M. and Goto, H., Tech. Dig. First Optoelectronics Conference (OEC'86), Tokyo, Japan, July, 1986, pp.192-193
[4] Fujiwara, M. et al., Tech. Dig. Topical Meeting on Photonic Switching, Incline Village, Nevada, March 15-20, 1987, pp.27-29, ThA4
[5] Fujiwara, M. et al., Trans. IEICE Japan E72, pp.55-62, 1989
[6] O'Mahony, M.J., IEEE J. Lightwave Technol. 6, pp.531-544, 1988
[7] Hodgkinson, T.G. et al., Br.Telecom Technol.J. 3, pp.5-18, 1985
[8] Bachus, E.-J.and Heydt, G., Tech. Dig. ECOC'86 , Barcelona, Spain, Sept.22-25, 1986, 2, pp.73-77
[9] Granestrand, P. et al., Electron. Lett., 22, pp.817-818,1987
[10] Kondo, M. et al., Tech. Dig. IECE Japan National Conference on Optical and Radio Wave Electronics, Tokyo, Oct., 1986, 2, pp.118, 289 (in Japanese)
[11] Hermes, T., et al., IEEE J. Lightwave Tech. LT-4, pp.467-471, 1985
[12] Suzuki, S. et al., Tech. Dig. OFC/IOOC'87, Reno, Jan.19-22, 1987, 3, p.146
[13] Clos, C., Bell System Tech. Jour. 32, pp. 407-424, 1953
[14] Emura, K. et al., Electron. Lett., 22, pp.1096-1097, 1986
[15] Yamazaki, S. et al., Electron. Lett. 21, pp.283-284, 1985
[16] Suzuki, S. and Nagashima, K., Tech. Dig. Topical Meeting on Photonic Switching, Nevada, March 18-20 1987, pp.21-23, ThA2
[17] Shimazu, Y. et al., IEEE J. Lightwave Technol. LT-5, pp.1742-1747, 1987
[18] Nishio, M. et al., Tech. Dig. ECOC'88, Brighton, UK, Sept. 11-15 1988, part 2, pp.49-52
[19] Goto, N and Miyazaki, Y., Tech. Dig. GLOBECOM'87, Tokyo, Nov. 15-18 1987, 2, pp. 1305-1309, 33.7
[20] Olsson, N.A. and Tsang, W.T., IEEE J. Quantum Electron. QE-20, p332, 1984.
[21] Kobrinski, H. , Electron. Lett. 23, pp. 974-976, 1987
[22] Hill, G.R., IEEE INFOCOM '88, March 27-31 1988, pp. 354-362, 4B.1
[23] Payne, D.B. and Stern, J.R., IEEE J. Lightwave Technol. LT-4, pp.864-869, 1986
[24] Kobrinski, H. et al., Electron. Lett. 23, pp.824-826, 1987
[25] Glance, B.S. et al., IEEE J. Lightwave Technol. 6, pp.67-72, 1988
[26] Fujiwara, M. et al., Tech. Dig. ECOC88, Brighton, UK, Sept. 11-15 1988, Part1, pp.139-142
[27] Murata, S. et al., Electron. Lett. 24, pp.577-579, 1988
[28] Emura, K. et al., Tech. Dig. OFC'88, New Orleans, 25-28 Jan. 1988, p.54, WC4
[29] Toba, H. et al., Electron. Lett. 23, pp.788-789, 1987
[30] Shibutani, M. et al., Tech. Dig. IOOC'89, Kobe, 18-21 July, 1989, 4, pp. 40-41, 21B4-5
[31] Shimosaka, N., et al, Tech. Dig. OFC'88, New Orleans, Jan. 25-28 1988, p168, ThG3
[32] Emura, K. et al., IEEE J. Lightwave Technol. LT-5, pp.469-477, 1987
[33] Eng, K.Y., Tech. Dig. GLOBECOME'87, Tokyo, Nov. 15-18 1987, pp.1861-1865, 47.2
[34] Arthurs, E. et al., IEEE J. Select. Areas Commun. 6, pp. 1500-1510, 1988
[35] Suzuki, S. et al., Tech. Dig. GLOBECOM'88, pp.933-937, 29.2
[36] Nishio, M. et al.; Tech. Dig. Topical Meeting on Photonic Switching, Salt Lake City, March 1-3, 1989, pp.98-100, ThE5
[37] Shimosaka, N. et al., Tech. Dig. ECOC'89, Gothenburg, September 10-14 1989, WeA15-2

AUTHOR INDEX